HERBALS AS NUTRACEUTICALS

Their Role in Healthcare

AAP Advances in Nutraceuticals

HERBALS AS NUTRACEUTICALS

Their Role in Healthcare

Edited by

Raj K. Keservani, PhD, MPharm

Rajeshwar Kamal Kant Arya, PhD, MPharm

Rajesh K. Kesharwani, PhD, MTech

AAP | APPLE ACADEMIC PRESS

First edition published 2025

Apple Academic Press Inc.
1265 Goldenrod Circle, NE,
Palm Bay, FL 32905 USA

4164 Lakeshore Road, Burlington,
ON, L7L 1A4 Canada

CRC Press
2385 NW Executive Center Drive,
Suite 320, Boca Raton FL 33431

4 Park Square, Milton Park,
Abingdon, Oxon, OX14 4RN UK

Apple Academic Press exclusively co-publishes with CRC Press, an imprint of Taylor & Francis Group, LLC

Library and Archives Canada Cataloguing in Publication

CIP data on file with Canada Library and Archives

Library of Congress Cataloging-in-Publication Data

CIP data on file with US Library of Congress

ISBN: 978-1-77491-614-8 (hbk)
ISBN: 978-1-77491-615-5 (pbk)
ISBN: 978-1-00348-839-2 (ebk)

ABOUT THE BOOK SERIES: AAP ADVANCES IN NUTRACEUTICALS

Series Editor:
Raj K. Keservani, PhD, MPharm
Faculty of B. Pharmacy, CSM Group of Institutions, Allahabad, India

Rajesh K. Kesharwani, PhD
Nehru Gram Bharati (Deemed to be University), Prayagraj, India

Anil K. Sharma, PhD, MPharm
Department of Pharmacy, School of Medical and Allied Sciences, GD Goenka University, Gurugram, India

In the modern era, mankind has witnessed a paradigm shift with respect to fundamental eating behavior. The lack of physical workouts and busy schedules at offices and in households have promoted the consumption of junk foods, which eventually results in numerous diseases and disorders since the nutritional content from fast food is inadequate. Plumpness and obesity have become a global health threat. The leading causes of death from most developing nations are noncommunicable diseases such as cardiovascular diseases, cancer, arthritis, osteoporosis, and liver toxicity. Patients suffering from such lifestyle ailments are a bit apprehensive towards prolonged use of costly modern therapeutics, which encourages instead the use of alternative approaches for management of such diseases and disorders.

The emerging sector of the nutraceutical industry encompasses products derived from nature, dietary supplements, and functional foods. Nutraceuticals are used for treatment and prevention of a broad range of diseases, such as the common cold, arthritis, sleep-related disorders, cancers, cardiovascular complications, metabolic disorders, and others.

The research on nutraceuticals is increasing day by day, considering the beneficial effects of food or food supplements in the management of diverse diseases. The issue of paramount concern is standardization and establishment of clinical efficacy of nutraceuticals, which is in fact a challenge for researchers around the globe.

This book series aims at realizing the significance of the variety of nutraceuticals for human well-being. The books in the series will also emphasize the role of dieticians and nutritionists for the prescription of judicious eating habits. The key food components such as carbohydrates, proteins, and lipids as well as micronutrients (vitamins, minerals) are demonstrated to maintain good health, obviating the need for medicines. Thus, nutraceuticals that are indeed derived from food ingredients are believed to be potential alternative therapeutics.

Asymptomatic diseases/disorders necessitate proper diagnosis. Stressed circumstances have been reported to cause weight loss. Workplace stress is a key variable in the genesis and propagation of metabolic disorders. The nuances of the underlying mechanisms are still to be deciphered. Poor lifestyle trends have been attributed along with the consumption of junk foods in several such instances. Sugar-rich carbonated beverages, the use of contraband drugs, and the consumption of liquor in excess are assigned as risk factors. This book series will cover advances and applications in addition to providing the basics of nutraceuticals.

The books in this series will be valuable resources for industry professionals to design and develop quality products for end users. In addition, the book series will be fortified by ideas of innovation, concept building, manufacturing aspects, quality control, and regulatory status of nutraceuticals.

Ostensibly, business has seen a rise in the nutraceutical consumption in recent years. Pure, safe, efficacious products are the need of the hour globally. There are a number of books on nutraceuticals in the market, yet these are only a few books presenting certain aspects of nutraceuticals therein. This book series aims to offer a holistic view of this promising strategy, which is picking up pace with time. This book series will be a unique endeavor to bring manufacturing, research and development, and marketing strategies under a single umbrella.

CURRENT BOOKS IN THE SERIES

- **Micronutrients and Macronutrients as Nutraceuticals**
  Editors: Prakash Chandra Gupta, PhD, Sayan Bhattacharyya, MBBS, MD, Nisha Sharma, PhD, MPharm, Rajesh K. Kesharwani, PhD, and Raj K. Keservani, PhD, MPharm

- **Formulations, Regulations, and Challenges of Nutraceuticals**
  Editors: Tingirikari Jagan Mohan Rao, PhD, Rajesh K. Kesharwani, PhD, Raj. K. Keservani, PhD, and Anil K. Sharma, PhD

- **Food Supplements and Dietary Fibers in Health and Disease**
  Editors: Bhushan R. Rane, PhD, MPharm, Raj K. Keservani, PhD, MPharm, Durgesh Singh, DPhil, Nayan A. Gujarathi, PhD, and Ashish S. Jain, PhD
- **Nutraceuticals in Cancer Prevention, Management, and Treatment**
  Editors: Raj K. Keservani, PhD, MPharm Bui Thanh Tung, PhD, Sippy Singh, DPhil, and Rajesh K. Kesharwani, PhD
- **Advances in Flavonoids for Human Health and Prevention of Diseases**
  Editors: Nisha Sharma, PhD, Deepika Saini, PhD, Rajesh K. Kesharwani, PhD, Prakash Chandra Gupta, and Raj. K. Keservani, PhD
- **The Flavonoids: Extraction and Applications**
  Editors: Deepika Saini, PhD, Rajesh K. Kesharwani, PhD, and Raj K. Keservani, PhD, MPharm
- **Immune-Boosting Nutraceuticals for Better Human Health: Novel Applications**
  Editors: Urmila Jarouliya, PhD, Raj K. Keservani, PhD, MPharm Rajesh K. Kesharwani, PhD, Virendra K. Patel, PhD, and Adi D. Bharti
- **Applications of Functional Foods in Disease Prevention**
  Editors: Raj K. Keservani, PhD, MPharm, and Eknath D. Ahire, MSPharm
- **Plant Metabolites and Vegetables as Nutraceuticals**
  Editors: Raj K. Keservani, PhD, MPharm, Bui Thanh Tung, PhD, Rajesh K. Kesharwani, PhD, MPharm, and Eknath D. Ahire
- **Preventive and Therapeutic Role of Vitamins as Nutraceuticals**
  Editors: Khemchand R. Surana, Eknath D. Ahire, Raj K. Keservani, PhD, MPharm, and Rajesh K. Kesharwani, PhD
- **Nutraceuticals and Bone Health**
  Editors: Deepak Sharma, PhD, Madan Mohan Gupta, PhD, Anil K. Sharma, PhD, MPharm, Raj K. Keservani, PhD, MPharm, and Rajesh K. Kesharwani, PhD
- **Flavonoids as Nutraceuticals**
  Editors: Rajesh K. Kesharwani, PhD, MTech, Deepika Saini, PhD, Raj K. Keservani, PhD, MPharm, and Anil Kumar Sharma, PhD
- **Herbals as Nutraceuticals: Their Role in Healthcare**
  Editors: Raj K. Keservani, PhD, MPharm, Rajeshwar Kamal Kant Arya, PhD, and Rajesh K. Kesharwani, PhD

- **Nutraceutical Fruits: Overview and Disease Prevention**
  Editors: Raj K. Keservani, PhD, MPharm
- **Nutrigenomics and Nutraceuticals**
  Editors: Raj K. Keservani, PhD, MPharm, Eknath D. Ahire, Shubham J. Khainar, PhD, Sanjay J. Kshirsagar, PhD, and Rajesh K. Kesharwani, PhD
- **Nutraceuticals in Cardiac Health Management**
  Editors: Raj K. Keservani, PhD, MPharm, Rajesh K. Kesharwani, PhD, Praveen Kumar Jain, PhD, Bhushan R. Rane, PhD, and Nayan A. Gujarathi, PhD
- **Nutraceuticals for the Treatment and Prevention of Sexual Disorders**
  Editors: Raj K. Keservani, PhD, MPharm, Sharangouda J. Patil, PhD, and Ivan Aranha, PhD
- **Nutraceuticals in Obesity Management and Control**
  Editors: Raj K. Keservani, PhD, MPharm, and Alka Lohani, PhD
- **Antioxidants as Nutraceuticals**
  Editors: Nayan A. Gujarathi, PhD, Raj K. Keservani, PhD, MPharm, Rajesh K. Kesharwani, PhD, Bhushan R. Rane, PhD, and Yogeeta Sameer Goyal, PhD
- **The Nature of Nutraceuticals: History, Properties, Sources, and Nanotechnology**
  Editors: Rajesh K. Kesharwani, PhD, Prashant Kumar, PhD, and Raj K. Keservani, PhD, MPharm
- **Nutraceuticals in Insomnia and Sleep Problems**
  Editors: Raj K. Keservani, PhD, MPharm, Sayan Bhattacharyya, PhD, and Rajesh K. Kesharwani, PhD
- **Nutraceuticals in Respiratory and Pulmonary Diseases**
  Editors: Deepika Saini, PhD, Rajesh K. Kesharwani, PhD, and Raj K. Keservani, PhD, MPharm
- **Nutraceuticals in Arthritis and Psoriasis**
  Editors: Meenakshi Jaiswal, PhD, Raj K. Keservani, PhD, MPharm Rajesh K. Kesharwani, PhD, and Swati G. Talele, PhD

ABOUT THE EDITORS

Raj K. Keservani, PhD, MPharm
Associate Professor, Faculty of B.Pharmacy,
CSM Group of Institutions, Prayagraj, India

Raj K. Keservani, PhD, MPharm, is working as Associate Professor in the faculty of B. Pharmacy at the CSM Group of Institutions, Prayagraj, India. He has more than 12 years of academic (teaching) experience from various institutes of India in pharmaceutical education. He has supervised more than 25 undergraduate students for their research work. He has published over 30 peer-reviewed papers in the field of pharmaceutical sciences in national and international journals, six patents, over 70 book chapters, three co-authored books, and over 40 edited books. He is also active as a reviewer for several international scientific journals. His research interests include nutraceutical and functional foods, novel drug delivery systems (NDDS), transdermal drug delivery/drug delivery, health science, cancer biology, and neurobiology. Dr. Keservani graduated with a pharmacy degree from the Department of Pharmacy, Kumaun University, Nainital, Uttarakhand, India. He received his Master of Pharmacy (MPharm) (specialization in pharmaceutics) from the School of Pharmaceutical Sciences, Rajiv Gandhi Proudyogiki Vishwavidyalaya, Bhopal, India. He earned his PhD degree from IKG Punjab Technical University, Jalandhar, Punjab, India. Dr. Keservani is a life member of the Society of Pharmaceutical Education and Research (SPER).

Rajeshwar Kamal Kant Arya, PhD, MPharm
Assistant Professor, Department of Pharmaceutical Sciences,
Sir J.C. Bose Technical Campus Bhimtal, Kumaun University,
Nainital, Uttarakhand, India

Rajeshwar Kamal Kant Arya, PhD, has been working as an Assistant Professor in the Department of Pharmaceutical Sciences, Sir J.C. Bose Technical Campus Bhimtal, Kumaun University Nainital, India, for the last 14 years. He has more than 25 research papers to his credit in national and international journals of repute. He has presented more than ten papers so far at various conferences, seminars, and workshops. He has also filed two Indian patents recently. He has authored many chapters in edited books in varied domains of pharmaceutical sciences. He is also serving on the editorial board of several prestigious national and international journals. He has guided more than 55 MPharm students and is currently supervising four PhD scholars. He completed his graduation and PhD from Kumaun University, Nainital and his master's from Bundelkhand University, Jhansi, India.

Rajesh K. Kesharwani, PhD, MTech
Associate Professor, Department of Computer Application,
Nehru Gram Bharati (Deemed to be University), Prayagraj, India

Rajesh K. Kesharwani, PhD, MTech, working as an Associate Professor in the Department of Computer Application at Nehru Gram Bharati (Deemed to be University), Prayagraj, India. He has more than 11 years of research and nine years of teaching experience in various institutes in India, imparting bioinformatics and biotechnology education. He has received several awards, including the NASI-Swarna Jayanti Puruskar by The National Academy of Sciences of India. He has supervised one PhD and more than 20 under- and postgraduate students for their research work and has authored over 49 peer-reviewed articles, 20 book chapters, and 14 edited books with international publishers. He has been a member of many scientific communities as well as a reviewer for many international journals. He has presented many papers at various national and international conferences. Dr. Kesharwani received his PhD from the Indian Institute of Information Technology, Allahabad, and worked at NIT Warangal for two-semester. He has been a recipient of the Ministry of Human Resource recipient of Development (India) Fellowship and a Senior Research Fellowship from the Indian Council of Medical Research, India. His research fields of interest are medical informatics, protein structure and function prediction, computer-aided drug designing, structural biology, drug delivery, cancer biology, nano-biotechnology, and biomedical sciences.

CONTENTS

CONTRIBUTORS

Mohit Agrawal
Department of Pharmacology, Rajiv Academy for Pharmacy, Mathura, Uttar Pradesh, India

Akash J. Amkar
Shri D. D. Vispute College of Pharmacy and Research Center, Panvel, Raigad, Maharashtra, India

Malti Arya
Chandra Shekhar Singh College of Pharmacy, Uttar Pradesh, India

Rajeshwar Kamal Kant Arya
Department of Pharmaceutical Sciences, Sir J.C. Bose Technical Campus Bhimtal, Kumaun University, Nainital, Uttarakhand, India

Anant Bhandarkar
Department of Pharmacognosy, Goa College of Pharmacy, Goa University, Goa, India

Samir Bhargava
Faculty of Pharmacy, DIT University, Dehradun, Uttarakhand, India

Dheeraj Bisht
Department of Pharmaceutical Sciences, Sir J.C. Bose Technical Campus Bhimtal, Kumaun University, Nainital, Uttarakhand, India

Komal Chandra
Department of Pharmacy, Sir J.C. Bose Technical Campus Bhimtal, Kumaun University, Bhimtal, Uttarakhand, India

Priyanka J. Chauhan
Shri D.D. Vispute College of Pharmacy and Research Center, Panvel, Raigad, Maharashtra, India

Pranaya Chodankar
Department of Pharmacognosy, Goa College of Pharmacy, Goa University, Goa, India

Punam S. Gadekar
Shri D.D. Vispute College of Pharmacy and Research Center, Panvel, Raigad, Maharashtra, India

Rupa Gupta
Amity Institute of Pharmacy, Amity University, Gurugram, Haryana, India
Department of Pharmacy, Sushant University, Gurugram, Haryana, India

Ashish S. Jain
Shri D.D. Vispute College of Pharmacy and Research Center, Panvel, Raigad, Maharashtra, India

Arun Joshi
Department of Pharmacognosy, Goa College of Pharmacy, Goa University, Goa, India

Himanshu Joshi
Department of Pharmacology, College of Pharmacy, Graphic Era Hill University, Bhimtal, Uttarakhand, India

Nidhi S. Kate
Shri D.D. Vispute College of Pharmacy and Research Center, Panvel, Raigad, Maharashtra, India

Ramandeep Kaur
Pharmacy Academy, IFTM University, Moradabad, Uttar Pradesh, India

Virender Kaur
Department of Pharmacy, Sir J.C. Bose Technical Campus Bhimtal, Kumaun University, Bhimtal, Uttarakhand, India

Meena Kausar
Department of Pharmaceutical Sciences, Sir J.C. Bose Technical Campus Bhimtal, Kumaun University, Nainital, Uttarakhand, India

Raj K. Keservani
Faculty of B. Pharmacy, CSM Group of Institutions, Prayagraj, Uttar Pradesh, India

Shahbaz Khan
Pharmacy Academy, IFTM University, Moradabad, Uttar Pradesh, India

Ankit Kumar
Department of Pharmaceutical Sciences, Sir J.C. Bose Technical Campus Bhimtal, Kumaun University, Nainital, Uttarakhand, India

Aadesh Kumar
Department of Pharmaceutical Sciences, Sir J.C. Bose Technical Campus Bhimtal, Kumaun University, Nainital, Uttarakhand, India

Bhavna Kumar
Faculty of Pharmacy, DIT University, Dehradun, Uttarakhand, India

Manish Kumar
Department of Pharmacy, Madhav University, Pindwara, Rajasthan, India

Suresh Kumar
Department of Biotechnology, Sir J.C. Bose Technical Campus Bhimtal, Kumaun University, Bhimtal, Uttarakhand, India

Sushil Kumar
School of Pharmaceutical Sciences, IFTM University, Moradabad, Uttar Pradesh, India

Tirath Kumar
Department of Pharmacy, Sir J.C. Bose Technical Campus Bhimtal, Kumaun University, Bhimtal, Uttarakhand, India

Alka Lohani
School of Pharmaceutical Sciences, IFTM University, Moradabad, Uttar Pradesh, India

Rekha Mehta
Amrapali Institute of Pharmacy and Sciences, Haldwani, Uttarakhand, India

Niharika Mishra
Department of Pharmaceutical Sciences, Sir JC Bose Technical Campus Bhimtal, Kumaun University, Naintal, Uttarakhand, India

Govind Raj Pal
Department of Pharmaceutical Sciences, Sir J.C. Bose Technical Campus Bhimtal, Kumaun University, Naintal, Uttarakhand, India

Virendra Kumar Patel
Faculty of Pharmacy, RKDF University, Bhopal, Madhya Pradesh, India

Amol D. Patil
Shri D.D. Vispute College of Pharmacy and Research Center, Panvel, Raigad, Maharashtra, India

Hritik A. Patil
Shri D.D. Vispute College of Pharmacy and Research Center, Panvel, Raigad, Maharashtra, India

Julie A. Patil
Shri D.D. Vispute College of Pharmacy and Research Center, Panvel, Raigad, Maharashtra, India

Nimisha N. Patil
Shri D.D. Vispute College of Pharmacy and Research Center, Panvel, Raigad, Maharashtra, India

Shivani S. Pawar
Shri D.D. Vispute College of Pharmacy and Research Center, Panvel, Raigad, Maharashtra, India

Sunil Kumar Prajapati
Institute of Pharmacy, Bundelkhand University, Jhansi, Uttar Pradesh, India

Vijay Singh Rana
Faculty of Pharmacy, DIT University, Dehradun, Uttarakhand, India

Bhushan R. Rane
Shri D.D. Vispute College of Pharmacy and Research Center, Panvel, Raigad, Maharashtra, India

Samali S. Raut
Shri D.D. Vispute College of Pharmacy and Research Center, Panvel, Raigad, Maharashtra, India

Gaurav Rawat
Department of Zoology, D.S.B. Campus, Kumaun University, Nainital, Uttarakhand, India

Neeraj Kumar Sethiya
Faculty of Pharmacy, DIT University, Uttarakhand, India

Amit Sharma
Department of Pharmaceutical Sciences, Sir JC Bose Technical Campus Bhimtal, Kumaun University, Naintal, Uttarakhand, India

Ajay Kumar Shukla
Institute of Pharmacy, Dr Ram Manohar Lohia Avadh University, Ayodhya, Uttar Pradesh, India

Anita Singh
Department of Pharmaceutical Sciences, Sir J.C. Bose Technical Campus Bhimtal, Kumaun University, Nainital, Uttarakhand, India

Arun Kumar Singh
Amrapali Institute of Pharmacy and Sciences, Haldwani, Uttarakhand, India

Manmohan Singhal
Faculty of Pharmacy, DIT University, Dehradun, Uttarakhand, India

Rahul Tiwari
Department of Pharmacology, Rajiv Academy for Pharmacy, Mathura, Uttar Pradesh, India

Ritesh Tiwari
Chandra Shekhar Singh College of Pharmacy, Uttar Pradesh, India

Ritu Tomar
Faculty of Pharmacy, DIT University, Dehradun, Uttarakhand, India

Navneet Verma
Pharmacy Academy, IFTM University, Moradabad, Uttar Pradesh, India

Vimal Kumar Yadav
Institute of Pharmacy, Dr. Ram Manohar Lohia Avadh University, Ayodhya, Uttar Pradesh, India

ABBREVIATIONS

AA	arachidonic acid
AAGR	average annual growth rate
AAPCC	American Association of Poison Control Centre
AD	Alzheimer's disease
ADHD	attention-deficit hyperactivity disorder
AMPK	AMP-activated protein kinase
APC	antigen-presenting cells
API	active pharmaceutical ingredients
ARGCM	Australian Regulatory Guidance for Complementary Medicine
ASSOCHAM	Association Chamber of Commerce and Industry of India
ATRA	all trans-retinoic corrosive
BBB	blood–brain barrier
BDNF	brain-derived neurotrophic factors
BMI	body mass index
BPH	benign prostatic hyperplasia
BW	body weight
CAGR	compound annual growth rate
CMA	cow's milk allergy
CMI	cell-mediated immune
CO_2	carbon dioxide
COPD	chronic obstructive pulmonary disease
COX	cyclooxygenase
CSIR	Council of Scientific and Industrial Research
CT	computed tomography
CVD	cardiovascular disease
DA	dopamine deficit
DBT	Department of Biotechnology
DM	diabetes mellitus
DSHEA	Dietary Supplements and Health Education Act
ED	erectile dysfunction
EGCG	epigallocatechin-3 gallate
EHD	electrohydrodynamic
EKG/ECG	electrocardiogram
EMA	European Medicines Agency
ENA	European Nutraceutical Association
FFC	foods with function claims
FIM	Foundation for Innovation in Medicine

FMD	fasting-mimicking diet
FMT	fecal microbiota transplant
FSSA	Food Safety and Standard Act
FTC	Federal Trade Commission
GAGs	glycosaminoglycans
GFR	glomerular filtration rate
GIT	gastrointestinal tract
GLP	glucagon-like peptide
HBV	hepatitis B virus
HC	Health Canada
HDL	high-density lipoproteins
HFC	high-fat content
HSDD	hypoactive sexual desire disorder
IBD	inflammatory bowel disease
IBS	irritable bowel syndrome
ICMR	Indian Council of Medical Research
IFN-γ	interferon-γ
IL	interleukin
IMID	immune-mediated inflammatory disease
LAB	lactic acid bacteria
LDL	low-density lipoproteins
LH	luteinizing hormone
LOX	lipooxygenase
MAO	monoamine oxidase
MAPK	mitogen-activated protein kinase
MC	*Momordica charantia*
MRI	magnetic resonance imaging
NCs	nutraceuticals
NDIN	New Dietary Ingredient Notification
NHANES	National Health and Nutrition Examination Survey
NHP	natural health product
NHPR	Natural Health Products Regulations
NO	nitric oxide
OA	osteoarthritis
OTC	over-the-counter
PCOS	polycystic ovarian syndrome
PD	Parkinson's disease
PDDAT	primary degenerative dementia of the Alzheimer's type
PE	premature ejaculation
PET	positron emission tomography
PGE2	prostaglandin E2
PGs	prostaglandins
PPD	protopanaxadiol group

PPT	protopanaxatriol
PUFA	polyunsaturated fatty acid
RA	rheumatoid arthritis
ROS	reactive oxygen species
SAC	S-allyl-cysteine
SAMC	S-allyl-mercapto cysteine
SARS-CoV-2	severe acute respiratory syndrome coronavirus 2
SC	stratum corneum
SPECT	single photon emission computed tomography
TEWL	trans-epidermal water loss
TGA	Therapeutic Goods Act
TLRs	toll-like receptors
TNF	tissue necrosis factor
UCC	urothelial cell carcinoma
US	ultrasound
UTI	urinary tract infections
UV	ultraviolet
WHO	World Health Organization

PREFACE

Nutraceuticals are food supplements used to diagnose, treat, cure, or prevent diseases and provide health benefits to human beings. Several herbal products have been used as nutraceuticals, and they possess many benefits over synthetic ones, for example, with minimal or no side effects due to their natural occurrence. Herbal nutraceuticals are economical and better tolerated by the patients. Nowadays, the whole world is aiming at herbal treatments, and herbal nutraceuticals are also called dietary supplements, nutraceuticals, herbals, or dietary fibers. A WHO survey report claimed that 80–85% of world's population depend on herbal treatment.

Nutraceuticals possess various vitamins, alkaloids, proteins, and fibers. Nutraceuticals have high antioxidant potential, which helps in the prevention of various diseases, such as cardiovascular diseases, hepatic disorders, neurological disorders, ophthalmology, and renal disorders. Nutraceuticals even play a pivotal role in the prevention of cancer. Various studies confirm the role of herbal nutraceuticals in cancer. Some commonly used herbal nutraceuticals in treating various diseases include amla, *Aloe vera*, Giloe, bitter gourd, and blackberry or jamun. Some important and most demanded nutraceuticals are Panax ginseng, cod liver oil, green tea, glucosamine, omega-3, folic acid, lutein, and others. The present market for nutraceuticals is growing day by day. A variety of herbal plant materials as well as nutraceuticals are commonly used for treating diabetes.

Herbal nutraceuticals have been found to detoxify environmental and ingested carcinogen agents by stimulating antioxidant enzymes, DNA repair pathways, and tumor-suppressor proteins. They prevent cancer progression by inhibiting epithelial–mesenchymal transition, angiogenesis, and metastasis and promoting apoptosis. Herbal nutraceuticals, in addition to their anticancer actions, have also been documented to boost therapeutic efficacy of chemotherapies against hypoxic malignancies and ensure that these cytotoxic treatments do not have harmful side effects. Herbal nutraceuticals are categorized into distinct classes, such as cannabinoids, saponins, flavonoids, carotenoids, sitosterols, anthocyanins, terpenes, and isothiocyanates, based on their chemical structures and functions.

The present book provides a deep insight into herbal nutraceuticals and their role in treating various diseases, their mechanisms of action, in vitro, in vivo, and clinical studies supporting the role of herbal nutraceuticals in various diseases.

This book, *Herbals as Nutraceuticals,* comprises 14 chapters in three parts and describes the role of herbal nutraceuticals in the management and prevention of different types of diseases.

PART I: OVERVIEW, CONCEPT, AND ADVANCES OF HERBAL NUTRACEUTICALS

Chapter 1, *Health Benefits of Herbal Nutraceuticals: An Overview,* written by Meena Kausar et al., focuses on the health benefits of nutraceuticals and the miraculous effects of herbal nutraceuticals in curing various diseases, such as cancer, diabetes, cardiovascular diseases, hepatic disorders, neurological disorders, ophthalmology, and renal disorders.

Chapter 2, *Nutraceutical Potential of Ayurvedic Herbals,* written by Aadesh Kumar et al., examines the physiological activities of numerous physiologically active ingredients found in commonly used herbs.

Chapter 3, *Current Concepts and Prospects of Herbal Nutraceuticals,* written by Rajeshwar Kamal Kant Arya and colleagues, focuses on the safety, efficacy, marketing, and regulatory parameters of herbal nutraceuticals.

PART II: ROLE OF HERBAL NUTRACEUTICALS IN COMBATTING DISEASE

Anant Bhandarkar and coworkers highlight the potential role of nutraceuticals in neurodegeneration, psychotic, cancer, obesity, and cardiovascular diseases in Chapter 4, *Nutraceuticals: A Potential Source in Treating Human Ailments.*

Komal Chandra and associates report and briefly discuss the antimicrobial properties of herbal nutraceuticals in Chapter 5, *Antimicrobial Properties of Herbal Nutraceuticals.*

Neeraj Kumar Sethiya et al. have tried to customize some of the selected recent plant-based nutraceutical research findings to be useful against CVD prevention in *Chapter 6, Nutraceutical Herbs in Cardiovascular Diseases.*

Alka Lohani and associates in Chapter 7, *Herbal Nutraceuticals: Safe and Potent Therapeutics in Combating Cancer* discuss that nutraceuticals are mostly made up of nutrients, herbals, and dietary supplements, which make them useful in maintaining health, fighting disease, and improving the overall quality of life.

Neeraj Kumar Sethiya and colleagues, in Chapter 8, *Nutraceutical Herbs as Immuno-Inflammatory,* discuss that nutraceutical products are recognized through various studies that they have positively controlled and prevented by influencing risk factors associated with immunoinflammation.

Chapter 9, *Nutraceuticals: An Emerging Concept for Skin Care,* written by Alka Lohani et al., describes nutraceuticals as promising supplements for preventing, delaying, or minimizing skin aging, as well as for treating some skin problems.

Chapter 10, *Nutraceutical Herbs in Obesity and Diabetes,* written by Arun Kumar Singh and coworkers, reports the potential of nutraceutical herbs against obesity and diabetes.

Pal et al., in Chapter 11, *Herbal Supplements*, discuss the in vitro, in vivo, and clinical studies with their possible mechanisms of action of various herbal supplements available in the market for treating various disorders, for example, weight gain, weight loss, sexual disorders. Herbal supplements also possess immunomodulatory properties.

PART III: MISCELLANEOUS

Chapter 12, *Herbal Supplements and Their Regulation*, written by Bhushan R. Rane and associates, highlights a large amount of discussion relevant to learn briefly about several herbal supplements that are typically used to promote human health and the status of herbal medications and supplements across various national legislatures.

Shukla et al. briefly report the recent advancement of herbal drugs for diseases in Chapter 13, *Recent Advancement of Herbal Nutraceuticals.*

Bhushan R. Rane and associates highlight the impact of probiotics on intestinal health as well as their potential as nutraceuticals in Chapter 14, *Probiotic Role of Nutraceuticals*.

PART I

OVERVIEW, CONCEPTS, AND ADVANCES IN HERBAL NUTRACEUTICALS

CHAPTER 1

HEALTH BENEFITS OF HERBAL NUTRACEUTICALS: AN OVERVIEW

MEENA KAUSAR[1], DHEERAJ BISHT[1], RAJ K. KESERVANI[2], and RAJESHWAR KAMAL KANT ARYA[1]

[1]*Department of Pharmaceutical Sciences, Sir J.C. Bose Technical Campus Bhimtal, Kumaun University, Nainital, India*

[2]*Faculty of B. Pharmacy, CSM Group of Institutions, Prayagraj, Uttar Pradesh, India*

ABSTRACT

Nutraceuticals are nutrients in the form of medicine; they are micronutrients that are abundantly available in fruits and vegetables. Nutraceuticals play a vital role in curing various diseases by fighting the disease-causing agents and maintaining the body's normal function. Nutraceuticals, called dietary supplements and functional food, possess different vitamins, alkaloids, proteins, and fibers. Nutraceuticals have high antioxidant potential, which helps in the prevention of various diseases like cardiac, hepatic, neurological, ophthalmic, and renal diseases, and even in cancer—nutraceuticals play a pivotal role in the prevention of cancer. Various studies confirm the role of herbal nutraceuticals in the treatment of cancer. There are some commonly used herbal nutraceuticals in the treatment of various diseases, for example, amla, aloe vera, giloe, bitter gourd, and blackberry or Jamun. Some of the important and most demanded nutraceuticals are *Panax ginseng*, green tea, glucosamine, omega-3, folic acid, and lutein. The present market for nutraceuticals is increasing day-by-day. This chapter is focused on the health benefits of nutraceuticals and the miraculous effects

Herbals As Nutraceuticals: Their Role in Healthcare.
Raj K. Keservani, Rajeshwar Kamal Kant Arya, and Rajesh K. Kesharwani (Eds.)

of herbal nutraceuticals in the cure of various diseases like cancer, diabetes, cardiovascular diseases, hepatic disorders, neurological disorders, ophthalmology, and renal disorders.

1.1 INTRODUCTION

The word "nutraceuticals" (nutrients + pharmaceuticals) was first used by *Stephen L. Defelice* (New Jersey, 1989) (Brower, 1998), which means nutrients in the form of medicine (Kalra, 2003; Keservani et al., 2010a; Keservani et al., 2010b; Keservani et al., 2017) that provide health benefits, fight against various disease-causing agents, delay aging, increase expectancy of life, and also help in maintaining normal body functions (Nasri, 2014; Trottier, 2010). *Hippocrates* (the father of medicine) has also mentioned nutraceuticals in his statement "Let Food Be Your Medicine," and many other philosophers had also mentioned nutraceuticals as dietary supplements (Keservani et al., 2020), functional, or multifunctional food (Kalra, 2003). The Dietary Supplements & Health Education Act (DSHEA), 1994, defined "Nutraceuticals are products used to supplement the diet, containing one or more ingredients like vitamins, minerals, proteins." Nutraceuticals are available in tablets, capsules, pills, or liquid formulations (USP, 2006) for daily consumption (Zeisel, 1999). According to the DSHEA act, manufacturers must ensure their safety before marketing any nutraceutical products (Nasri et al., 2014). Some of the important and most demanded nutraceuticals are ginseng, cod liver oil, green tea, glucosamine, omega-3, folic acid, and lutein, which possess many therapeutic values (Nasri et al., 2014).

A market survey in 2018 reveals how demand for nutraceuticals is expanding daily (Hardy, 2000). As of now, herbal (Sharma et al., 2020; Rane et al., 2020a; Rane et al., 2020b; Rane et al., 2020c) nutraceuticals have been utilized against various major diseases like tumors (Shirzad et al., 2011), atherosclerosis (Madihi et al., 2013), diabetes (Baradaran et al., 2013; Nasri, 2013), different types of cardiovascular diseases (Khosravi-Boroujeni et al., 2013), and CNS (Roohafza et al., 2013) diseases.

In this chapter, we have discussed the miraculous effects of herbal nutraceuticals in curing various diseases like cancer, diabetes, cardiovascular diseases, hepatic disorders, neurological disorders, ophthalmology, renal disorders, and the recent advancement in the use of nutraceuticals in various diseases.

1.2 ROLE OF HERBAL NUTRACEUTICALS IN IMMUNE FUNCTIONS

Herbal nutraceuticals promote health, regulate immune functions (Keservani et al., 2015), and help eliminate different types of immunological diseases (Santini et al., 2018). Mushrooms, orange juice, cereals, oatmeal, and salmon are rich sources of vitamin D. Vitamin D inhibits excessive synthesis and release of IL-1β and IL-6 in sepsis conditions (Sassi et al., 2018) and alters all the immune system-related consequences (Marino and Misra, 2017). Vitamin D also inhibits the activation of Th1 (T-helper cells type-1), Th17 (Dankers et al., 2017), and subsequent release of pro-inflammatory cytokines, which include interferon-γ (IFN-γ), IL-12, and IL-17 (Medrano et al., 2017). Grapes consist of polyphenolic compound resveratrol, which exhibits immunomodulatory properties by inhibiting SIRT1 (a protein involved in various signaling pathways), down-regulating the differentiation and proliferation of immune cells, and also inhibiting the COX enzyme (Malaguarnera, 2019). Studies have revealed that resveratrol down-regulates the expression of Toll-like receptors (TLRs) on the surface of immune cells (innate immunity types) (Jardim et al., 2018), decreases the activity of mitogen-activated protein kinase (MAPK) (Xia et al., 2017) involved in various types of viral infections and immunological diseases, for example, systemic lupus erythematosus (Kavas et al., 2013; Yahfoufi et al., 2018). Quercetin, a polyphenolic compound (flavonoid) found in apples, berries, green tea, and onions, has been shown to have a direct inhibitory effect on NF-kB, (Li et al., 2016) release of pro-inflammatory cytokines, COX-2, and lipid peroxidation (Rivera et al., 2008). Studies also found that quercetin enhances Th-1 (Nair et al., 2002) activity while decreasing the synthesis of Th-2 cytokines (Catanzaro and Corsini, 2018). Curcumin (*Curcuma longa*), a non-flavonoid, also exhibits functions similar to quercetin by (1) decreasing the synthesis of reactive oxygen species (ROS) and nitric oxide (NO), (2) releasing pro-inflammatory cytokines (Burge et al., 2019), and (3) down-regulating the expression of Toll-like receptors (TLRs) and endothelial adhesion molecules; with the abovementioned mechanism curcumin prevents the initiation of immune system-mediated diseases (Sellami et al., 2018).

1.3 ROLE OF HERBAL NUTRACEUTICALS IN INFLAMMATION

Inflammation may be defined as the responsiveness of the body to an injury (irritation), having common manifestations including pain, redness, swelling, and heat. Nutraceuticals have been found to be very effective in inflammation. Nutraceuticals obtained from soybean, ginger, glucosamine,

S-adenosylmethionine, and chondroitin have been proven beneficial in osteoarthritis. Micronutrients, including vitamin C and D, also come under the category of anti-inflammatory. Potent anti-inflammatory alkaloids and oxindole have also been extracted from cat's claw (Rafieian-Kopaei, 2014). Chondroitin and glucosamine are mostly employed as nutraceuticals in osteoarthritis because they control the nitrous oxide (NO) and prostaglandin E2 (PGE2) synthesis and regulate the gene expression involved in the pathophysiology of osteoarthritis (Chan et al., 2005). Pineapple contains a proteolytic enzyme, bromelain, that is effective against inflammation (Pawan et al., 2012). Many fruits such as *Vaccinium myrtillus, Vaccinium corymbosum, Vaccinium ashei,* and *Vaccinium angustifolium* contains resveratrol which exhibits functions similar to sirtuin (inhibition of COX-1 enzyme), inhibits prostaglandins, leukotrienes synthesis, and results in anti-inflammatory and antifungal activities (Jang and Cai, 1997). Nuts, green vegetables, and vegetable oils (e.g., seed oil) contain γ-linolenic acid in small amounts (Kaur et al., 2014); this acid is also synthesized inside the body, and after metabolism, γ-linolenic acid is converted into dihomo- γ-linolenic acid. Dihomo-γ-linolenic acid gets oxidized in the presence of lipooxygenase (LOX) and cyclooxygenase (COX) enzymes and produces eicosanoids which exhibit anti-inflammatory action (Wang et al., 2012).

1.4 ROLE OF HERBAL NUTRACEUTICALS IN ALLERGY

Allergy is an autoimmune body response to an allergen mediated by IgE antibodies. Generally considered as a type-I hypersensitivity reaction (Uzzaman and Cho, 2012), nutraceuticals have shown anti-allergic properties. Quercetin is an herbal nutraceutical that is found in various food that prevents the LDL-C-(low-density lipoprotein-C) mediated allergic reactions, and also possesses free radical scavenging activity, and acts as an antioxidant that also minimizes the chances of cardiovascular diseases by preventing vascular damage associated with LDL-C (Ho et al., 2014).

1.5 ROLE OF HERBAL NUTRACEUTICALS IN CARDIOVASCULAR DISEASES

Cardiovascular disorder is a broad term that includes many cardiac and blood vessels-associated diseases such as congestive heart failure, angina pectoris, hypertension, and peripheral vascular disease (Rafieian-Kopaei,

2012; Keservani et al., 2016a). Due to the rapid increase in the number of cardiovascular patients, research is also increasing in this field (Behradmanesh and Nasri, 2012; Nasri, 2012; Asgary et al., 2013). Generally, vegetables and fruit-rich diets are recommended for patients having any type of cardiovascular disease (Hu and Willett, 2002). Fruits contain flavonoids, such as flavones, flavonols, and flavanones (Khosravi-Boroujeni et al., 2013), that help to overcome cardiac complications. Flavonoids block the major enzyme involved in the pathophysiology of hypertension (angiotensin-converting enzyme), and they inhibit platelet aggregations increase blood flow as well as also possess COX (cyclooxygenase) enzyme inhibiting property, and prevent PGs (Prostaglandins) breakdown (Nasri et al., 2014; Asgary, 2013). Flavonoids are widely found in apples, black grapes, red wine, onions, cruciferous, cherries, and grapefruit (Kelishadi et al., 2013; Gharipour et al., 2013). A study shows that the consumption of flavonoids on a regular basis reduces the chances of myocardial infarction and coronary artery diseases in older persons (Rafieian-Kopaei et al., 2013). Citrus fruits (orange, lemon) contains flavonoid hesperidin, a flavanone glycoside, and this flavonoid is recommended in different types of vascular (especially associated with veins) abnormalities and hemorrhoids (Garg et al., 2001). Recently, the rhizomes of ginger (*Zingiber Officinalis*) have also been suggested for tachycardia and cardiac arrhythmias (Ghayur et al., 2005). Many green vegetables contain phytosterols that reduce the probability of cardiovascular diseases by interacting with dietary cholesterol, reducing its uptake, and increasing its excretion from the body (Gita, 2004; Khosravi-Boroujeni, 2012). Grains, fruits, and leaves contain octacosanol, which has a lipid-reducing quality (Sidhu, 2003; Heidarian et al., 2013). Other plant derivatives such as anthocyanins, proanthocyanidins, serotonin, stilbenes, tetrahydro-β-carbolines, and dietary indoleamines are also beneficial against various cardiovascular diseases (Iriti and Faoro, 2006).

1.6 ROLE OF HERBAL NUTRACEUTICALS IN CANCER

The World Cancer Report, 2020, reveals that about 15 million fresh cancer cases are registered, mostly in developing countries (Nasri, 2013). Healthy and nutritious food, vegetables, and fruits such as guava, grapefruit, watermelon, and tomatoes can reduce the probability of cancer (Stahl and Sies, 2005).

Lycopene, a carotenoid, possesses antioxidant as well as anti-cancer properties. It decreases cellular DNA damage. This information attracts the researcher's attention for further exploration (Willis and Wians, 2003). Lycopene is an unsaturated hydrocarbon (chemical formula $C_{40}H_{56}$) and quenches the single oxygen atom. It mainly accumulates in the skin, adrenal gland, testes, and prostate and protects against cancer (Willis and Wians, 2003; Shirzad et al., 2013). Daidzein from soybean (*Glycine max*), biochanin from red clover (*Trifolium pretense*), and genistein, like other isoflavones from different legumes species, are considered to inhibit the proliferation of prostate-cancerous cells (Kruger et al., 2002). Carotene is found in carrots, sweet potatoes, and apricots and is responsible for reddish-purple appearances. β-carotene possesses the highest antioxidant and anti-cancer properties, α-carotene exhibits about 51–55%, and ε-carotene results in 43–49% of total antioxidant activities (Stahl and Sies, 2005).

Fruits, vegetables, and herbs have high demand as anti-cancer and antimutagenics compounds because they contain chemoprotective constituents with no side effects. Various phytoconstituents exhibit hormonal properties (also known as phytoestrogens) and are effective against breast and prostate cancer (Limer and Speirs, 2004). Saponin, tannins, and pectin found abundantly in fruit and vegetables also have anti-proliferative, anti-cancer, and antimutagenic possessing (Li et al., 2003; Pizzi, 2021; Wikiera et al., 2014). Saponins extracted from soybean, spinach, tomatoes, clover, potatoes, and commercially utilized herbs like Yucca *schidigera* and Quillaja *saponaria* show anti-proliferative, anti-cancer, and antimutagenic potential. Tannins found in tea, grapes, cranberries, blue, and blackberries have free radical scavenging and detoxifying properties (Li et al., 2003). Pectin (Malus *Domestica*) have found to be effective against prostate cancer by reducing the blood cholesterol level and preventing the adhesion of cancerous cells with normal cells (Nasri et al., 2014). It has been proved that the intake of vitamin E, vitamin C, cysteine, lycopene, selenium, and glutathione-rich fruits and vegetables with antioxidant properties are beneficial for tumors (Rafieian-Kopaei and Nasri, 2013). A study has revealed that retinol, N-acetylcysteine, α-tocopherol, β-carotene, and retinyl palmitate do not show any cancer-protective effect in smokers (Brouns, 2002).

1.7 ROLE OF HERBAL NUTRACEUTICALS IN DIABETES

Various nutraceuticals of herbal origin have also shown remarkable antidiabetic properties. It is a well-known fact about diabetes that diabetes-insipidus is a kind of polyuria and causes fluid imbalance inside the body while diabetes mellitus is a type of disorder related to glucose metabolism and also affects lipid metabolism (Sharma et al., 2016; Sharma et al., 2015). It affects the affected person's physical and mental health (Sharma et al., 2016). Indian Council of Medical Research (ICMR), Council of Scientific and Industrial Research (CSIR), and Department of Biotechnology (DBT) have come to a step ahead in formulating herbal-based anti-diabetic formulation, which will not only help in maintaining blood glucose levels within the range but also provide nutrition to the affected patients (Raut et al., 2013; Chopra et al., 2011). The leaves of the horseradish tree (*Moringa oleifera*) have shown potent hypoglycemic properties in many clinical studies and has been proved as a better nutraceutical because they possess high calcium level (17 times more as compared to milk), iron (25 times more as compared to spinach), protein (9 times more as compared to yogurt), vitamins (10 times more as compared to carrot), and potassium (15 times more as compared to bananas) (Mbikay, 2012; Jaiswal et al., 2009; Kumari, 2010; Giridhari et al., 2011). *Emblica officinalis* (Amla), *Aloe barbadensis* (Aloe vera), *Tinospora cordifolia* (Giloe), *Momordica charantia* (Bitter gourd), and *Eugenia jambolana* (Blackberry or Jamun) are considered to be rich in minerals including Mn, Cr, Mg, Zn, and are beneficial for diabetic patients. *Emblica officinalis* and *Aloe barbadensis* consist of a high amount of vitamin C (ascorbic acid), which is believed to act as a good reducing agent (Virdi et al., 2003; Vikrant et al., 2001).

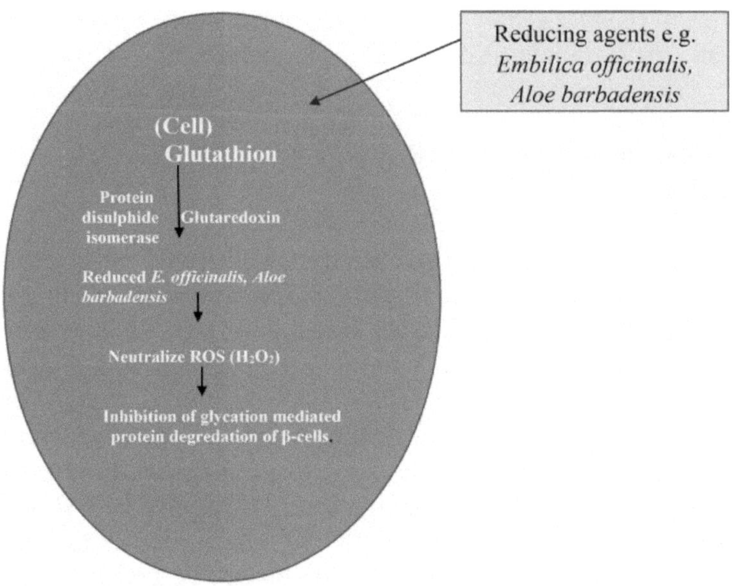

FIGURE 1.1 Role of reducing agents in β-cells protection

Chromium administration, along with vitamin B7 (biotin), enhances the release of insulin and decreases blood glucose level (9.7%) and HbA1C (0.54%) within 2 h of administration (Geohas et al., 2007). Chromium shows anti-diabetic activity because it (1) increases the total number of insulin receptors; (2) increases insulin binding with insulin receptors; and (3) regulate the phosphorylation or dephosphorylation mechanism of the insulin-releasing pathway by interacting with insulin receptors (tyrosine kinase) (Kota et al., 2013; Hua et al., 2012). Chromium has also been shown to have an inhibitory effect on tyrosine phosphatase, including in the down-regulation of the insulin receptor (Srivastava et al., 2012). Other micronutrients like magnesium also increase the insulin secretion from β-cells of pancreatic islets because Mg^{++} acts as a cofactor in the oxidation pathway of glucose and regulates the transportation of glucose to the cell membrane (Srivastava et al., 2012). α-Lipoic acid, an organosulfur obtained from natural sources (Salehi et al., 2019), is an antioxidant that decreases complications probabilities associated with diabetic neuropathy such as vascular diseases, cataracts, and polyneuropathies (Packer et al., 2001; Straface et al., 2002). Diabetes gives rise to various types of cardiovascular abnormalities and nutraceuticals such as ascorbic acid, omega-3 polyunsaturated fatty acid, vitamin E, L-arginine, and α-lipoic acid, which are believed to reduce such types of complications (Baldi, 2013). Extracts of many plants, including felty germander (*Teucrium polium*), bitter melon (*Momordica charantia*), and dalchini (*Cinnamomum zeylanicum*), proved to be effective in diabetes (Momeni, 2012; Rafieian-Kopaei, 2013; Kazemi et al., 2010).

1.8 ROLE OF HERBAL NUTRACEUTICALS IN HEPATIC DISORDERS

Many Phyllanthus genus species are effective in acute and chronic hepatic diseases. The extract obtained from *Phyllantus amarus* (Bhui Amla) was found to be effective against hepatitis B virus (HBV) (Thyagarajan et al., 1988) due to inhibition of DNA polymerase (hepatitis causing virus) (Lee et al., 1996); it also inhibits viral expression and cAMP depended on protein kinase in woodchuck (Polya et al., 1995). Alcoholic extract of *Silybum marianum* (milk thistle) seeds contains three isomers flavonolignans (known as silymarin), which include silybin, silychrstine, and silydianin (Luper, 1998). Silymarin decreases the generation of free radical species as well as lipid peroxidation (Pietrangelo et al., 1995), possesses anti-fibrotic property

(Boigk et al., 1997), and also prevents the toxins binding to cell-membrane receptors present in hepatocytes (Tuchweber et al., 1979; Faulstich et al., 1980). The root powder extract of Glycyrrhiza *glabra* consists of major glycyrrhizin, which is supposed to be effective in in-vivo hepatoprotective studies (Yoshikawa et al., 1997). It contains anti-inflammatory activities and inhibits tissue necrosis factor (TNF), and $CD4^{+}$ T-cell mediated cytotoxicity. An ayurvedic formulation consists of many herbs marketed under the name Liv-52. This preparation consists of *Tamarix gallica, Cassia occidentalis, Solanum nigrum, Cichorium intybus, Capparis spinosa, Terminalia arjuna,* and *Achillea millefolium,* and is used in viral hepatitis, liver cirrhosis, anorexia, and radiation/chemotherapy-mediated liver injury (Radha and Chawla, 2005).

1.9 ROLE OF HERBAL NUTRACEUTICALS IN NEUROLOGICAL DISORDERS

1.9.1 ALZHEIMER'S DISEASE

Alzheimer's (memory loss) is a neurodegenerative disorder (Nunomura et al., 2001) related to aging and caused by a low intake of antioxidants in the diet. Therefore, oxidative stress increases the disease progression (Nunomura et al., 1999). Studies revealed that intake of polyunsaturated, saturated, and trans-fatty acids-rich diets decreases neurodegeneration. Additionally, the intake of antioxidants slows the disease progression (Glevitzky et al., 2019). Flavonoids (Keservani and Sharma, 2014; Keservani et al., 2016b) are polyphenolic compounds, and they are supposed to interact with the neuronal-glial signaling pathway and improve neuronal function. Flavonoids are found in vegetables, fruits, and drinks (Glevitzky et al., 2019). For example, catechin, epicatechin, epigallocatechin, and epigallocatechin gallate are very effective in Alzheimer's diseases. It has been reported that Alzheimer's patients have a low percentage of major carotenoids such as β-carotene, lutein, and carotenoids and they are found in pigments (mainly deep yellow, orange, and red) of fruits and vegetables (Bungau et al., 2019). *Vaccinium oxycoccus* (Cranberries) and *Fragaria ananassa* (Strawberries) contain a major constituent cyanidin (anthocyanidins) which inhibits the pro-inflammatory cytokines and exhibits potent anti-inflammatory and neuroprotective properties (Sawmiller et al., 2014; Wang et al., 2016). Ginseng and other ginsenosides from *Panax ginseng* also exhibit neuroprotective properties (Li et al., 2021).

1.9.2 PARKINSONISM

Parkinson's is a neurodegenerative disorder related to the regeneration of dopaminergic neurons in substantia nigra, and subsequent decrease in dopamine (Mythri et al., 2015). *Camellia sinensis* (Tea) consist of epigallocatechin-3-gallate, a polyphenolic compound that can cross BBB (Smith et al., 2011) and show a beneficial effect on Parkinson's (Malar et al., 2020). Antiparkinsonism property of vincamine and its derivative vinpocetine (*Vinca rosea,* alkaloid) has been proven. Vincamine causes vasodilatation, dilates the neurons' capillaries, and increases the blood supply to the brain. Increase in generation of ATPs through a citric acid cycle results in increased glucose and nutrient supply to the brain (Navarro and Boveris, 2009; Lama et al., 2020). Vitamin E, creatine, and glutathione also proved beneficial in Parkinson's disease (Anwar et al., 2007).

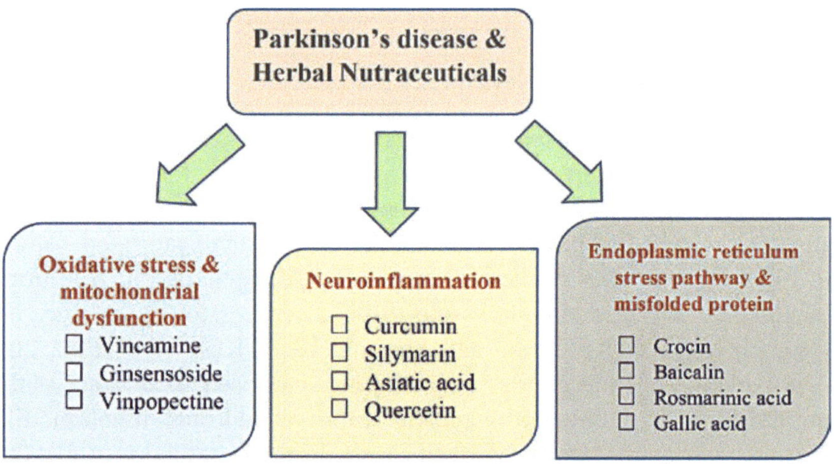

FIGURE 1.2 Three different mechanisms of herbal nutraceuticals in Parkinsonism.

1.9.3 DEPRESSION

Depression is a bipolar disorder characterized by a sad mood, anorexia, insomnia, and decreased physical activities, and it also affects social interaction (Ceskova and Silhan, 2018). Intake of macro- and micronutrient-rich diet including zinc, folic acid, and omega-3 fatty acid such as whole-grain improve brain functioning (Van der Burg et al., 2019).

1.9.4 PSYCHOTIC DISORDERS

Apart from exhibiting different functions in various parts of the nervous system, nutraceuticals also exhibit an important role in the management of psychotic disorders, including schizophrenia (Cloutier et al., 2016). Generally, omega-3 fatty acids and vitamins are recommended for antipsychotic therapy. The human body consists of two main polyunsaturated fatty acids, arachidonic acid (linoleic acid derivative) and eicosapentaenoic acid, and less important polyunsaturated fatty acids such as docosahexaenoic acid (Brown and Roffman, 2016). All these polyunsaturated fatty acids are essen-tial for the integrity of cell membranes made up of phospholipids and can be uptaken by diet because they are not synthesized in the body. They help maintain noradrenaline, dopamine, and 5HT availability on postsynaptic neurons and reduce their presynaptic availability. These polyunsaturated fatty acids also reduce oxidative stress-induced consequences in the brain (Sarris et al., 2015). It has been reported that nutrients along with antioxidant, anti-inflammatory anticholinesterase, and antihypercholesterolemic properties help improve the brain's cognitive functions (Preethi Pallavi et al., 2018). When effective nutraceuticals are used to manage neurodegenerative disor-ders (Keservani et al., 2016b), they can replace other synthetic drugs like rivastigmine, donepezil, tacrine, and galantamine (Lenaz, 2001). *Withania somnifera* (ashwagandha) possesses withanine, a steroidal alkaloid, which was reported to have memory and neurocognitive-enhancing activities (Farooqui et al., 2018).

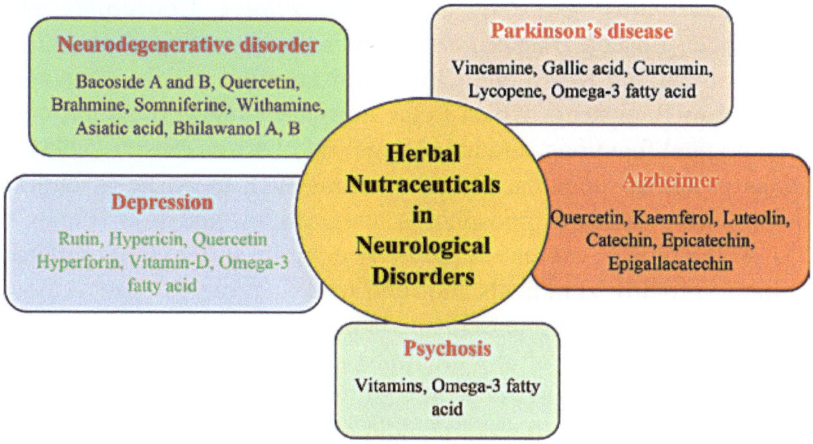

FIGURE 1.3 Herbal nutraceuticals for CNS disorders.

1.10 ROLE OF NUTRACEUTICALS IN OPHTHALMOLOGY

Synthetic ophthalmic preparations have several adverse effects. For this, the herbal preparations are marketed to minimize the risk associated with synthetic formulations. Several studies revealed that lutein and zeaxanthin are very beneficial for the eye. Fruits, vegetables such as carrots, tomatoes, green leafy vegetables (including collards), and sweet potatoes possess lutein. Zeaxanthin is another carotenoid pigment found in egg yolks, cabbage, beans, peas, kiwi, spinach, and corn. Both carotenoids have good ophthalmic properties and are employed against different types of visual disturbances (Abdel-Aal el-SM et al., 2013) and can be used to inhibit cataracts, eye-blinding problems, and muscular dystrophy (Khan et al., 2020). A study suggests that Tagetes *erecta* (marigold) contains about 86% of lutein and zeaxanthin (Kruger et al., 2002), and the Tagetes *erecta* is very beneficial in maintaining eye health (Zhang et al., 2020).

1.11 ROLE OF HERBAL NUTRACEUTICALS IN SPORTS

The utilization of herbal supplements is increasing daily for different purposes such as medicine, to increase body mass or physical strength (Singh et al., 2017a; Singh et al., 2017b). Athletes have used many herbs to improve physical strength (Sellami et al., 2018). Various herbal nutraceuticals are currently prescribed for enhancing physical activity, including Panax ginseng, Camilla sinensis, Coffea Arabica, Ephedra sinica, Zingiber officinale, Tribulus *terrestris,* and Ginko *biloba.* Concerning physical performance, ginseng (Panax *ginseng*) is a widely utilized plant in Asian countries, including Korea and China. Ginsenosides are the main chemical compound in P. ginseng, which helps normalize blood pressure, enhances immunological functions, and helps to get rid of stress (Bucci, 1993). It also contains proteins, saponins, fibers, vitamins such as A, B, C, and D, and minerals like potassium, phosphorus, magnesium, and iron (Popov et al., 1975). At the time of exercise, *P. ginseng* exhibits antioxidant properties via inhibition of hydroxyl radicals and lipid peroxidation and also assists mitochondrial activity (Zhong and Jiang, 1972). The root extract of *P. ginseng* is used to improve sexual abnormalities such as erectile dysfunctions and enhance libido in men (Lee et al., 2021). Unlike other supplements, *P. ginseng* has some side effects, including sleeplessness, headache, palpitation, blood pressure, and other abdominal disturbances. Vaginal bleeding and breast tenderness have been shown in females taking *P. ginseng*. Interactions of *P. ginseng* have also been observed with digoxin, insulin, anticoagulants, and monoamine oxidase inhibitors (MAO inhibitors). Generally, it is contraindicated in tachycardia (Siegel, 1980).

Caffeine is the most important chemical constituent (alkaloid) of *C. arabica* (Coffee). Immuno-stimulation effect of coffee beans on a cellular level has also been studied (Rafiul Haque et al., 2013; Bucci, 2000). It was revealed in a study that before 1 h of physical work or exercise, consumption of caffeine in a small amount (approximately 2–8 mg/kg of body weight) (Kovacs et al., 1998) enhances the serum level of catecholamine and also boosts immune functions (Senchina et al., 2014). High levels of catecholamine in blood improve aerobic as well as anaerobic functions. It was observed that consumption of 9 mg/kg (according to body weight) of caffeine before running (Graham et al., 1991) and swimming as well enhances the performance (Collomp et al., 1992). Due to its strong antioxidant properties, caffeine exhibits an ergogenic effect (increases muscle contraction, fat metabolism, and aerobic endurance) (Kamat et al., 2000). A high dose or prolonged utilization of caffeine supplements can exhibit severe consequences, including anxiety, tachycardia, insomnia, diarrhea, and dehydration (Lieberman et al., 2002).

Paullinia cupana, commonly known as Guarana, consists of chemical constituents such as caffeine, tannins, saponins, theophylline, and theobromine. It contains approximately two times more caffeine than *Coffea arabica* (Hsu, 2002). Due to a higher percentage of caffeine, P. cupana exhibits a stimulation effect on the central nervous system and increases body mass (Burke, 2008). The seeds of *P. cupana* have been employed against UTIs (urinary tract infections), diarrhea, headache, and paralysis. The interactions of *P. cupana* with other caffeine supplements, monoamine oxidase inhibitors, clozapine, paracetamol, adenosine, and lithium have also been seen (Siegel, 1980). Some major side effects of *P. cupana* are anorexia, insomnia, increased heart rate, and abdominal disturbances.

The supplements of green tea extracts have been shown to enhance endurance capability and increase plasma adrenergic and glycerol levels, antioxidant properties, and muscular lipid oxidation (Nakagawa et al., 1999). The *C. sinensis* (green tea) is increasing energy expenditure and free fatty acids utilization (Martin et al., 2018; Gahreman et al., 2016). The extract also decreases exercise-induced oxidative damage of DNA (Chacko et al., 2010).

Ephedrine is one of the most important alkaloids of the plant *E. sinica*. Several studies have been conducted on this plant which supports the major properties of *E. sinica,* including ergogenic property, decrease in body mass (Shekelle et al., 2003), and improved physical strength (Bell et al., 1998). Medicinally, during exercise, *E. sinica* enhances the reaction time, and also increases mental alertness and aerobic capacitance, and reduces fatigue (Powers, 2001). The combined use of ephedrine and caffeine in runners and athletes showed enhanced muscle performance, helps in weight loss, and exhibits mild adverse effects (Molnár et al., 2000). Hallucination, headache, insomnia, hypertension, and anorexia are the most common side effects associated with the consumption of ephedrine (Greenway, 2001; USFDA, 2014). Several chemical constituents are present in ginger. Some studies have also revealed that *Z. officinale* exhibits potent analgesic activity, and in athletes, gingerols have shown fatigue resistance. At the same time, some researchers also support that ginger does not affect athletes' metabolic rate, body composition, oxygen consumption, and muscular strength (Wilson, 2015).

Gokharu (*T. terrestris*) is a flowering plant found all around the world. Protodioscin and the saponins such as dioscine and diosgenin are the main chemical constituents of *T. terrestris*, which are supposed to improve physical performance and libido as well (Semerdjieva and Zheljazkov, 2019). In Atlanta, Summer Olympics, a Bulgarian athlete (1996), gave credit to *T. terrestris* for medal-winning. This plant attracts the population's attention to the utilization of *T. terrestris* in sports to improve physical fitness (Ivanova et al., 2016). The supplements of *T. terrestris* extract have also been shown to enhance the production of luteinizing hormone (LH) and increase the growth of muscles.

After considering all the beneficial effects of *T. terrestris* extract, it can be concluded that the extract can be used as a potent enhancer of performance by increasing testosterone and decreasing oxidative muscular damage and inflammation. It has also been seen that the extract of *T. terrestris* can be utilized as an alternative treatment for some types of diseases, including hypoactive sexual desire disorder (HSDD) in postmenopausal women and cardiovascular diseases (Zhu et al., 2017). It was also seen in athletes from throwing, sprints, and weight lifting disciplines that the use of *T. terrestris* enhances physical performance and muscular mass (Pokrywka et al., 2014). Despite all the positive effects in sportspersons, *T. terrestris* in high doses (more than 1000 mg/kg body weight) may result in hypertension, fatigue, insomnia, palpitation, and cardiac arrhythmia (Ryan et al., 2015; Qureshi et al., 2014).

Rhodiola *rosea* is commonly known as golden root or roseroot (Family-Crassulaceae), a traditionally used medicinal plant in Asia and Europe. The chemical constituents of *R. rosea* root are rosin, rosiridin, tyrosol, rosine rosarines, polyphenols minerals, vitamins, gallic acid, chiorogenic acid, and tannins. The ingestion of root extract of *R. rosea* is supposed to stimulate the production of serotonin, acetylcholine, noradrenaline, and dopamine. These four biomolecules directly stimulate the cerebral cortex, improve memory, increase concentration, attention, and physical performance, and enhance fatigue resistance (Cui et al., 2015). Endogenous steroids such as 4-androstene-3,17-dione, dehydro-epiandrosterone, and pseudoendogenous steroids including 1,4-androstadiene are present in the root and rhizome extract of *R. rosea* (Walpurgis et al., 2016). It was observed that the consumption of *R. rosea* extracts in a dose of 3 mg/kg slows down the heart rate to maximize the exercise and also decreases the perception of high effort during intense exercise in recreationally fit women (Noreen et al., 2013).

Codyceps *sinensis* is an entomopathogenic fungus, commonly known as caterpillar fungus (Family- Clavicipitaceae), which consists of stearic acid, amino acid, mycose, palmitic acid, D-mannitol, amino acid, cholesterol, and 5α-8α-epidioxy-5α-ergosta-6,22-dien-3β-ol (Zhang, 2011). Consumption of 100–500 mg/kg *C. sinensis* for 5 days raises the hemoglobin level and aerobic capacity (Li et al., 1993). The supplements of *C. sinensis* extract increase the production of lactic acid, which ultimately raise the chances of muscle fatigue, and increase heart rate and blood pressure (Nagata et al., 2006). Furthermore, a drink of *C. sinensis* (240 mg) improves the functioning of the cardiovascular system in runners (Nagata and Tajima, 2000). It was seen in well-trained athletes that the consumption of *C. sinensis* in combination with *Rhodiola crenulata* enhances aerobic performance (Chen et al., 2014). Abdominal disturbances and diarrhea are the most common adverse effect of *C. sinensis* associated with the higher dose.

Ginkgo *biloba* or maidenhair tree is a tree native to China and consists of biologically active flavonoids and terpenoids (Curtis-Prio et al., 1999). It was also seen that *G. Biloba* stimulates the production and release of relaxing factors from endothelium. This relaxing factor ameliorates microcirculation and increases muscle blood flow (Cott, 1995). It also enhances endurance performance (Schneider, 1992). The use of *G. Biloba* in combination with *R. Rosea* for 7 weeks enhances endurance performance in young and healthy athletes (Baker et al., 2014; Nash and Shah, 2015).

1.12 ROLE OF HERBAL NUTRACEUTICALS IN RENAL DISORDERS

A survey found that about 8–16% of the world population is affected by chronic kidney diseases (Nash and Shah, 2015; Coresh et al., 2007). Ultimately, the risk of kidney-associated cardiovascular disorders such as hypertension is also enhanced. Kidney Disease Outcomes Quality Initiative defines chronic kidney disease as the damage to the kidney confirmed by the structural or functional abnormalities associated with the kidney or glomerular filtration rate (GFR) <60 mL/min/1.73 m^2 for about 3 months (Nash and Shah, 2015).

Various valuable herbal compounds for nephrology have been found through in-vivo studies, for example, curcumin, resveratrol, catechins, epigallocatechin-3-gallate, polyunsaturated fatty acids, and linolenic acid. Curcumin is a polyphenolic compound found in rhizomes of *C. longa* (Keith et al., 2004). Curcumin exhibits several activities such as anti-inflammatory, anti-cancer, antiulcer, and antioxidant (Metzler et al., 2013). Apart from these activities, curcumin also proved beneficial in patients with chronic kidney diseases (Deng et al., 2018; Hernández-Reséndiz et al., 2015), acute kidney injury (Jiménez-Osorio et al., 2016), kidney transplantation (Fan et al., 2017), renal cells carcinoma in animal, as well as in human studies (Shoskes et al., 2005).

Resveratrol is a potent antioxidant polyphenol abundantly found in grapes and red wine (Singh et al., 2015). Resveratrol is effective against oxidative stress-induced cellular damage, inflammation, and diabetes (Kruse, 2019) and enhances the production of nitric oxide (Pandey and Rizvi, 2009) but diminishes the production of ROS mediated by nitric oxide. Resveratrol protects mesenchymal stem cells from degradation and apoptosis induced by TNF-β. Resveratrol also inhibits NF-kβ in chondrocytes and reduces the secretion of IL-6 (Singh et al., 2015). In human studies, it was found that resveratrol exhibits beneficiary effects on vascular functions in persons affected with chronic kidney diseases (Saldanha et al., 2013). In National Health and Nutrition Examination Survey (NHANES), it was found that the consumption of a small amount of wine by patients affected with chronic kidney disease reduced the probability of cardiovascular disorders associated with kidney disease (Jespersen et al., 2018).

Catechins are the most important polyphenolic (flavonoids) compounds of green tea which are potent antioxidant and anti-inflammatory compounds. Studies show that consumption of green tea or coffee in moderate amount by patients with renal impairment show a beneficial impact on renal functions due to a decrease in the production of ROS, NF- kβ translocation, intracellular adhesion molecule-1 (ICAM-1) expression, and activated protein-1, catechins exhibit pro-inflammatory, and pro-apoptotic oxidative stress (Chen et al., 2004). Human studies revealed that the daily consumption of 450 mg of catechins (from decaffeinated green tea) decreases plasma level of hypochlorous acid enhanced by hemodialysis, diminishes H_2O_2 activity before dialysis, slows down ROS production, decreases pro-inflammatory cytokines synthesis, and phosphatidylcholine H_2O_2 (Hsu et al., 2007).

Epigallocatechin-3-gallate is a compound obtained from green tea exhibit reno-protective potential in unilateral ureteral obstructive animal models (Wang et al., 2015). Caffeine is another important chemical compound found in significant amounts in coffee, which also exhibits a beneficial effect on kidney functions (Cosola et al., 2018). Consumption of one cup of coffee (caffeine) daily by a healthy volunteer slightly enhances the glomerulus filtration rate (GFR) but does not cause glomerulus hyperfiltration (Herber-Gast et al., 2016). Moreover, animal studies indicate some types of adverse effects due to consumption of caffeine in a higher amount, such as proteinuria, higher-renin concentration-induced hypertension, and another type of renal dysfunctions (Tofovic et al., 2007).

Plants and meat-derived polyunsaturated fatty acid (PUFA) ω-3PUFA, and marine-derived PUFA is ω-6PUFA.150 possess anti-inflammatory properties. PUFA protects from kidney damage (Calder and Grimble, 2002; Ajay et al., 2019). The human body cannot synthesize PUFA, but the supplementation of PUFA proved beneficial in the treatment of a chronic type of kidney disease (Aguila et al., 2005). Furthermore, the administration of ω-3PUFA reduces the chances of cardiovascular disorders occurring due to blood pressure, triglycerides, low-density lipoproteins (LDL), and high-density lipoproteins (HDL) (Aguila et al., 2005).

Butyrivibrio fibrisolvens, a fermented bacterium, synthesizes conjugated linolenic acid in grass-fed ruminants such as goats, sheep, and cattle (Ogborn, 2003). Chemically, conjugated linolenic acid is the mixture of linolenic acid isomers with conjugated double (=) bonds. Many plants such as catalpa, bitter ground, snake ground, pot marigold, trichosanthes, jacaranda, and tung seeds contain conjugated linolenic acid (Aguila et al., 2005). Anti-inflammatory, anti-diabetic, and anti-atherosclerotic properties of conjugated linolenic acid have been revealed in in-vivo studies due to the reduction in PGE2 renal synthesis; conjugated linolenic acid exhibits anti-fibrotic, and renal anti-inflammatory activity (Cicero et al., 2013).

1.13 CONCLUSION

Fruit and vegetables are good sources of nutraceuticals. The herbal nutraceuticals found in fruits, vegetables, and herbs show marvelous effects on various diseases. Herbal nutraceuticals play an important role in the prevention of various diseases and treatment as well. Herbal nutraceuticals contain micro- and macro-elements with flavonoids, saponins, fibers, alkaloids, and polyphenolic compounds such as resveratrol. Various studies exhibited the health benefits of nutraceuticals. Nutraceuticals are very effective in treating or managing various diseases because of their antioxidant and anti-inflammatory potential. From this study, we can conclude that fruits and herbal nutraceuticals must be included in the daily diet for the prevention and management of various diseases.

KEYWORDS

- **herbal immunomodulator**
- ***Curcuma longa***
- **flavonoids**
- **caffeine**
- **ginko biloba**
- **resveratrol**

REFERENCES

Aguila, M. B.; Pinheiro, A. R.; Aquino, J. C.; Gomes, A. P.; Mandarim-de-Lacerda, C. A. Different Edible Oil Beneficial Effects (Canola Oil, Fish Oil, Palm Oil, Olive Oil, and Soybean Oil) on Spontaneously Hypertensive Rat Glomerular Enlargement and Glomeruli Number. *Prostaglandins Other Lipid Mediat.* **2005,** *76*, 74–85.

Ajay, A. K.; Vig, S.; Sabbisetti, V. Mechanism of Action of Functional Lipids and Metabolites for Patients with Chronic Kidney Disease. *Funct. Foods Health Dis.* **2019,** *9*, 412–429.

Anwar, F.; Latif, S.; Ashraf, M.; Gilani, A. H. Moringa oleifera: A Food Plant with Multiple Medicinal Uses. *Phytother Res.* **2007,** *21*, 17–25.

Asgary, S.; Kelishadi, R.; Rafieian-Kopaei, M.; Najafi, S.; Najafi, M.; Sahebkar, A. Investigation of the Lipid-Modifying and Anti-inflammatory Effects of Cornus mas L. Supplementation on Dyslipidemic Children and Adolescents. *Pediatr. Cardiol.* **2013,** *34*, 1729–1735.

Asgary, S.; Keshvari, M.; Sahebkar, A.; Hashemi, M.; Rafieian-Kopaei, M. Clinical Investigation of the Acute Effects of Pomegranate Juice on Blood Pressure and Endothelial Function in Hypertensive Individuals. *ARYA Atheroscler.* **2013,** *9*, 326–331.

Baker, L. B.; Nuccio, R. P.; Jeukendrup, A. E. Acute Effects of Dietary Constituents on Motor Skill and Cognitive Performance in Athletes. *Nutr. Rev.* **2014,** *72*, 790–802.

Baradaran, A.; Madihi, Y.; Merrikhi, A.; Rafieian-Kopaei, M.; Nasri, H. Serum lipoprotein (a) in Diabetic Patients with Various Renal Function not yet on Dialysis. *Pak. J. Med. Sci.* **2013,** *29* (Suppl), 354–357.

Behradmanesh, S.; Nasri, P. Serum Cholesterol and LDL-C in Association with Level of Diastolic Blood Pressure in Type 2 Diabetic Patients. *J. Renal Inj. Prev.* **2012,** *1*, 23–26.

Bell, D. G.; Jacobs, I.; Zamecnik, J. Effects of Caffeine, Ephedrine and their Combination on Time to Exhaustion During High-Intensity Exercise. *Eur. J. Appl. Physiol.* **1998,** *77*, 427–433.

Boigk, G.; Stroedter, I.; Herbst, H.; Waldschmidt, J.; Riecken, E. O.; Schuppan, D. Silymarin Retards Collagen Accumulation in Early and Advanced Biliary Fibrosis Secondary to Complete Bile Duct Obliteration in Rats. *Hepatology* **1997,** *26*, 643–649.

Brouns, F. Soya Isoflavones: A New and Promising Ingredient for the Health Foods Sector. *Food Res. Int.* **2002,** *35*, 187–193.

Brower, V. Nutraceuticals: Poised for a Healthy Slice of the Healthcare Market? *Nat. Biotechnol.* **1998,** *16*, 728–731.

Brown, H. E.; Roffman, J. L. Emerging Treatments in Schizophrenia: Highlights from Recent Supplementation and Prevention Trials. *Harv. Rev. Psychiatry* **2016,** *24*, 1–7.

Bucci, L. In *Nutrients as Ergogenic Aids for Sports and Exercise*; Boca Raton: CRC Press, 1993.

Bucci, L. R. Selected Herbals and Human Exercise Performance. *Am. J. Clin. Nutr.* **2000,** *72* (2 Suppl), 624S–636S.

Bungau, S.; Abdel-Daim, M. M.; Tit, D. M.; Ghanem, E.; Sato, S.; Maruyama-Inoue, M.; Yamane, S.; Kadonosono, K. Health Benefits of Polyphenols and Carotenoids in Age-Related Eye Diseases. *Oxid. Med. Cell. Longev.* **2019**.

Burge, K.; Gunasekaran, A.; Eckert, J.; Chaaban, H. Curcumin and Intestinal Inflammatory Diseases: Molecular Mechanisms of Protection. *Int. J. Mol. Sci.* **2019,** *20*, 1912.

Burke, L. M. Caffeine and Sports Performance. *Appl. Physiol. Nutr. Metab.* **2008,** *33*, 1319–1334.

Calder, P. C.; Grimble, R. F. Polyunsaturated Fatty Acids, Inflammation and Immunity. *Eur. J. Clin. Nutr.* **2002,** *56*, S14–S19.

Catanzaro, M.; Corsini, E. Immunomodulators Inspired by Nature: A Review on Curcumin and Echinacea. *Molecules* **2018,** *23*, 2778.

Ceskova, E.; Silhan, P. Novel Treatment Options in Depression and Psychosis. *Neuropsychiatr. Dis. Treat.* **2018,** *14*, 741.

Chacko, S. M.; Thambi, P. T.; Kuttan, R.; Nishigaki, I. Beneficial Effects of Green Tea: A Literature Review. *Chin. Med.* **2010,** *5*, 13.

Chen, W. C.; Hayakawa, S.; Shimizu, K.; Chien, C. T.; Lai, M. K. Catechins Prevents Substance P-Induced Hyperactive Bladder in Rats Via the Down Regulation of ICAM and ROS. *Neurosci. Lett.* **2004,** *367*, 213–217.

Chen, C. Y.; Hou, C. W.; Bernard, J. R.; Chen, C. C.; Hung, T. C.; Cheng, L. L.; et al. Rhodiola crenulata- and Cordyceps sinensis-Based Supplement Boosts Aerobic Exercise Performance After Short-Term High Altitude Training. *High Alt. Med. Biol.* **2014,** 15, 371–379.

Chopra, A.; Saluja, M.; Tillu, G.; Venugopalan, A.; Sarmukaddam, S.; Raut, A. K.; et al. A Randomized Controlled Exploratory Evaluation of Standardized Ayurvedic Formulations in Symptomatic Osteoarthritis Knees: A Government of India NMITLI Project. *Evid. Based Complement. Alternat. Med.* **2011,** *2011*, 724291.

Cicero, A. Effect of a Lipid-Lowering Nutraceutical on Pulse-Wave-Velocity in Hypercholesterolemic Patients With or Without Chronic Kidney Disease. *Open Hypertens. J.* **2013,** *5*, 18–22.

Cloutier, M.; Aigbogun, M. S.; Guerin, A.; Nitulescu, R.; Ramanakumar, A. V.; Kamat, S. A.; DeLucia, M.; Duty, R.; Legacy, S. N.; Henderson, C. The Economic Burden of Schizophrenia in the United States in 2013. *J. Clin. Psychiatry* **2016,** *77*, 764–771.

Collomp, K.; Ahmaidi, S.; Chatard, J. C.; Audran, M.; Préfaut, C. Benefits of Caffeine Ingestion on Sprint Performance in Trained and Untrained Swimmers. *Eur. J. Appl. Physiol. Occup. Physiol.* **1992,** *64*, 377–380.

Coresh, J.; Selvin, E.; Stevens, L. A.; Manzi, J.; Kusek, J. W.; Eggers, P.; Van Lente, F.; Levey, A. S. Prevalence of Chronic Kidney Disease in the United States. *JAMA* **2007,** *298*, 2038–2047.

Cosola, C.; Sabatino, A.; di Bari, I.; Fiaccadori, E.; Gesualdo, L. Nutrients, Nutraceuticals, and Xenobiotics Affecting Renal Health. *Nutrients* **2018,** *10*.

Cott, J. NCDEU Update. Natural Product Formulations Available in Europe for Psychotropic Indications. *Psychopharmacol. Bull.* **1995,** *31*, 745–751.

Cui, J. L.; Guo, T. T.; Ren, Z. X.; Zhang, N. S.; Wang, M. L. Diversity and Antioxidant Activity of Culturable Endophytic Fungi from Alpine Plants of Rhodiola Crenulata, R. Angusta, and R. Sachalinensis. *PLoS One* **2015,** *10*, e0118204.

Curtis-Prior, P.; Vere, D.; Fray, P. Therapeutic Value of Ginkgo Biloba in Reducing Symptoms of Decline in Mental Function. *J. Pharm. Pharmacol.* **1999,** *51*, 535–541.

Dankers, W.; Colin, E. M.; van Hamburg, J. P.; Lubberts, E. Vitamin D in Autoimmunity: Molecular Mechanisms and Therapeutic Potential. *Front. Immunol.* **2017,** *7*, 697.

Deng, Q.; Liang, L.; Liu, Q.; Duan, W.; Jiang, Y.; Zhang, L. Autophagy is a Major Mechanism for the Dual Effects of Curcumin on Renal Cell Carcinoma Cells. *Eur. J. Pharmacol.* **2018,** *826*, 24–30.

Fan, Y.; Chen, H.; Peng, H.; Huang, F.; Zhong, J.; Zhou, J. Molecular Mechanisms of Curcumin Renoprotection in Experimental Acute Renal Injury. *Front. Pharmacol.* **2017,** *8*, 912.

Faulstich, H.; Jahn, W.; Wieland, T. Silybin Inhibition of Amatoxin Uptake in the perfused rat liver. *Arzneimittelforschung* **1980,** *30*, 452–454.

FOOD US. Drug Administration Code of Federal Regulations Title 21. Department of Health and Human Services, ed. 21CFR20157; US Food and Drug Administration: Washington, 2014.

Gahreman, D. E.; Boutcher, Y. N.; Bustamante, S.; Boutcher, S. H. The Combined Effect of Green Tea and Acute Interval Sprinting Exercise on Fat Oxidation of Trained and Untrained Males. *J. Exerc. Nutr. Biochem.* **2016,** *20*,1–8.

Garg, A.; Garg, S.; Zaneveld, L. J.; Singla, A. K. Chemistry and Pharmacology of the Citrus Bioflavonoid Hesperidin. *Phytother. Res.* **2001,** *15*, 655–669.

Geohas, J.; Daly, A.; Juturu, V.; Finch, M.; Komorowski, J. R. Chromium Picolinate and Biotin Combination Reduces Atherogenic Index of Plasma in Patients With Type Two Diabetes Mellitus: A Placebo-Controlled, Double Blinded, Randomized Clinical Trial. *Am. J. Med. Sci.* **2007,** *333*, 145–153.

Gharipour, M.; Ramezani, M. A.; Sadeghi, M.; Khosravi, A.; Masjedi, M.; Khosravi-Boroujeni, H.; et al. Sex Based Levels of C-Reactive Protein and White Blood Cell Count in Subjects with Metabolic Syndrome: Isfahan Healthy Heart Program. *J. Res. Med. Sci.* **2013,** *18,* 467–472.

Ghayur, M. N.; Gilani, A. H.; Afridi, M. B.; Houghton, P. J. Cardiovascular Effects of Ginger Aqueous Extract and its Phenolic Constituents are Mediated Through Multiple Pathways. *Vascul. Pharmacol.* **2005,** *43,* 234–241.

Giridhari, V. V.; Malhati, D.; Geetha, K. Anti-Diabetic Properties of Drumstick (Moringa oleifera) Leaf Tablets. *Int. J. Health Nutr.* **2011,** *2,* 1–5.

Gita, C. Functional Food Attributes of n-3 Polyunsaturated and Conjugated Linoleic Acid Enriched Chicken Eggs. *Curr. Top. Nutraceutical. Res.* **2004,** *2,* 113–121.

Glevitzky, I.; Dumitrel, G. A.; Glevitzky, M.; Pasca, B.; Otrisal, P.; Bungau, S.; Cioca, G.; Pantis, C.; Popa, M. Statistical Analysis of the Relationship Between Antioxidant Activity and the Structure of Flavonoid Compounds. *Rev. Chim.* **2019,** *70,* 3103–3107.

Graham, T. E.; Spriet, L. L. Performance and Metabolic Responses to a High Caffeine Dose During Prolonged Exercise. *J. Appl. Physiol.* **1991,** *71,* 2292–2298.

Greenway, F. L. The Safety and Efficacy of Pharmaceutical and Herbal Caffeine and Ephedrine use as a Weight Loss Agent. *Obes Rev.* **2001,** *2,* 199–211.

Hardy, G. Neurtraceuticals and Functional Foods: Introduction and Meaning. *Nutrition* **2000,** *16,* 688–689.

Heidarian, E.; Rafieian-Kopaei, M.; Ashrafi, K. The Effect of Hydroalcoholic Extract of Allium Latifolium on the Liver Phosphatidate Phosphatase and Serum Lipid Profile in Hyperlipidemic Rats. *J. Babol. Univ. Med. Sci.* **2013,** *15,* 37–46.

Herber-Gast, G. C.; Van Essen, H.; Verschuren, W. M.; Stehouwer, C. D.; Gansevoort, R. T.; Bakker, S. J.; Spijkerman, A. M. Coffee and Tea Consumption in Relation to Estimated Glomerular Filtration Rate: Results from the Population-Based Longitudinal Doetinchem Cohort Study. *Am. J. Clin. Nutr.* **2016,** *103,* 1370–1377.

Hernández-Reséndiz, S.; Correa, F.; García-Niño, W. R.; Buelna-Chontal, M.; Roldán, F. J.; Ramírez-Camacho, I.; Delgado-Toral, C.; Carbó, R.; Pedraza-Chaverrí, J.; Tapia, E. Cardioprotection by Curcumin Post-Treatment in Rats with Established Chronic Kidney Disease. *Cardiovasc. Drugs Ther.* **2015,** *29,* 111–120.

Ho, M. H.; Wong, W. H.; Chang, C. Clinical Spectrum of Food Allergies: A Comprehensive Review. *Clin. Rev. Allergy Immunol.* **2014,** *46* (3), 225–240.

Hsu, P. P. Natural Medicines Comprehensive Database. *J. Med. Lib. Assoc.* **2002,** *90* (1), 114.

Hsu, S. P.; Wu, M. S.; Yang, C. C.; Huang, K. C.; Liou, S. Y.; Hsu, S. M.; Chien, C. T. Chronic Green Tea Extract Supplementation Reduces Hemodialysis-Enhanced Production of Hydrogen Peroxide and Hypochlorous Acid, Atherosclerotic Factors, and Pro-inflammatory Cytokines. *Am. J. Clin. Nutr.* **2007,** *86,* 1539–1547.

Hu, F. B.; Willett, W. C. Optimal Diets for Prevention of Coronary Heart Disease. *JAMA* **2002,** *288,* 2569–2578.

Iriti, M.; Faoro, F. Grape Phytochemicals: A Bouquet of Old and New Nutraceuticals for Human Health. *Med. Hypotheses.* **2006,** *67,* 833–838.

Ivanova, S.; Ivanov, K.; Mladenov, R.; Papanov, S.; Ivanova, S.; Obreshkova, D.; Atanasov, P. P. V. Food Supplements with Anabolic and Androgenic Activity UHPLC Analysis of Food Additives, Containing Tribulusterrestris Extract. *World J. Pharma. Res.* **2016,** *5,* 6–13.

Jaiswal, D.; Kumar Rai, P.; Kumar, A.; Mehta, S.; Watal, G. Effect of Moringa oleifera Lam. Leaves Aqueous Extract Therapy on Hyperglycemic Rats. *J. Ethnopharmacol.* **2009,** *123*, 392–396.

Jang, M.; Cai, L.; Udeani, G. O.; Slowing, K. V.; Thomas, C. F.; Beecher, C. W.; et al. Cancer Chemopreventive Activity of Resveratrol, A Natural Product Derived from Grapes. *Science* **1997,** *275*, 218–220.

Jardim, F. R.; de Rossi, F. T.; Nascimento, M. X.; da Silva Barros, R. G.; Borges, P. A.; Prescilio, I. C.; De Oliveira, M. R. Resveratrol and Brain Mitochondria: A Review. *Mol. Neurobiol.* **2018,** *55*, 2085–2101.

Jespersen, T.; Kruse, N.; Mehta, T.; Kuwabara, M.; Noureddine, L.; Jalal, D. Light Wine Consumption is Associated with a Lower Odd for Cardiovascular Disease in Chronic Kidney Disease. *Nutr. Metab. Cardiovasc. Dis.* **2018,** *28*, 1133–1139.

Jiménez-Osorio, A. S.; García-Niño, W. R.; González-Reyes, S.; Álvarez-Mejía, A. E.; Guerra-León, S.; Salazar-Segovia, J.; Falcón, I.; Montes de Oca-Solano, H.; Madero, M.; Pedraza-Chaverri, J. The Effect of Dietary Supplementation with Curcumin on Redox Status and nrf2 Activation in Patients with Nondiabetic or Diabetic Proteinuric Chronic Kidney Disease: A Pilot Study. *J. Ren. Nutr.* **2016,** *26*, 237–244.

Kalra, E. K. Nutraceutical – Definition and Introduction. *AAPS Pharm Sci.* **2003,** *5*, E25.

Kamat, J. P.; Boloor, K. K.; Devasagayam, T. P.; Jayashree, B.; Kesavan, P. C. Differential Modification by Caffeine of Oxygen-Dependent and Independent Effects of Gamma-Irradiation on Rat Liver Mitochondria. *Int. J. Radiat. Biol.* **2000,** *76*,1281–1288.

Kaur, N.; Chugh, V.; Gupta, A. K. Essential Fatty Acids as Functional Components of Foods- A Review. *J. Food Sci. Technol.* **2014,** *51* (10), 2289–2303. DOI: 10.1007/s13197-012-0677-0

Kavas, G. O.; Ayral, P. A.; Elhan, A. H. The Effects of Resveratrol on Oxidant/Antioxidant Systems and their Cofactors in Rats. *Adv. Clin. Exp. Med.* **2013,** *22*,151–155.

Kazemi, S.; Asgary, S.; Moshtaghian, J.; Rafieian, M.; Adelnia, A.; Shamsi, F. Liver-Protective Effects of Hydroalcoholic Extract of *Allium hirtifolium* boiss. In Rats with Alloxan-Induced Diabetes Mellitus. *ARYA Atheroscler.* **2010,** *6*, 11–15.

Keith, D. S.; Nichols, G. A.; Gullion, C. M.; Brown, J. B.; Smith, D. H. Longitudinal Follow-up and Outcomes Among a Population with Chronic Kidney Disease in a Large Managed Care Organization. *Arch. Intern. Med.* **2004,** *164*, 659–663.

Keservani, R. K.; Kesharwani, R. K.; Sharma, A. K.; Vyas, N.; Chadoker, A. Nutritional Supplements: An Overview. *Int. J. Curr. Pharm. Rev. Res.* **2010b,** *1* (1), 59–75.

Keservani, R. K.; Kesharwani, R. K.; Sharma, A. K.; Jarouliya, U. Dietary Supplements, Nutraceutical and Functional Foods in Immune Response (Immunomodulators), Chapter 20. In *Nutraceutical and Functional Foods in Human Life and Disease Prevention*; Bagchi, D., Preuss, H. G., Swaroop, A., Ed.; CRC Press, Taylor and Francis: Boca Raton, FL, 2015; pp 343–358. ISBN: 9781482237214.

Keservani, R. K.; Kesharwani, R. K.; Vyas, N.; Jain, S.; Raghuvanshi, R.; Sharma, A. K. Nutraceutical and Functional Food as Future Food: A Review. *Der Pharmacia Lettre* **2010a,** *2* (1), 106–116.

Keservani, R. K.; Sharma, A. K. Flavonoids: Emerging Trends and Potential Health Benefits. *J. Chin. Pharm. Sci.* **2014,** *23* (12), 815.

Keservani, R. K.; Sharma, A. K.; Kesharwani, R. K. Nutraceutical and Functional Foods for Cardiovascular Health. In *Food Process Engineering*, pp 291–312; Apple Academic Press, CRC Press, 2016a; pp 257–278. ISBN: 9781771884020.

Keservani, R. K.; Sharma, A. K.; Kesharwani, R. K. Medicinal Effect of Nutraceutical Fruits for the Cognition and Brain Health. *Scientifica* **2016b,** *3109254,* 10. DOI: 10.1155/2016/3109254.

Keservani, R. K.; Sharma, A. K.; Kesharwani, R. K. An Overview and Therapeutic Applications of Nutraceutical and Functional Foods. In *Recent Advances in Drug Delivery Technology*, 2017; pp 160–201. ISBN: 9781522507543.

Keservani, R. K.; Sharma, A. K.; Kesharwani, R. K. Eds.; In *Nutraceuticals and Dietary Supplements: Applications in Health Improvement and Disease Management*; CRC Press, 2020. ISBN: 9781771888738.

Khosravi-Boroujeni, H.; Mohammadifard, N.; Sarrafzadegan, N.; Sajjadi, F.; Maghroun, M.; Khosravi, A.; et al. Potato Consumption and Cardiovascular Disease Risk Factors Among Iranian Population. *Int. J. Food Sci. Nutr.* **2012,** *63*, 913–920.

Khosravi-Boroujeni, H.; Sarrafzadegan, N.; Mohammadifard, N.; Sajjadi, F.; Maghroun, M.; Asgari, S.; et al. White Rice Consumption and CVD Risk Factors Among Iranian Population. *J. Health Popul. Nutr.* **2013,** *31*, 252–261.

Kidney Disease: Improving Global Outcomes (KDIGO) CKD-MBD Update Work Group. KDIGO 2017 Clinical Practice Guideline Update for the Diagnosis, Evaluation, Prevention and Treatment of Chronic Kidney Disease-Mineral and Bone Disorder (CKD-MBD). *Kidney Int. Suppl.* **2017,** *7*, 1–59.

Kota, S. K.; Jammula, S.; Kota, S. K.; Krishna, S. V.; Meher, L. K.; et al. Nutraceuticals in Dyslipidemia Management. *J. Med. Nutr. Nutraceut.* **2013,** *2*, 26–40.

Kovacs, E. M.; Stegen, J. H. C. H.; Brouns, F. Effect of Caffeinated Drinks on Substrate Metabolism, Caffeine Excretion, and Performance. *J. Appl. Physiol.* **1998,** *85*, 709–715.

Kruger, C. L.; Murphy, M.; DeFreitas, Z.; Pfannkuch, F.; Heimbach, J. An Innovative Approach to the Determination of Safety for a Dietary Ingredient Derived from a New Source: Case Study Using a Crystalline Lutein Product. *Food Chem. Toxicol.* **2002,** *40*, 1535–1549.

Kruse, N. T. Nutraceuticals as a Potential Adjunct Therapy Toward Improving Vascular Health in CKD. *Am. J. Physiol. Regul. Integr. Comp. Physiol.* **2019,** *317*, 719–732.

Kumari, D. J. Hypoglycemic Effect of Moringa oleifera and Azadirachta Indica in Type-2 Diabetes. *Bioscan* **2010,** *5*, 211–214.

Lama, A.; Pirozzi, C.; Avagliano, C.; Annunziata, C.; Mollica, M. P.; Calignano, A.; Meli, R.; Mattace Raso, G. Nutraceuticals: An Integrative Approach to Starve Parkinson's Disease. *Brain Behav. Immun. Health* **2020,** *2*, 100037.

Lee, H. W.; Lee, M. S.; Kim, T. H.; et al. Ginseng for Erectile Dysfunction. *Cochrane Database Syst. Rev.* **2021,** *4* (4), CD012654. Published Apr 19, 2021. DOI: 10.1002/14651858. CD012654.pub2

Lee, C. D.; Ott, M.; Thyagarajan, S. P.; Sharfritz, D. A.; Burk, R. D.; Gupta, S. Phyllanthus Amarus Down Regulates Hepatitis B Virus mRNA Transcription and Replication. *Eur. J. Clin. Invest.* **1996,** *24*,161–168.

Lenaz, G. The Mitochondrial Production of Reactive Oxygen Species: Mechanisms and Implications in Human Pathology. *IUBMB Life* **2001,** *52*, 159–164.

Li, Y.; Chen, G. Z.; Jiang, D. Z. Effect of Cordyceps Sinensis on Erythropoiesis in Mouse Bone Marrow. *Chin. Med. J.* **1993,** *106*, 313–316.

Li, J.; Huang, Q.; Chen, J.; et al. Neuroprotective Potentials of Panax Ginseng Against Alzheimer's Disease: A Review of Preclinical and Clinical Evidence. *Front. Pharmacol.* **2021,** *12*. https://www.frontiersin.org/article/10.3389/fphar.2021.688490

Li, H.; Wang, Z.; Liu, Y. Review in the Studies on Tannins Activity of Cancer Prevention and Anti-Cancer. *Zhong Yao Cai.* **2003,** *26*, 444–448.

Li, Y.; Yao, J.; Han, C.; Yang, J.; Chaudhry, M. T.; Wang, S.; Liu, H.; Yin, Y. Quercetin, Inflammation and Immunity. *Nutrients* **2016,** *8*, 167.

Lieberman, H. R.; Tharion, W. J.; Shukitt-Hale, B.; Speckman, K. L.; Tulley, R. Effects of Caffeine, Sleep Loss, and Stress on Cognitive Performance and Mood During U.S. Navy SEAL Training. Sea-air-land. *Psychopharmacology* **2002,** *164*, 250–261.

Limer, J. L.; Speirs, V. Phyto-oestrogens and Breast Cancer Chemoprevention. *Breast Cancer Res.* **2004,** *6*, 119–127.

Luper, S. A Review of Plants Used in the Treatment of Liver Disease: Part 1. *Altern. Med. Rev.* **1998,** *3*, 410–421.

Madihi, Y.; Merrikhi, A.; Baradaran, A.; Rafieian-kopaei, M.; Shahinfard, N.; Ansari, R.; et al. Impact of Sumac on Postprandial High-Fat Oxidative Stress. *Pak. J. Med. Sci.* **2013,** *29*, 340–345.

Malaguarnera, L. Influence of Resveratrol on the Immune Response. *Nutrients* **2019,** *11*, 946.

Marino, R.; Misra, M. Extra-Skeletal Effects of Vitamin D. *Nutrients* **2019,** *11*, 1460.

Martin, B. J.; MacInnis, M. J.; Gillen, J. B.; Skelly, L. E.; Gibala, M. J. Short-Term Green Tea Extract Supplementation Attenuates the Postprandial Blood Glucose and Insulin Response Following Exercise in Overweight Men. *Appl. Physiol. Nutr. Metab.* **2016,** *41*, 1–7.

Mbikay, M. Therapeutic Potential of Moringa oleifera Leaves in Chronic Hyperglycemia and Dyslipidemia: A Review. *Front Pharmacol.* **2012,** *3*, 1–12.

Medrano, M.; Carrillo-Cruz, E.; Montero, I.; Perez-Simon, J. A. Vitamin D: Effect on Haematopoiesis and Immune System and Clinical Applications. *Int. J. Mol. Sci.* **2018,** *19*, 2663.

Metzler, M.; Pfeiffer, E.; Schulz, S. I.; Dempe, J. S. Curcumin Uptake and Metabolism. *Biofactors* **2013,** *39*, 14–20.

Molnár, D.; Török, K.; Erhardt, E.; Jeges, S. Safety and Efficacy of Treatment With an Ephedrine/Caffeine Mixture. The First Double-Blind Placebo-Controlled Pilot Study in Adolescents. *Int. J. Obes. Relat. Metab. Disord.* **2000,** *24*, 1573–1578.

Momeni, A. Serum Uric Acid and Diabetic Nephropathy. *J. Renal Inj. Prev.* **2012,** *1*, 37–38.

Mythri, R.; Kumar, A.; Mms, B. Nutraceuticals and Other Natural Products in Parkinson's Disease Therapy: Focus on Clinical Applications in Bioactive Nutraceuticals and Dietary Supplements in Neurological and Brain Disease; Academic Press: Cambridge, MA, USA, 2015; pp 421–431.

Nagata, A.; Tajima, T. Anti-Fatigue Effectiveness of Cordyceps Sinensis Extract by the Double-Blind Method. *Hiro to Kyuyo no Kagaku.* **2000,** *17*, 89–97.

Nagata, A.; Tajima, T.; Uchida, M. Supplemental Anti-Fatigue Effects of Cordyceps sinensis (Tochu-Kaso) Extract Powder During Three Stepwise Exercise of Human. *Japanese J. Phys. Fit Sport Med.* **2006,** *55* Supplement, S145–S152.

Nair, M. P.; Kandaswami, C.; Mahajan, S.; Chadha, K. C.; Chawda, R.; Nair, H.; Kumar, N.; Nair, R. E.; Schwartz, S. A. The Flavonoid, Quercetin, Differentially Regulates Th-1 (IFNgamma) and Th-2 (IL4) Cytokine Gene Expression by Normal Peripheral Blood Mononuclear Cells. *Biochim. Biophys. Acta* **2002,** *1593*, 29–36.

Nakagawa, K.; Ninomiya, M.; Okubo, T.; Aoi, N.; Juneja, L. R.; Kim, M.; et al. Tea Catechin Supplementation Increases Antioxidant Capacity and Prevents Phospholipid Hydroperoxidation in Plasma of Humans. *J. Agric. Food Chem.* **1999,** *47*, 3967–3973.

Nash, K. M.; Shah, Z. A. Current Perspectives on the Beneficial Role of Ginkgo biloba in Neurological and Cerebrovascular Disorders. *Integr. Med. Insights* **2015,** *10*, 1–9.
Nasri, H. Comment on: Serum Cholesterol and LDL-C in Association with Level of Diastolic Blood Pressure in Type 2 Diabetic Patients. *J. Renal Inj. Prev.* **2012,** *1*, 13–14.
Nasri, H. Impact of Diabetes Mellitus on Parathyroid Hormone in Hemodialysis Patients. *J. Parathyr. Dis.* **2013,** *1*, 9–11.
Nasri, H.; Ardalan, M. R.; Rafieian-Kopaei, R. On the Occasion of World Hypertension Day 2014. *J. Parathyr. Dis.* **2014,** *2*, 19–20.
Nasri, H.; Baradaran, A.; Shirzad, H.; Rafieian-Kopaei, M. New Concepts in Nutraceuticals as Alternative for Pharmaceuticals. *Int. J. Prev. Med.* **2014,** *5* (12), 1487–1499.
Nasri, H.; Motamedi, P.; Dehghani, N.; Nasri, P.; Taheri, Z.; Kinani, F.; et al. Vitamin D and Immune System. *J. Renal Endocrinol.* **2014,** *1*, 5–7.
Nasri, H.; Rafieian-Kopaei, M. Herbal Medicine and Diabetic Kidney Disease. *J. Nephropharmacol.* **2013,** *2*, 1–2.
Navarro, A.; Boveris, A. Brain Mitochondrial Dysfunction and Oxidative Damage in Parkinson's Disease. *J. Bioenerg. Biomembr.* **2009,** *41*, 517–521.
Noreen, E. E.; Buckley, J. G.; Lewis, S. L.; Brandauer, J.; Stuempfle, K. J. The Effects of an Acute Dose of Rhodiola rosea on Endurance Exercise Performance. *J. Strength Cond. Res.* **2013,** *27*, 839–847.
Nunomura, A.; Perry, G.; Aliev, G.; Hirai, K.; Takeda, A.; Balraj, E. K.; Jones, P. K.; Ghanbari, H.; Wataya, T.; Shimohama, S.; et al. Oxidative Damage is the Earliest Event in Alzheimer Disease. *J. Neuropathol. Exp. Neurol.* **2001,** *60*, 759–767.
Nunomura, A.; Perry, G.; Pappolla, M. A.; Wade, R.; Hirai, K.; Chiba, S.; Smith, M. A. RNA Oxidation is a Prominent Feature of Vulnerable Neurons in Alzheimer's Disease. *J. Neurosci.* **1999,** *19*, 1959–1964.
Ogborn, M. R.; Nitschmann, E.; Bankovic-Calic, N.; Weiler, H. A.; Fitzpatrick-Wong, S.; Aukema, H. M. Dietary Conjugated Linoleic Acid Reduces PGE2 Release and Interstitial Injury in Rat Polycystic Kidney Disease. *Kid. Int.* **2003,** *64*, 1214–1221.
Packer, L.; Kraemer, K.; Rimbach, G. Molecular Aspects of Alpha Lipoic Acid in the Prevention of Diabetes Complications. *Nutrition* **2001,** *17*, 888–895.
Pandey, K. B.; Rizvi, S. I. Plant Polyphenols as Dietary Antioxidants in Human Health and Disease. *Oxid. Med. Cell. Longev.* **2009,** *2*, 270–278.
Pavan, R.; Jain, S.; Shraddha; Kumar A. Properties and Therapeutic Application of Bromelain: A Review. *Biotechnol. Res. Int.* **2012,** *2012*, 976203. DOI: 10.1155/2012/976203
Pietrangelo, A.; Borrella, F.; Casalgraudi, G.; et al. Antioxidantactivity of Slilybin In Vivo During Long-Term Iron Overload in Rats. *Gastroenterology* **1995,** *109*, 1941–1949.
Pokrywka, A.; Obmiński, Z.; Malczewska-Lenczowska, J.; Fijatek, Z.; Turek-Lepa, E.; Grucza, R. Insights into Supplements with Tribulus Terrestris used by Athletes. *J. Hum. Kinet.* **2014,** *41*, 99–105.
Polya, G. M.; Wang, B. H.; Foo, L. Y. Inhibition of Signal Regulated Protein Kinases by Plant Derived Hydrolyzable Tannins. *Phytochemistry* **1995,** *38*, 307–314.
Popov, I. M.; Goldwag, W. J. A Review of the Properties and Clinical Effects of Ginseng. *Am. J. Chin. Med.* **1973,** *1*, 263–270.
Powers, M. E. Ephedra and its Application to Sport Performance: Another Concern for the Athletic Trainer? *J. Athl. Train.* **2001,** *36*, 420–424.
Preethi Pallavi, M. C.; Sampath Kumar, H. M. Nutraceuticals in Prophylaxis and Therapy of Neurodegenerative Diseases. In *Discovery and Development of Neuroprotective Agents*

from Natural Products; Brahmachari, G., Ed.; Elsevier: Amsterdam, The Netherlands, 2018; pp 359–376.

Qureshi, A.; Naughton, D. P.; Petroczi, A. A Systematic Review on the Herbal Extract Tribulus Terrestris and the Roots of its Putative Aphrodisiac and Performance Enhancing Effect. *J. Diet Suppl.* **2014,** *11*, 64–79.

Radha, K.; Chawla, Y. K. Herbal Medicines for Liver Diseases. *Dis. Dig. Sci.* **2005,** *10*, 1807–1812.

Rafieian-Kopaei, M. Medicinal Plants and the Human Needs. *J. Herb. Med. Plarmacol.* **2012,** *1*, 1–2.

Rafieian-Kopaei, M. Identification of Medicinal Plants Affecting on Headaches and Migraines in Lorestan Province, West of Iran. *Asian Pac. J. Trop. Med.* **2014,** *7*, 376–379.

Rafieian-Kopaei, M.; Baradaran, A.; Rafieian, M. Plants Antioxidants: From Laboratory to Clinic. *J. Nephropathol.* **2013,** *2*, 152–153.

Rafieian-Kopaei, M.; Nasri, H. Ginger and Diabetic Nephropathy. *J. Renal. Inj. Prev.* **2013,** *2*, 9–10.

Rafiul Haque, M.; Ansari, S. H.; Rashikh, A. Coffea Arabica Seed Extract Stimulate the Cellular Immune Function and Cyclophosphamide-Induced Immunosuppression in Mice. *Iran J. Pharm. Res.* **2013,** *12*, 101–108.

Rane, B. R.; Bharath, M. S.; Patil, R. R.; Keservani, R. K.; Jain, A. S. Novel Approaches in Nutraceuticals, Chapter 11. In *Enhancing the Therapeutic Efficacy of Herbal Formulations through Novel Drug Delivery Systems*; Kesharwani, R. K., Keservani, R. K., Sharma, A. K., Ed.; IGI Global International Publisher: Pennsylvania, USA, 2020b; pp 241–266. ISBN13: 9781799844532.

Rane, B. R.; Patil, A. S.; Keservani, R. K.; Jain, A. S. Novel Approaches in Herbal Formulation, Chapter 2. In *Enhancing the Therapeutic Efficacy of Herbal Formulations through Novel Drug Delivery Systems*; Kesharwani, R. K., Keservani, R. K., Sharma, A. K., Ed.; IGI Global International Publisher: Pennsylvania, USA, 2020a; pp 43–68. ISBN13: 9781799844532.

Rane, B. R.; Tadavi, S. A.; Keservani, R. K. Naturopathy, Chapter 12. In *Herbal Product Development*; Sharma, A. K., Keservani, R. K., Gautam, S. P., Ed.; Apple Academic Press, CRC Press, Taylor & Francis Group, 2020c; pp 321–347. ISBN: 9781771888776.

Raut, A.; Bichile, L.; Chopra, A.; Patwardhan, B.; Vaidya, A. Comparative Study of Amrutbhallataka and Glucosamine Sulphate in Osteoarthritis: Six Months Open Label Randomized Controlled Clinical Trial. *J. Ayurveda Integr. Med.* **2013,** *4*, 229–236.

Rivera, L.; Morón, R.; Sánchez, M.; Zarzuelo, A.; Galisteo, M. Quercetin Ameliorates Metabolic Syndrome and Improves the Inflammatory Status in Obese Zucker Rats. *Obesity* **2008,** *16*, 2081–2087.

Roohafza, H.; Sarrafzadegan, N.; Sadeghi, M.; Rafieian-Kopaei, M.; Sajjadi, F.; Khosravi-Boroujeni, H. The Association Between Stress Levels and Food Consumption Among Iranian Population. *Arch. Iran Med.* **2013,** *16*, 145–148.

Ryan, M.; Lazar, I.; Nadasdy, G. M.; Nadasdy, T.; Satoskar, A. A. Acute Kidney Injury and Hyperbilirubinemia in a Young Male After Ingestion of Tribulus Terrestris. *Clin. Nephrol.* **2015,** *83* (2015), 177–183.

Saldanha, S. N.; Tollefsbol, T. O. The Role of Nutraceuticals in Chemoprevention and Chemotherapy and Their Clinical Outcomes. *J. Oncol.* **2012,** *19*, 24–64.

Santini, A.; Cammarata, S. M.; Capone, G.; Ianaro, A.; Tenore, G. C.; Pani, L.; Novellino, E. Nutraceuticals: Opening the Debate for a Regulatory Framework. *Br. J. Chin. Pharmacol.* **2018,** *84*, 659–672.

Sarris, J.; Logan, A. C.; Akbaraly, T. N.; Amminger, G. P.; Balanzá-Martínez, V.; Freeman, M. P.; Hibbeln, J.; Matsuoka, Y.; Mischoulon, D.; Mizoue, T.; et al. Nutritional Medicine as Mainstream in Psychiatry. *Lancet Psychiatry* **2015,** *2*, 271–274.

Sassi, F.; Tamone, C.; D'Amelio, P. Vitamin D: Nutrient, Hormone, and Immunomodulator. *Nutrients* **2018,** *10*, 1656.

Sawmiller, D.; Li, S.; Shahaduzzaman, M.; Smith, A. J.; Obregon, D.; Giunta, B.; Borlongan, C. V.; Sanberg, P. R.; Tan. J. Luteolin Reduces Alzheimer's Disease Pathologies Induced by Traumatic Brain Injury. *Int. J. Mol. Sci.* **2014,** *15*, 895–904.

Schneider, B. Biloba Extract in Peripheral Arterial Diseases: Meta-Analysis of Controlled Clinical Studies. *Arzneimittelforschung* 1992, *42*, 428–436.

Sellami, M.; Slimeni, O.; Pokrywka, A.; et al. Herbal Medicine for Sports: A Review. *Null* **2018,** *15* (1), 14. DOI: 10.1186/s12970-018-0218-y

Senchina, D. S.; Hallam, J. E.; Kohut, M. L.; Nguyen, N. A.; Perera, M. A. Alkaloids and Athlete Immune Function: Caffeine, Theophylline, Gingerol, Ephedrine, and their Congeners. *Exerc. Immunol. Rev.* **2014,** *20*, 68–93.

Sharma, R.; Amin, H.; Prajapati, P. K. Comparative Lipid Profiles in Non-diabetic Obese and Type-2 Diabetic Obese. *Astrocyte* **2015,** *2*, 99–100.

Sharma, R.; Amin, H.; Prajapati, P. K. Comparative Lipid Profile of Type 2 Obese Diabetics and Obese Non-diabetics: A Hospital Based Study from Hilly Terrains of Mandi, Himachal Pradesh. *Int. J. Health Allied Sci.* **2016,** *5*, 63–64.

Sharma, R.; Amin, H.; Prajapati, P. K. Influence of Psychiatric Comorbidity on the Treatment Process of Type 2 Diabetic Patient. *Indian J. Soc. Psychiatry* **2016,** *32*, 177–178.

Sharma, A. K.; Keservani, R. K.; Gautam, S. P. Herbal Product Development, Apple Academic Press, CRC Press, Taylor & Francis Group, 2020. ISBN: 9781771888776.

Shekelle, P. G.; Hardy, M. L.; Morton, S. C.; Maglione, M.; Mojica, W. A.; Suttorp, M. J.; et al. Efficacy and Safety of Ephedra and Ephedrine for Weight Loss and Athletic Performance: A Meta-Analysis. *JAMA J. Am. Med. Assoc.* **2003,** *289*, 1537–1545.

Shirzad, H.; Burton, R. C.; Smart, Y. C.; Rafieian-kopaei, M.; Shirzad, M. Natural Cytotoxicity of NC-2+Cells Against the Growth and Metastasis of WEHI-164 Fibrosarcoma. *Scand. J. Immunol.* **2011,** *73*, 85–90.

Shirzad, H.; Kiani, M.; Shirzad, M. Impacts of Tomato Extract on the Mice Fibrosarcoma Cells. *J. Herb. Med. Pharmacol.* **2013,** *2*, 13–16.

Shoskes, D.; Lapierre, C.; Cruz-Correa, M.; Muruve, N.; Rosario, R.; Fromkin, B.; Braun, M.; Copley, J. Beneficial Effects of the Bioflavonoids Curcumin and Quercetin on Early Function in Cadaveric Renal Transplantation: A Randomized Placebo Controlled Trial. *Transplantation* **2005,** *80*, 1556–1559.

Sidhu, K. S. Health Benefits and Potential Risks Related to Consumption of Fish or Fish Oil. *Regul. Toxicol. Pharmacol.* **2003,** *38*, 336–344.

Siegel, R. K. Ginseng and High Blood Pressure. *JAMA* **1980,** *243* (1), 32–32.

Singh, C. K.; Liu, X.; Ahmad, N. Resveratrol, in its Natural Combination in Whole Grape, for Health Promotion and Disease Management. *Ann. N. Y. Acad. Sci.* **2015,** *1348* (1), 150–60. DOI: 10.1111/nyas.12798.

Singh, P.; Kesharwani, R. K.; Keservani, R. K. Antioxidants and Vitamins: Roles in Cellular Function and Metabolism. In *Sustained Energy for Enhanced Human Functions and Activity*; Academic Press, 2017a; pp 385–407. ISBN: 9780128054130.

Singh, P.; Kesharwani, R. K.; Keservani, R. K. Protein, Carbohydrates, and Fats: Energy Metabolism. In *Sustained Energy for Enhanced Human Functions and Activity*; Academic Press, 2017b; pp 103–115. ISBN:9780128054130.

Srivastava, N.; Tiwari, G.; Tiwari, R.; Bhati, L. K.; Rai, A. K. Neutraceutical Approaches to Control Diabetes: A Natural Requisite Approach. *J. Nat. Sci. Biol. Med.* **2012,** *3*, 168–176.

Stahl, W.; Sies, H. Bioactivity and Protective Effects of Natural Carotenoids. *Biochim. Biophys. Acta* **2005,** *1740*, 101–107.

Straface, E.; Rivabene, R.; Masella, R.; Santulli, M.; Paganelli, R.; Malorni, W. Structural Changes of the Erythrocyte as a Marker of Noninsulin-Dependent Diabetes: Protective Effects of N-acetylcysteine. *Biochem. Biophys. Res. Commun.* **2002,** *290*, 1393–1398.

Thyagarajan, S. P.; Subramanian, S.; Thirunaksundari, T.; Venkateswaran, P. S.; Blumberg, B. S. Effects of *Phyllanthus amarus* on Chronic Carriers of Hepatitis B Virus. *Lancet* **1988,** *2*, 764–766.

Tofovic, S. P.; Jackson, E. K. Effects of Long-Term Caffeine Consumption on Renal Function in Spontaneously Hypertensive Heart Failure Prone Rats. *J. Cardiovasc. Pharmacol.* **1999,** *33*, 360–366.

Trottier, G.; Boström, P. J.; Lawrentschuk, N.; Fleshner, N. E. Nutraceuticals and Prostate Cancer Prevention: A Current Review. *Nat. Rev. Urol.* **2010,** *7*, 21–30.

Tuchweber, B.; Sieck, R.; Trost, W. Prevention of Silybin of Phalloidin Induced Acute Hepatoxicity. *Toxicol. Appl. Pharmacol.* **1979,** *51*, 265–275.

United State Pharmacopeia USP. The Regulation of Dietary Supplements, 2006.

Uzzaman, A.; Cho, S. H. Classification of Hypersensitivity Reactions. *Allergy Asthma Proc.* **2012,** *33* (1), 96–99.

Van der Burg, K. P.; Cribb, L.; Firth, J.; Karmacoska, D.; Sarris, J. Nutrient and Genetic Biomarkers of Nutraceutical Treatment Response in Mood and Psychotic Disorders: A Systematic Review. *Nutr. Neurosci.* **2019,** 1–17.

Vikrant, V.; Grover, J. K.; Tandon, N. Treatment with Extracts of Momordica Charantia and Eugenia Jambolana Prevents Hyperglycemia and Hyperinsulinaemia in Fructose Fed Rats. *J. Ethnopharmacol.* **2001,** *76*, 13943.

Virdi, J.; Sivakami, S.; Shahani, S. Antihyperglycemic Effects of Three Extracts from Momordica Charantia. *J. Ethnopharmacol.* **2003,** *88*, 107–111.

Walpurgis, K.; Schultze, G.; Mareck, U.; Geyer, H.; Schänzer, W.; Thevis, M. Recent Advances in Doping Analysis, 2016.

Wang, H.; Cheng, H.; Che, Z. Ameliorating Effect of Luteolin on Memory Impairment in an Alzheimer's Disease Model. *Mol. Med. Rep.* **2016,** *13*, 4215–4220.

Wang, X.; Lin, H.; Gu, Y. Multiple Roles of Dihomo-γ-linolenic Acid Against Proliferation Diseases. *Lipids Health Dis.* **2012,** *11*, 25. Published 2012 Feb 14. DOI: 10.1186/1476-511X-11-25

Wang, Y.; Wang, B.; Du, F.; Su, X.; Sun, G.; Zhou, G.; Bian, X.; Liu, N. Epigallocatechin-3-Gallate Attenuates Oxidative Stress and Inflammation in Obstructive Nephropathy via NF-κB and Nrf2/HO-1 Signalling Pathway Regulation. *Basic Clin. Pharmacol. Toxicol.* **2015,** *117*, 164–172.

Willis, M. S.; Wians, F. H. The Role of Nutrition in Preventing Prostate Cancer: A Review of the Proposed Mechanism of Action of Various Dietary Substances. *Clin. Chim. Acta* **2003,** *330*, 57–83.

Wilson, P. B. Ginger (Zingiber officinale) as an Analgesic and Ergogenic Aid in Sport. *J. Strength Cond. Res.* **2015,** *29*, 2980–2995.

Xia, N.; Daiber, A.; Förstermann, U.; Li, H. Antioxidant Effects of Resveratrol in the Cardiovascular System. *Br. J. Pharmacol.* **2017,** *174*, 1633–1646.

Yahfoufi, N.; Alsadi, N.; Jambi, M.; Matar, C. The Immunomodulatory and Anti-Inflammatory Role of Polyphenols. *Nutrients* **2018,** *10*, 1618.

Yoshikawa, M.; Matsui, Y.; Kawamoto, H.; et al. Effects of Glycyrrhizin on Immune Mediated Cytotoxicity. *J. Gastroenterol. Hepatol.* **1997,** *12*, 243– 247.

Zeisel, S. H. Regulation of "Nutraceuticals". *Science* **1999,** *285*, 1853–1855.

Zhang, Z.; Wang, X.; Zhang, Y.; Ye, G. Effect of Cordyceps Sinensis on Renal Function of Patients with Chronic Allograft Nephropathy. *Urol. Int.* **2011,** *86*, 298–301.

Zhang, H.; Zhang, S.; Zhang, H.; et al. Carotenoid Metabolite and Transcriptome Dynamics Underlying Flower Color in Marigold (Tagetes erecta L.). *Sci. Rep.* **2020,** *10* (1), 16835. DOI: 10.1038/s41598-020-73859-7

Zhong, G.; Jiang, Y. Calcium Channel Blockage and Anti-Free-Radical Actions of Ginsenosides. *Chin. Med. J.* **1997,** *110*, 28–29.

Zhu, W.; Du, Y.; Meng, H.; Dong, Y.; Li, L. A Review of Traditional Pharmacological Uses, Phytochemistry, and Pharmacological Activities of Tribulus Terrestris. *Chem. Cent. J.* **2017,** *11*, 60.

CHAPTER 2

NUTRACEUTICAL POTENTIAL OF AYURVEDIC HERBALS

ANKIT KUMAR[1], AADESH KUMAR[1], RAJESHWAR KAMAL KANT ARYA[1], and GAURAV RAWAT[2]

[1]*Department of Pharmaceutical Sciences, Sir J.C. Bose Technical Campus Bhimtal, Kumaun University, Nainital, India*

[2]*Department of Zoology, D.S.B. Campus, Kumaun University, Nainital, India*

ABSTRACT

Herbs and natural sources are being used as natural remedies for a wide range of physiological disorders since prehistoric times. They were used in traditional medicine because they were a natural gift that may help with recuperation. These herbs are rare and have substantial nutritional and therapeutic properties, making them highly sought after by the pharmaceutical, cosmetic, and herbal-based industries. This chapter examines the physiological activities of numerous physiologically active ingredients found in commonly used herbs, including their chemical constituents such as minerals, vitamins, flavonoids, alkaloids, micronutrients, fatty acids, essential oils, and fibers, among others. It also describes a few important plants, such as *Allium sativum, Gingko biloba, Panax ginseng, Curcuma longa, Allium sepa, Asperagus racemosa,* and *Withania somnifera.* This chapter looks at some of the Indian Himalayan herbs and aromatic plants that have a lot of potential for use as nutraceuticals. The current phase also explains the market worth of nutrition values and their market projection. In the year 2021, the value of the Global Nutraceutical Market was anticipated to be US Dollar 45,600 crore,

Herbals As Nutraceuticals: Their Role in Healthcare.
Raj K. Keservani, Rajeshwar Kamal Kant Arya, and Rajesh K. Kesharwani (Eds.)

and it is predicted to hike with a compound annual growth rate (CAGR) of 9.0 % by the end of this decade.

2.1 INTRODUCTION

Nutraceuticals are therapeutic foods that promote health by modifying immunity and thereby preventing and curing specific diseases (Keservani et al., 2010a; Keservani et al., 2010b; Keservani et al., 2017; Keservani et al., 2020). They are oral dietary components that are naturally occurring, bio-active or chemical compounds that, in addition to serving a nutritional function, have health-promoting, disease-curing, or preventative qualities (Souyoul et al., 2018; Sachdeva et al., 2020). Throughout the prehistoric era, plants, and other sources, were used to treat a variety of diseases throughout the world. To trace the history of nutraceuticals, it is necessary to establish linkages between unconventional medicines such as allopathy, herbalism, ethnopharmacology, and phytotherapy. When active compounds taken from plants and animals source are concentrated and supplied in an appropriate dosage form, they can generate a most useful toolkit beneficial for preventing and supporting the treatment of certain pathologic disorders (Santini et al., 2016; Makkar et al., 2020). Carbohydrates (Singh et al., 2017b), alkaloids, lipids, terpenes, polyphenols, sterols, and steroids are the primary pharmacologically active herbal components (Sharma et al., 2020; Rane et al., 2020a; Rane et al., 2020b; Rane et al., 2020c). The claimed nutraceutical compounds derived from these sources omega-3 fatty acids, isoflavones, allyl sulfur compounds, phytosterols, tocotrienols, quercetin, carotenoids, and fiber are a few of the more well-known chemical constituents (Andrew and Izzo, 2017). As a result of their natural nature, there is an increasing need for nutraceuticals, which blur the line between medications and food, thereby assisting farmers in diversifying their crops and encouraging research and new ideas. However, varied rules for different countries, security, and proof of nutritional benefits are the main problems that nutraceuticals have to deal with (Santini and Novellino, 2017). Nutraceuticals are based on a prevention-focused mindset. Most of the time, it can be applied to dietary supplements and functional foods (Fig. 2.1).

2.1.1 NUTRITION IN DIET

This is the way by which we introduce more nutritional value to the food. Such supplements consist of one/more nutritional components listed and are designed to enrich the diet with minerals, vitamins, amino acids, and enzymes. These all were obtained from herbs or other natural sources, metabolites, and other ingredients.

2.1.2 FUNCTIONAL FOODS

Functional foods are any ready-to-eat components that supply a health improvement in addition to the usual minerals it comprises. Multifunctional meals include therapeutically active ingredients originating from plants or animals (Nwosu and Ubaoji, 2020).

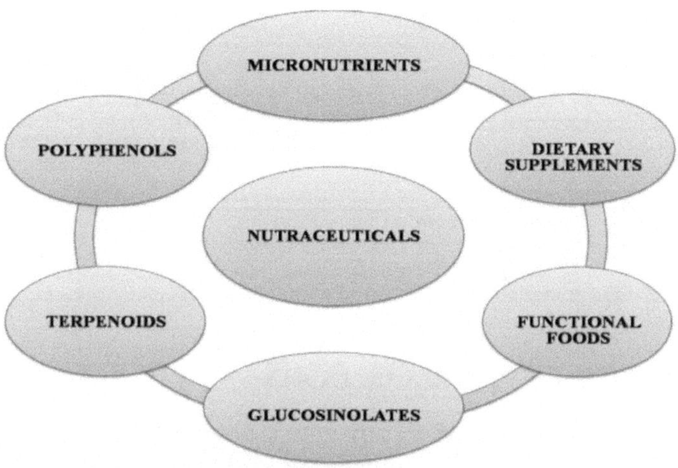

FIGURE 2.1 Nutraceutical uses in different forms.

2.2 CLASSIFICATION OF NUTRACEUTICALS

Nutraceuticals and functional foods are classified according to their source: natural or traditional versus unnatural or non-traditional.

2.2.1 TRADITIONAL NUTRACEUTICALS

They are natural items that do not alter the meal in any way. They contain a range of natural ingredients that provide benefits other than nutrition, including omega-3 fatty acids, minerals, micronutrients, and vitamins present in salmon, saponins, phytosterols, phytic acids, soya and lycopene, vitamin C, and carotenes in tomatoes.

2.2.2 NON-CLASSICAL NUTRITION VALUES

These are synthetic foods that have been biotechnologically produced. Bioactive components in food samples are developed to create human-wellness goods. The classification of nutraceuticals are summarized in Figure 2.2 (Alamgir, 2017).

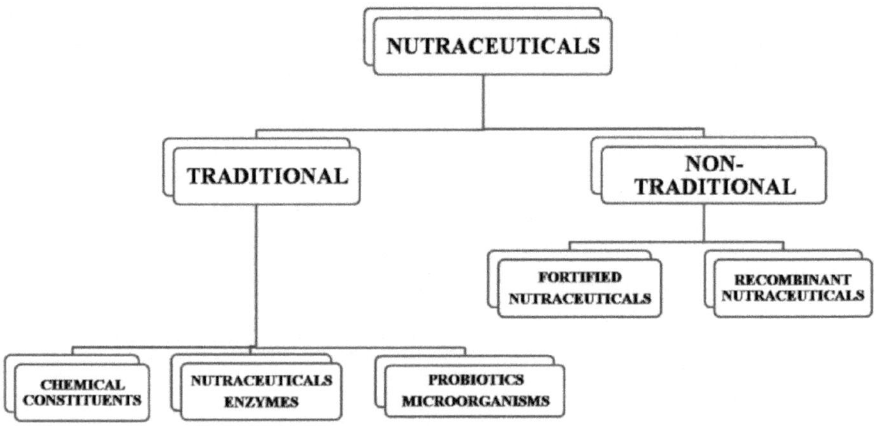

FIGURE 2.2 Diagrammatic representation of nutraceutical classification.

2.3 SOURCES OF NUTRACEUTICALS

Nutraceuticals are now better known as a class of chemicals that are both architecturally and utilitarian diversified bioactive compounds. In addition to nutritional or unambiguous pharmacological consequences, they give long-term herbal cures as well as corporal perquisite which distinguish nutraceuticals from fully functional commodities and medications. The source of nutraceuticals can be discussed as follows:

i. Botanical Source: in the form of phytochemicals and vitamins
ii. Zoological Source: cellulose-containing moieties and conjugated linoleic acid. Microbes containing amino acids; and
iii. Aquatic Sources: like glucosamine and chitosan.

Diverse herbal origin was used for their therapeutic and fortitude-recovery properties, including cinnamon (dalchini), saffron (kesar), honey (sehad), garlic (lehsun), ginger (adrak), pomegranate (anar), mint (pudina), etc., (Wang et al., 2016; Alali et al., 2021). To date, the majority of oceanic microflora was unknown, since marine scientists focused on either big appealing organisms or commercially important food species available in the marine environment (Dewapriya and Se-Kwon, 2014). The most frequent way to introduce these organisms into the gastrointestinal tract (GIT) is through fermented milk products, especially yogurts, drinks, and cryo-dried bacteria capsules. Current findings describe that Lactobacillus species derived from the sea have numerous benefits over traditional probiotic varieties. *Lactobacillus paracasei*, a strain obtained from oysters that are resistant to stress, was found to have a substantial antagonistic effect on Vibro spp. fish pathogens like *V. alginolyticus* and *V. proteolyticus* (Dewapriya and Se-Kwon, 2014; Simat et al., 2020). Sources of nutraceuticals are also summarized (Fig. 2.3).

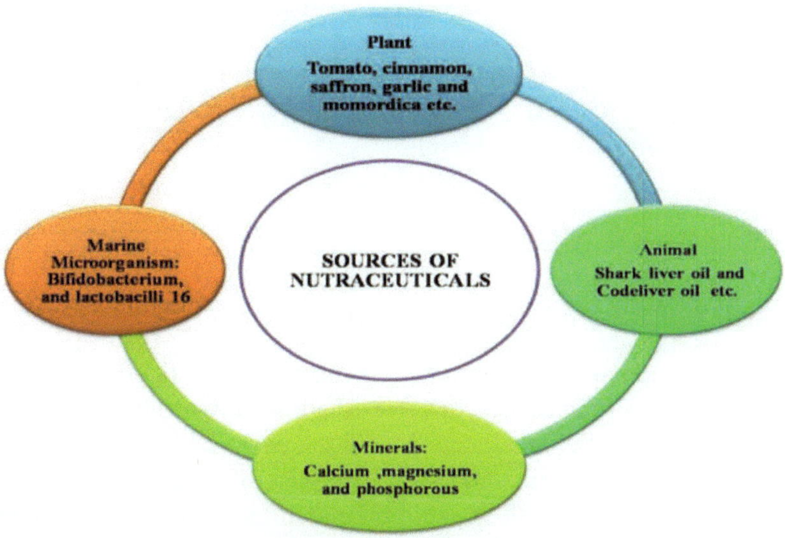

FIGURE 2.3 Nutraceuticals sources and some examples.

Probiotics are utilized for their alleged benefits on the GIT as well as integral organs such as the liver, brain, vaginal canal, and bloodstream. Most chronic diseases are associated with elevated levels of free radicals; antioxidants (Singh et al., 2017a) are essential in the treatment of virtually all diseases. The aging effect along with a lack of potent antioxidants accelerates oxidative stress. Many studies have found that flavonoids have anti-cancer capabilities due to their high antioxidant content. The beneficial effect of nutraceuticals is summarized in Figure 2.4 (Chintale et al., 2013; Keservani and Sharma, 2014; Keservani et al., 2016a; Keservani et al., 2016b; Kumar and Kumar, 2015; Chui et al., 2020).

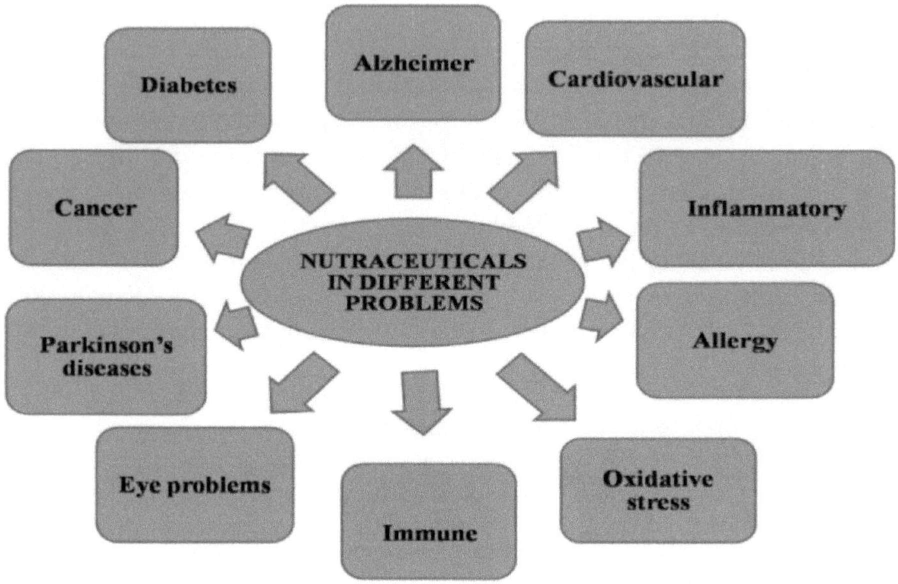

FIGURE 2.4 Role of nutraceuticals in various problems.

2.4 HISTORY OF NUTRACEUTICALS

The idea of nutraceuticals dates back to 3000 years. Hippocrates (460–377 B.C.), the father of medicines, discovered a link between consuming healthy meals and their medicinal advantages. In the early days of the 20th-century, bakery owners in America had started the use of little amounts of iodine, which were added with salt to avoid the risk of goiter disease. Theophrastus (370–285 BC), the father of modern Botany, his Plant Etiology is concerned with physiology and structure, while his inquiry into plants is concerned with description and categorization. Schwan, in 1837, used microbes (fungus or yeast) for product creation via the alcoholic fermentation process since 2000 BC. Funk (1884–1967), known as the "Father of Vitamin Therapy" pioneered a new line of scientific research on vitamins, establishing them as food ingredients required for optimal health maintenance. Nutrition has already been practiced from the beginning of eras. Diet was mostly governed by the availability and consumer acceptability of foods, as well as teachings; skills for gathering and preparing food were developed through trial and error, as well as a tremendous capacity for human ingenuity. In 1989, the man behind the term "nutraceutical" was Dr. Stephen Defelice who merged the two terms "nutrition" and "pharmaceuticals" (Singh and Sinha, 2012;

Souyoul et al., 2018; Nwosu and Ubaoji, 2020; Trüeb, 2020). Due to the prevalence of adverse effects caused by various pharmaceutical therapies, nutraceutical admixtures have received interest as a second choice therapy with a preventive attitude, as well as the benefits of being extra cost-effective and accessible (Alali et al., 2021).

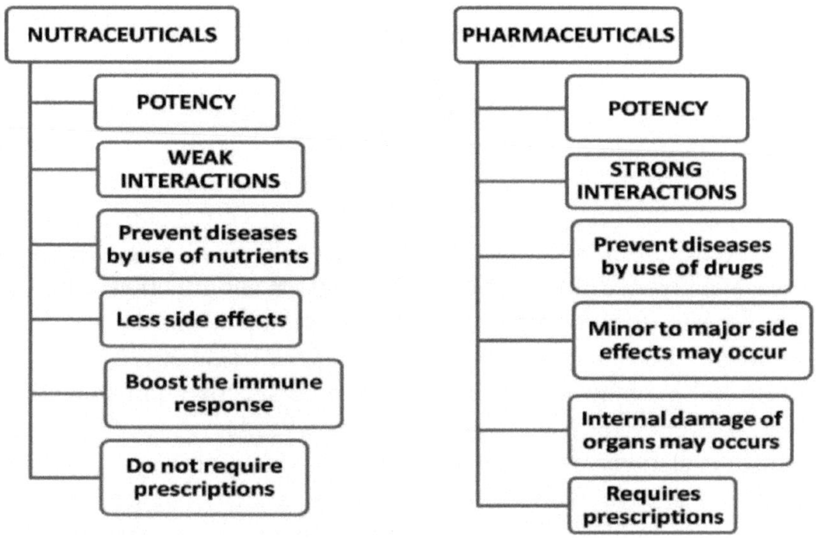

FIGURE 2.5 Difference between nutraceuticals and pharmaceuticals.

2.5 AYURVEDIC HERBALS

One of the world's oldest medicinal systems Ayurveda dates back to thousands of years. Its origins can be traced back thousands of years to the Vedic period in India and neighboring nations. These ancient manuscripts contain thorough descriptions of over 700 plants and 6000 formulas, as well as explanations of numerous ailments, diagnostic methods, nutrition, and lifestyle advice. The three fundamental principles, known as doshas (Vata, pitta, and Kapha), are drawn from five elements of Indian philosophy. Food, daily routine activity, the environment, and stress conditions can disrupt or imbalance these functions. Plant-based remedies are essential in the ayurvedic treatment regimen (Kumar et al., 2017; Gyawali et al., 2021). Ayurveda explains the idea of Samsarjana Karma Dravya, which deals with different methods of food processing to provide the desired nutritional content of specific foods (Gupta et al., 2018; Chavda et al., 2022). Ayurvedic herbs have enough promise for COVID-19 prevention and treatment. Traditional uses and scientific support for these plants and their phytochemicals in vitro and in vivo models against infectious diseases, with a focus on COVID-19, may be useful for generating a unique source of natural goods. Some of the helpful herbs used for the therapy include Indian Gooseberry, cinnamon, Nepalese pepper, Drumstick tree, Ashwagandha, garlic, turmeric, ginger, basil, chirata, licorice, and tinospora (Gyawali et al., 2020). Some of the biologically useful herbs are listed in Table 2.1.

2.5.1 MARKET SHARE OF NUTRACEUTICALS

The worldwide nutraceutical market was estimated at USD 454.55 billion in 2021, and it is expected to develop at a 9.0% compound annual growth rate (CAGR) from 2021 to 2030.This was calculated based on the present increase in demand for nutritional supplements and functional foods, which are major variables for the market forecast period of this decade. The use of nutritional values and functional foods for the treatment of cardiovascular disease and malnutrition brings a positive perspective on medical nutrition to the forefront (Nutraceuticals Market Size, 2021).

India is also known as the home of Ayurveda, and it accounts for 3% of total global sales of nutraceuticals, according to a combined study conducted by Assocham and RNCOS. Beverages, functional foods, and dietary supplements are all part of the Indian nutraceuticals business. Breakfast cereals and enriched flour are examples of utilitarian foods. Fortified juices, sports drinks, and glucose are some examples of functional beverages. Additives in food include plant base and non-herbal extracts containing macronutrients. This represents the trend of more than 65% of the Indian nutraceutical market (Chokhani, 2021).

2.6 CONCLUSION

Human body is made up of 100% natural bioactive chemical components that play a different role in daily routine. From wakening from bed to sleep, there is the involvement of energy to perform all-day activities. That energy comes from the food, diet, and nutrition, consumed by us to feed our bodies on demand. The nutritional value is in the form of carbohydrates, proteins, vitamins, minerals, anthraquinole, enzymes, and micronutrients. Somehow when our food is a deficit and it is needed to provide the required energy value then there is the need to consume some value other than food that

TABLE 2.1 Some of the Herbs Used as a Source of Nutraceuticals in Various Forms.

S. No.	Botanical name	Common name	Part used	Component present	Form used	Used for treatment	Reference
1.	*Allium sativum*	Garlic	Dried bulbs	Alliin and allicin	Fresh dried cloves, capsules, odorless tablets, tinctures, and garlic extracts	Chemopreventive, cancer, diabetes, arteriosclerosis, lowering cholesterol, respiratory infections, and cardiovascular conditions	Touloupakis and Ghanotakis, 2010; Pandey et al., 2011; Chauhan et al., 2013; Williamson et al., 2019
2.	*Ginkgo biloba*	Maidenhair tree	Leaves	Ginkgolide and bilobalide	Dried leaf tea	Age-related memory, Restore function in cognitively impaired individuals	Pandey et al., 2011; Chauhan et al., 2013; Pitt and Leung, 2016
3.	*Zingiber officinale*	Ginger	Rhizomes	Zingiberene and gingerols	Fresh and dried roots, capsules, tablets, and tinctures	Indigestion, motion sickness, nausea, antioxidant, cholesterol control, gastrointestinal, inflammation	Pandey et al., 2011; Chauhan et al., 2013; Williamson et al., 2019
4.	*Echinacea purpurea*	Echinacea	Dried herb	Alkylamide and echinacoside	Dried raw herbs and roots, capsules, freshly juice of flowering plants, pills, and tinctures are all available	Cold, flu, minor infections, immunostimulant	Pandey et al., 2011; Chauhan et al., 2013
5.	*Panax ginseng*	Ginseng	Dried root	Ginsenosides, Panaxosides, maltol, ferulic acid, vanillic acid, p-coumaric acid, salicylic acid, p-hydroxybenzoic acid cinnamic acid, and gentisic acid	Dried roots, steamed roots, as capsules, extracts, tablets, tinctures, and decoction	Convalescence, tiredness, diabetes, cholesterol control, concentration improvement, an aphrodisiac	Chauhan et al., 2013; Dasgupta, 2019; Truong and Jeong, 2022

TABLE 2.1 *(Continued)*

S. No.	Botanical name	Common name	Part used	Component present	Form used	Used for treatment	Reference
6.	*Glycyrrhiza glabra*	Licorice	Dried root	Glycyrrhizin, Isoliquiritin, liquirtin, Liquiritin apioside, Glycyrrhetic acid, 18β-Glycyrrhetinic acid, Licochalcone A, Liquiritigenin, and Glabridin	Gum, capsules, tablets, powder, liquid extracts, and syrups	It shows anti-inflammatory, anti-allergic, expectorant, anti-ulcerative, antimicrobial, hepatoprotective, anticarcinogenic, antimutagenic, neuroprotective, antidepressive, antihyperglycemic activity, and memory enhancer	Chauhan et al., 2013; Kaur et al., 2013; Pastorino et al., 2018; Martini, 2020; Tiwari et al., 2021
7.	*Hypericum perforatum*	St. John's wort	Dried aerial part	Hypericin and hyperforin	Flowering tops in the form of decoction, oil for external use, capsules, pills, and in the form of tinctures	Epilepsy (mild to moderate), cuts and abrasions, depression	Pandey et al., 2011; Chauhan et al., 2013
8.	*Curcuma longa*	Turmeric	Rhizome	Curcumin	Dried root, whole, and powder	Reducing inflammation, indigestion, antioxidants, and liver problem	Pandey et al., 2011; Chauhan et al., 2013
9.	*Allium cepa*	Onion	Dried onion bulbs	Allicin quercetin-4-glucoside, cycloalliin, methiin, and alliin	Tablets, capsules, syrup, tinctures, and oil.	Antibiotic antiatherosclerosis, Tooth disorders, heart disease, skin disorders, cholera, anticancer, antimicrobial activity, wound healing, antiplatelet activity, antioxidant property, gallstone treatment, antidepressant, insecticidal, anti-inflammatory, hepatoprotective	Singh and Singh, 2008; Kumar et al., 2010; Chauhan et al., 2013; Chakraborty et al., 2022

TABLE 2.1 *(Continued)*

S. No.	Botanical name	Common name	Part used	Component present	Form used	Used for treatment	Reference
10.	*Valeriana officinalis Linn.*	Valeriana	Dried root	Valerenic acid and valerate	Root, powder, tea, capsules, tablets, tinctures, extracts	Anxiety, insomnia, hypertension	Pandey et al., 2011; Chauhan et al., 2013
11.	*Aloe barbadensis Mill.*	Aloe vera/ Ghritkumari	Dried juice	Aloins and aloesin	Gel, juice, Sunscreen, skin creams, lotions, and oral intake	Wounds healing, inflammation, constipation, insomnia, pain, hemorrhoids, itching, headache, hair loss, kidney disease, blisters, sunburn, ulcers, and immunomodulatory	Pandey et al., 2011; Hussain et al., 2015; Williamson et al., 2019
12.	*Hydrastis canadensis*	Goldenseal	Dried root	Hydrastine, canadine, hydrastinine berberine, sideroxylin, chlorogenic acid, and palmatine	Tablet, as a tincture	As astringent, inflammation, antimicrobial, hypolipidemic, anti-obesity, antitumor, immunomodulator, anti-arthritis, anticonvulsant, and free radical scavenging	Chauhan et al., 2013; Mandal et al., 2020
13.	*Cassia angustifolia*	Senna	Dried leaves	Sennosides, tinnevellin glycoside, apigenin-6,8-di-C-glycoside(III), aloe emodin and kaempferol	Tablet, syrup, and tea infusion	Indigestion, jaundice, malaria, constipation, loss of appetite, splenomegaly, hepatomegaly and aspurgative	Momin, 2005; Chauhan et al., 2013; Absar and Singh, 2018
14.	*Ferula assafoetida L.*	Asafoetida	Oleo gum resin	Rhamnose Glucose, 1-arabinose,galactose, glucuronic acid, glycoproteins Ferulic acid, asaresinotannols, farnesiferols, umbellic acid, Asacoumarin A, Epi-conferdione, Lehmferin, Ligupersin A, Microlobin, Feselol, terpenes, and many sulfur-containing compounds	Resin and its fluid extract and powder	Stimulant, carminative, expectorant, anti-inflammatory, anticarcinogenic, hepatoprotective, neuroprotective, memory enhancing, antioxidant, antispasmodic, antimicrobial, antiobesity, and anthelmintic	Mahendra and Bisht, 2012; Chauhan et al., 2013; Amalraj and Gopi, 2017

TABLE 2.1 *(Continued)*

S. No.	Botanical name	Common name	Part used	Component present	Form used	Used for treatment	Reference
15.	*Aegle marmelos Corr.*	Bael	Fruits, leaves, roots, bark, seeds	Marmelosin, alkaloids, mermesinin, rutin, phenylethyl cinnamides, anhydromarmeline, and aegelinosides sterols, essential oils, coumarin, psoralen, xanthotoxin, coumarins, tembamide, mermin, and skimmianine	Powder, tablets, capsules, Decoction, oil, syrup and dry extract	Digestive, treatment of diarrhoea, appetiser and dysentery treatment as anticancer, anti-inflammatory, antidiabetic and antimicrobial	Chauhan et al., 2013; Nigam and Nambiar, 2017; Pathirana, 2020
16.	*Asparagus racemosus*	Satawar/ Satmuli	Root	Shatavarin IV, racemofuran, asparagamine, racemosol, alkaloids, sarsasapogenin, and kaempferol	Powder, tablets, and syrup	The hypocholesterolemic effect, antioxidant, phytoestrogenic, diarrhea, dysentery, gastritis, antidepressant, antidiarrhoeal, analgesic, female tonic, and reproductive system	Hussain et al., 2015; Hasan et al., 2016; Sharma et al., 2017
17.	*Withania somnifera*	Ashwagandha	Root	Sitoindoside, Steroidal lactones, with anolides, Glycosides, alkaloids (isopelletierine and anaferine), saponins	Root powder, standardized extracts, and tinctures	Immunomodulatory effect, anti-inflammation, hypoglycemic, diuretic, and hypocholesterolemic effects, anti-inflammation	Pandey et al., 2011; Hussain et al., 2015; Bharti et al., 2016
18.	*Bacopa monniera*	Brahmi	Herb	Alkaloids, saponins, sterols, betulic acid, stigmasterol, beta-sitosterol, bacosides, and bacopasaponins	Extract	Enhance learning, anxiolytic properties, anticonvulsive, hepatoprotective, antiulcer, muscle relaxant and neurosis, restore function in cognitively impaired individuals	Hussain et al., 2015; Pitt and Leung, 2016

TABLE 2.1 *(Continued)*

S. No.	Botanical name	Common name	Part used	Component present	Form used	Used for treatment	Reference
19	*Pueraria tuberosa*	Vidari	Flowers, roots, tubers	Alkaloids, volatile oils, lupinoside, anthocyanin, coumarin, and isoflavonoids	Powder	Cooling agent, aphrodisiac, mucoprotective agent, galactagogue, antidiabetic and immunomodulatory activity, eases bowel movement, useful in relieving constipation, used in skin diseases	Pandey et al., 2011; Hussain et al., 2015
20.	*Emblica officinalis*	Amla	Fruit, leaves, bark	Zeatin, phyllembin, gallic acid, ascorbic, music acid, and superoxide dismutase	Fruit and powder	High vitamin C, antiaging, treats hemorrhage, diarrhea, and dysentery, antimicrobial, antioxidant, hepatoprotective, analgesic and antipyretic, stress, diuretic, liver function, diabetes, cardiovascular diseases, metabolic syndrome, cancer, gastrointestinal disorders, deficiency diseases	Pandey et al., 2011; Hussain et al., 2015; Ramadan et al, 2019
21.	*Terminalia chebula*	Haritaki/ harada	Fruit	Amino acids, glucose, succinic acid, contanoic acid, β-sitosterol, palmitic acid, teaflavin A, arjunolic acid, punicalagin, arjungenin, chebupentol and terminoic acid	Powder and fruit pulp	Effective in various problems like sore throat, asthma vomiting, bleeding piles, diarrhea, gout, heart diseases, antiulcerogenic effect, and anticancer activity	Hussain et al., 2015; Pandey et al., 2011

could provide proper nutrition to the body. These types of nutrition values are known as nutraceuticals. In the current scenario of COVID-19, the demand for nutraceuticals was observed, to increase due to the high antioxidant capacity of nutraceuticals. Antioxidants are helpful in immunities and the defense mechanism of the human body. The worldwide demand for nutraceuticals was reported as USD 45,600 crore (approx.) with an increase of 9% CAGR annually. This increase in the demand for nutraceuticals represents the expansion of the national as well as the international market of nutraceuticals.

KEYWORDS

- **nutraceuticals**
- **therapeutic**
- **ayurvedic herbals**
- **COVID-19**

REFERENCES

Absar, N.; Singh, S. Senna (Cassia angustifoliaVahl.): Recent Advances in Pharmacognosy and Prospects of Cultivation in India. *Bioved* **2018,** *29* (2), 399–408.

AlAli, M.; Alqubaisy, M.; Aljaafari, M. N.; AlAli, A. O.; Baqais, L.; Molouki, A.; ... Lim, S.-H. E. Nutraceuticals: Transformation of Conventional Foods into Health Promoters/ Disease Preventers and Safety Considerations. *Molecules* **2021,** *26* (9), 2540. DOI: 10.3390/ molecules26092540

Alamgir, A. N. M. Classification of Drugs, Nutraceuticals, Functional Food, and Cosmeceuticals; Proteins, Peptides, and Enzymes as Drugs. In *Therapeutic Use of Medicinal Plants and Their Extracts*: *Volume 1*; Progress in Drug Research; Springer: Cham, 2017; vol *73*. https://doi.org/10.1007/978-3-319-63862-1_5.

Amalraj, A.; Gopi, S. Biological Activities and Medicinal Properties of Asafoetida: A Review. *J. Tradit. Complement. Med.* **2017,** *7* (3), 347–359.

Andrew, R.; Izzo, A. A. Principles of Pharmacological Research of Nutraceuticals. *Br. J. Pharmacol.* **2017,** *174* (11), 1177–1194. DOI: 10.1111/bph.13779

Bharti, V. K.; Malik, J. K.; Gupta, R. C. Ashwagandha: Multiple Health Benefits. In *Nutraceuticals*; Academic Press, 2016; pp 717–733. https://doi.org/10.1016/ B978-0-12-802147-7.00052-8

Chakraborty, A. J.; Uddin, T. M.; Zidan, M.; Redwan, B. M.; Mitra, S.; Das, R.; ... Emran, T. B. Allium cepa: A Treasure of Bioactive Phytochemicals with Prospective Health Benefits. *Evid. Based Complement. Altern. Med.* **2022,** *2022*.

Chauhan, B.; Kumar, G.; Kalam, N.; Ansari, S. H. Current Concepts and Prospects of Herbal Nutraceutical: A Review. *J. Adv. Pharm. Technol. Res.* **2013,** *4* (1), 4. https://dx.doi.org/10.4103%2F2231-4040.107494

Chavda, V. P.; Patel, A. B.; Vihol, D.; Vaghasiya, D. D.; Ahmed, K. M. S. B.; Trivedi, K. U.; Dave, D. J. Herbal Remedies, Nutraceuticals, and Dietary Supplements for COVID-19 Management: An update. *Clin. Complement. Med. Pharmacol.* **2022,** *2* (1), 100021.

Chintale Ashwini, G.; Kadam Vaishali, S.; Sakhare Ram, S.; Birajdar Ganesh, O.; Nalwad Digambar, N. Role of Nutraceuticals in Various Diseases: A Comprehensive Review. *Int. J. Res. Pharm. Chem.* **2013,** *3*, 290–299.

Chiu, H. F.; Venkatakrishnan, K.; Wang, C. K. The Role of Nutraceuticals as a Complementary Therapy Against Various Neurodegenerative Diseases: A Mini-Review. *J. Tradit. Complement. Med.* **2020,** *10* (5), 434–439.

Chokhani, R. Why Nutraceutical Industry is Booming in India, Published on 18 August 2021. https://www.indianretailer.com/article/whats-hot/retail-trends/why-nutraceutical-industry-is-booming-in-india.a7302/ (assessed Apr 26, 2022).

Dasgupta, A. Antiinflammatory Herbal Supplements. In *Translational Inflammation*; Academic Press, 2019; pp 69–91.

Dewapriya, P.; Kim, S. Marine Microorganisms: An Emerging Avenue in Modern Nutraceuticals and Functional Foods. *Food Res. Int.* **2014,** *56*, 115–125. DOI: 10.1016/j.foodres.2013.12.022

Gupta, R. K.; Soni, P.; Singh, R.; Shrivastava, J.; Rajput, P. Health Benefits of Natural Nutrients: Ayurvedadravya as Nutraceuticals. *WJPMR* **2018,** *4* (4), 253–255.

Gyawali, R.; Paudel, P. N.; Basyal, D.; Setzer, W. N.; Lamichhane, S.; Paudel, M. K.; Khanal, P. A Review on Ayurvedic Medicinal Herbs as Remedial Perspective for COVID-19. *J. Karnali Acad. Health Sci.* **2020,** *3*.

Gyawali, D.; Vohra, R.; Orme-Johnson, D.; Ramaratnam, S.; Schneider, R. H. A Systematic Review and Meta-Analysis of Ayurvedic Herbal Preparations for Hypercholesterolemia. *Medicina* **2021,** *57* (6), 546.

Hasan, N.; Ahmad, N.; Zohrameena, S.; Khalid, M.; Akhtar, J. Asparagus Racemosus: For Medicinal Uses & Pharmacological Actions. *Int. J. Adv. Res.* **2016,** *4* (3), 259–267.

Hussain, S. A.; Panjagari, N. R.; Singh, R. R. B.; Patil, G. R. Potential Herbs and Herbal Nutraceuticals: Food Applications and Their Interactions with Food Components. *Crit. Rev. Food Sci. Nutrit.* **2015,** *55* (1), 94–122. DOI: 10.1080/10408398.2011.649148

Kaur, R.; Kaur, H.; Dhindsa, A. S. Glycyrrhizaglabra: A Phytopharmacological Review. *Int. J. Pharm. Sci. Res.* **2013,** *4* (7), 2470.

Keservani, R. K.; Kesharwani, R. K.; Sharma, A. K.; Vyas, N.; Chadoker, A. Nutritional Supplements: An Overview. *Int. J. Curr. Pharm. Rev. Res.* **2010b,** *1* (1), 59–75.

Keservani, R. K.; Kesharwani, R. K.; Vyas, N.; Jain, S.; Raghuvanshi, R.; Sharma, A. K. Nutraceutical and Functional Food as Future Food: A Review. *Der Pharmacia Lettre* **2010a,** *2* (1), 106–116.

Keservani, R. K.; Sharma, A. K. Flavonoids: Emerging Trends and Potential Health Benefits. *J. Chinese Pharm. Sci.* **2014,** *23* (12), 815.

Keservani, R. K.; Sharma, A. K.; Kesharwani, R. K. Nutraceutical and Functional Foods for Cardiovascular Health. In *Food Process Engineering*; pp 291–312; Apple Academic Press, CRC Press, 2016a; pp 257–278. ISBN: 9781771884020.

Keservani, R. K.; Sharma, A. K.; Kesharwani, R. K. Medicinal Effect of Nutraceutical Fruits for the Cognition and Brain Health. *Scientifica* **2016b,** 3109254, 10. DOI: 10.1155/2016/3109254.

Keservani, R. K.; Sharma, A. K.; Kesharwani, R. K. An Overview and Therapeutic Applications of Nutraceutical and Functional Foods. In *Recent Advances in Drug Delivery Technology*, 2017; pp 160–201. ISBN: 9781522507543.

Keservani, R. K.; Sharma, A. K.; Kesharwani, R. K. Eds.; In *Nutraceuticals and Dietary Supplements: Applications in Health Improvement and Disease Management*; CRC Press, 2020. ISBN: 9781771888738.

Kumar, K. S.; Debjit, B.; Pankaj, T. Allium cepa: A Traditional Medicinal Herb and Its Health Benefits. *J. Chem. Pharm. Res.* **2010,** *2* (1), 283–291.

Kumar, S.; Dobos, G. J.; Rampp, T. The Significance of Ayurvedic Medicinal Plants. *J. Evid. Based Complement. Alternat. Med.* **2017,** *22* (3), 494–501.

Kumar, K.; Kumar, S. Role of Nutraceuticals in Health and Disease Prevention: A Review. *South Asian J. Food Technol. Environ.* **2015,** *1* (2), 116–121.

Mahendra, P.; Bisht, S. Ferula asafoetida: Traditional Uses and Pharmacological Activity. *Pharmacogn. Rev.* **2012,** *6* (12), 141.

Makkar, R.; Behl, T.; Bungau, S.; Zengin, G.; Mehta, V.; Kumar, A.; Uddin, M. S.; Ashraf, G. M.; Abdel-Daim, M. M.; Arora, S.; Oancea, R. Nutraceuticals in Neurological Disorders. *Int. J. Mol. Sci.* **2020,** *21* (12), 4424. https://doi.org/10.3390/ijms21124424

Mandal, S. K.; Maji, A. K.; Mishra, S. K.; Ishfaq, P. M.; Devkota, H. P.; Silva, A. S.; Das, N. Goldenseal (Hydrastiscanadensis L.) and its Active Constituents: A Critical Review of their Efficacy and Toxicological Issues. *Pharmacol. Res.* **2020**. DOI: https://doi.org/10.1016/j.phrs.2020.105085

Martini, N. Liquorice. *J. Primary Health Care* **2020,** *12* (4), 397–398.

Momin, M. Studies in Development and Evaluation of Sennoside Formulations. *Indian J. Pharm. Sci.* **2005,** *67* (4), 458.

Nigam, V. G.; Nambiar, V. S. Knowledge, Practice and Use Of Aegle Marmelos (L.) Correa Leaves Among Naturopathy And Ayurvedic Practitioners Of Vadodara City and Desk Review On Various Commercial Formulations available In Health And Disease Specially Diabetes. *Int. J. Phytomed.* 2017, *9* (3), 451–460.

Nutraceuticals Market Size, Share & Trends Analysis Report By Product (Dietary Supplements, Functional Food, Functional Beverages), By Region (North America, Europe, APAC, CSA, MEA), And Segment Forecasts, 2021 – 2030; Published in Jan, 2022. **https://www.grandviewresearch.com/industry-analysis/nutraceuticals-market (accessed Apr 26, 2022**)

Nwosu, O. K.; Ubaoji, K. I. Nutraceuticals: History, Classification and Market Demand. In *Functional Foods and Nutraceuticals*; Springer: Cham, 2020; pp 13–22.

Pandey, N.; Meena, R. P.; Rai, S. K.; Pandey-Rai, S. Medicinal Plants Derived Nutraceuticals: A Re-emerging Health Aid. *Int. J. Pharma Bio Sci.* **2011,** *2* (4), 420–441.

Pastorino, G.; Cornara, L.; Soares, S.; Rodrigues, F.; Oliveira, M. B. P. Liquorice (Glycyrrhizaglabra): A Phytochemical and Pharmacological Review. *Phytother. Res.* **2018,** *32* (12), 2323–2339.

Pathirana, C. K.; Madhujith, T.; Eeswara, J. Bael (Aegle Marmelos L. Corrêa), A Medicinal Tree with Immense Economic Potentials. *Adv. Agric.* **2020,** *2020.*

Pitt, J.; Leung, Y. Cognitive Effects of Nutraceuticals. *Nutraceuticals* **2016,** 29–48. https://doi.org/10.1016/B978-0-12-802147-7.00003-6

Ramadan, M. F. Ed.; In *Fruit Oils: Chemistry and Functionality*; Switzerland: Springer, 2019; pp 765–788. *10.1007/978-3-030-12473-1*

Rane, B. R.; Bharath, M. S.; Patil, R. R.; Keservani, R. K.; Jain, A. S. Novel Approaches in Nutraceuticals, Chapter 11. In *Enhancing the Therapeutic Efficacy of Herbal Formulations through Novel Drug Delivery Systems*; Kesharwani, R. K., Keservani, R. K., Sharma, A. K., Ed.; IGI Global International Publisher: Pennsylvania, USA, 2020b; pp 241–266. ISBN13: 9781799844532.

Rane, B. R.; Patil, A. S.; Keservani, R. K.; Jain, A. S. Novel Approaches in Herbal Formulation, Chapter 2. In *Enhancing the Therapeutic Efficacy of Herbal Formulations through Novel Drug Delivery Systems*; Kesharwani, R. K., Keservani, R. K., Sharma, A. K., Ed.; IGI Global International Publisher: Pennsylvania, USA, 2020a; pp 43–68. ISBN13: 9781799844532.

Rane, B. R.; Tadavi, S. A.; Keservani, R. K. Naturopathy, Chapter 12. In *Herbal Product Development*; Sharma, A. K., Keservani, R. K., Gautam, S. P., Ed.; Apple Academic Press, CRC Press, Taylor & Francis Group, 2020c; pp 321–347. ISBN: 9781771888776.

Sachdeva, V.; Roy, A.; Bharadvaja, N. Current Prospects of Nutraceuticals: A Review. *Curr. Pharm. Biotechnol.* **2020,** *21* (10), 884–896. DOI: 10.2174/1389201021666200130113441

Santini, A.; Novellino, E. To Nutraceuticals and Back: Rethinking a Concept. *Foods* **2017,** *6* (9), 74. https://doi.org/10.3390/foods6090074

Santini, A.; Tenore, G. C.; Novellino, E. Nutraceuticals: A Paradigm of Proactive Medicine. *Eur. J. Pharm. Sci.* **2016,** *96*, 53–61. DOI: 10.1016/j.ejps.2016.09.003

Sharma, A. K.; Keservani, R. K.; Gautam, S. P. Herbal Product Development, Apple Academic Press, CRC Press, Taylor & Francis Group, 2020. ISBN: 9781771888776.

Sharma, A. K.; Tripathi, M. K.; Kumar, S. Multidimensional Efficacy of Asparagus racemosusWilld: A Medicinal Herb; Lenin Media Private Limited: Delhi, India, 2017.

Simat, V.; Elabed, N.; Kulawik, P.; Ceylan, Z.; Jamroz, E.; Yazgan, H.; ÄŒagalj, M.; Regenstein, J. M.; Fatih, Ã. Recent Advances in Marine-Based Nutraceuticals and Their Health Benefits. *Mar. Drugs* **2020,** *18* (12), 627. DOI: 10.3390/md18120627

Singh, P.; Kesharwani, R. K.; Keservani, R. K. Antioxidants and Vitamins: Roles in Cellular Function and Metabolism. In *Sustained Energy for Enhanced Human Functions and Activity*; Academic Press, 2017a; pp 385–407. ISBN: 9780128054130.

Singh, P.; Kesharwani, R. K.; Keservani, R. K. Protein, Carbohydrates, and Fats: Energy Metabolism. In *Sustained Energy for Enhanced Human Functions and Activity*; Academic Press, 2017b; pp 103–115. ISBN: 9780128054130.

Singh, D. K.; Singh, V. K. Pharmacological Effects of Garlic (Allium SativumL.). *Annu. Rev. Biomed. Sci.* **2008,** *10*, 6–26.

Singh, J.; Sinha, S. Classification, Regulatory Acts and Applications of Nutraceuticals for Health. *Int. J. Pharma. Bio. Sci.* **2012,** *2* (1), 177–187.

Souyoul, S. A.; Saussy, K. P.; Lupo, M. P. Nutraceuticals: A Review. *Dermatol. Ther. (Heidelb)* **2018,** *8*, 5–16. https://doi.org/10.1007/s13555-018-0221-x

Tiwari, R. K.; Singh, L.; Verma, S.; Sharma, V. Formulation and Evaluation of Isabgol and Liquorice-Based Nutraceuticals Floating Tablets for Management of Gastric Ulcer, 2021.

Touloupakis, E.; Ghanotakis, D. F. Nutraceutical Use of Garlic Sulfur-Containing Compounds. In *Bio-Farms for Nutraceuticals*; Springer: Boston, MA, 2010; pp 110–121.

Trüeb, R. M. Brief History of Human Nutrition. In *Nutrition for Healthy Hair*; Springer: Cham, 2020. https://doi.org/10.1007/978-3-030-59920-1_2

Wang, J.; Guleria, S.; Mattheos Ag K.; Yan, Y. Microbial Production of Value-Added Nutraceuticals. *Curr. Opin. Biotechnol.* **2016,** *37*, 97–104. DOI: 10.1016/j.copbio.2015.11.003

Wildman, R. E. C.; Wildman, R.; Wallace, T. C. Eds.; In *Handbook of Nutraceuticals and Functional Foods*, 2nd Ed.; CRC Press, 2006. https://doi.org/10.1201/9781420006186

Williamson, E. M.; Liu, X.; Izzo, A. A. Trends in Use, Pharmacology, and Clinical Applications of Emerging Herbal Nutraceuticals. *Br. J. Pharmacol.* **2020,** *177* (6), 1227–1240. DOI: 10.1111/bph.14943.

CHAPTER 3

CURRENT CONCEPTS AND PROSPECTS OF HERBAL NUTRACEUTICALS

MEENA KAUSAR[1], RAJESHWAR KAMAL KANT ARYA[1], DHEERAJ BISHT[1], and RAJ K. KESERVANI[2]

[1]*Department of Pharmaceutical Sciences, Sir J.C. Bose Technical Campus Bhimtal, Kumaun University, Nainital, India*

[2]*Faculty of B. Pharmacy, CSM Group of Institutions, Prayagraj, Uttar Pradesh, India*

ABSTRACT

Nutraceuticals are food supplements used to diagnose, treat, cure, or prevent diseases and also provide health benefits. Several herbal products have been used as nutraceuticals and they possess many benefits over the synthetic ones, for example, minimal or no side effects and due to natural occurrence, herbal nutraceuticals are economical and better tolerated by the patients. Nowadays, developed as well as developing countries, both are aiming at herbal treatments. A variety of herbs are found in India; Indian traditional systems of medicine such as Ayurveda also depend on herbs or medicinal plants. According to natural sources, categories, and chemical groups, herbal nutraceuticals are also called as- dietary supplements, nutraceuticals herbals, and dietary fibers. Dietary Supplements, Health, and Education Act (DSHEA, 1994) introduced by FDA, mentioned that dietary supplements must contain more than one content such as vitamins, minerals, amino acids, etc. A WHO survey report claimed that 80–85% world's population depends on herbal treatment. In 2007, the global sales of nutraceuticals were $74.6 billion at an

Herbals As Nutraceuticals: Their Role in Healthcare.
Raj K. Keservani, Rajeshwar Kamal Kant Arya, and Rajesh K. Kesharwani (Eds.)

average annual growth rate (AAGR) of 9.9% but the present global market of nutraceuticals is estimated at around $117 billion. In the United States, DSHEA, 1994, regulates the importing, manufacturing, packaging, and marketing of food or dietary supplements and other nutraceuticals. To ensure the safety and efficacy of herbal nutraceuticals processed by nutraceuticals industries, the Indian government passed Food Safety and Standard Act (FSSA) in 2006. This review focused on the safety, efficacy, marketing, and regulatory parameters of herbal nutraceuticals.

3.1 INTRODUCTION

Since ancient times, several herbs, and plants have been explored for their therapeutic values due to possession of abundant phytochemical contents. These phytochemicals either may be beneficial or result in serious secondary reactions known as adverse effects (Wink, 2010). Following the traditional system, the demand for herbal (Sharma et al., 2020; Rane et al., 2020a; Rane et al., 2020b; Rane et al., 2020c) drugs is still increasing in cosmetics, therapeutics, and food supplements, because most plant-derived products are less toxic as compared to synthetic products (Evans, 1994). Egyptian and Chinese papyrus narrates the use of plant-derived drugs, since 3000 BC. In America and Africa, herbal materials were used for healing therapies. In India, the traditional system of medicines in Ayurveda, Siddha, and Unani in which plant-based therapies are used (Ampofo et al., 2012). World Health Organization (WHO) defined the herbal drug as the total sum of practice, skill as well as knowledge gained by theories and experiences employed for the health and also in diagnosis, improvement, treatment, or prevention of physical and mental health (WHO, 2014). Although, plant-derived drugs or herbal drugs are naturally occurring substances that required no or less industrial processing and are used against a variety of diseases (Tilburt and Kaptchuk, 2008). Approximately, 70–80% of health-related issues can be treated with herbal medicines (Pal and Shukla, 2003).

WHO classified the plant-derived medicines in the following categories based on their evolution and current existence forms:

1. **Indigenous herbal products**
   These are plant-derived medicinal products which used in a particular community or society for many years. These products are also easily accessible in these communities.

2. **Herbal products in systems**
   These are the herbal medicines that have been in practice for many years and their particular theories and concepts are also well described.

3. **Modified herbal products**
   These are the herbal products modified in different shapes, and dosage forms by the pharmaceutical industries, so that their mechanism of action, route of administration or application, etc., can be modified (WHO, 2005).

4. **Imported herbal products**
   Many herbal drugs and their raw materials are imported and marketed in different countries. These types of products must be registered in the name of the countries of their origin 8. The different traditional system of medicines is there in which herbal drugs are used to diagnose, heal, treat, or prevent diseases but the availability of these herbal drugs is affected by the environmental and geographical conditions. Therefore, some herbal medicines are restricted only in some particular areas and they are imported by the other countries for their utilization (Conboy et al., 2007).

3.2 HISTORY OF HERBAL PRODUCTS

Traditionally, either a part of plants or whole plants is utilized by a human for their health-related benefits (Raghavendra et al., 2009). It is very difficult to describe the exact year, since when herbal medicines have been utilized. About 5000 years earlier, the handwritten document was found on a Sumerian clay slab in Nagpur, which described 12 drug preparation methods from plants in which approximately 250 plants have been described including henbane, mandrake, and opium poppy, etc. (Kelly, 2009). Indian Vedas also describe the use of herbal medicines. A Chinese writer, Emperor Shen Nung Circa, wrote a book in Chinese (2500 BC) in which approximately 365 medicinal plants and their parts have been described, which are still in use (Bottcher, 1965; Wiart, 2006). Theophrastus (371 BC) classified approximately 500 medicinal plants in his books namely—"De CausisPlantarium"—Plant Etiology and "De HistoriaPlantarium"—Plant History, respectively (Pelagic, 1970; Katic, 1958).

Every country possesses its documents for the utilization of plants, such as India has Ayurveda (Goyal et al., 2012). Charak Samhita (1000 BC) and Sushrut Shamhita (100 AD) describe 1500 plants and 10,000 herbal formulations. Madhav Nidan (800 AD) describes diagnostic methods for about 5000 diseases or disorders. Following are the eight branches of Ayurvedic studies:

1. Kaya chikista (General medicine)
2. Bhuthavidhya (Psychological treatment)
3. Shalya (Surgery)
4. Rasayana (Rejuvenation therapy)
5. Kaumarabhruthya (Pediatrics)
6. Vajeekarana (Sexual vitality)
7. Salakya (ENT, Dentistry, and Ophthalmology)
8. Agadatantra (Toxicology) (WHO, 2005).

The father of medicines, Hippocrates, wrote about 60 medicinal books focused on human anatomy and physiology (Rogers, 2010). He suggested the humoral theory, which describes that the human body is made up of four humors—blood, phlegm, yellow, and black bile, which manage the healthy body as well as diseases in the human body (Israili, 1981). In 19th century, many alkaloids such as morphine, and codeine were isolated from the Opium poppy. Similarly, various glycosides were also isolated such as Digitoxin from *Digitalis pupurea* (Dervendzi, 1992).

3.3 CURRENT SCENARIO

Herbal sources are playing an important role in the evolution of novel drugs. WHO estimated, that about 25–30% of new drugs have been prepared from herbal sources, and WHO also stated that herbal products play a vital role in the primary health care of human beings (Leslie, 2000). Recently, many herbal products have been successfully employed against severe diseases, for example, vincristine, and vinblastine obtained from *Catharanthusroseus* have been used against choriocarcinoma, Hodgkins and non-Hodgkins lymphomas, testicular cancer, throat cancer, leukemia in children, etc. (Farnsworth and Bingel, 1977). Podophyllotoxin (*Podophyllumemodi*) is effective against lungs, testicular carcinomas, and lymphomas. Taxol (*Taxusbrevifolia*) is effectively used in lung and ovarian cancer. Serpentine was isolated from *Rauwolfia serpentine* in 1953, effective against hypertension and tachycardia (Hasan et al., 2009).

About 90% of the African population, 70% of the Indian population, and 40% of the Chinese population depend on herbal products for primary health care (Ernst et al., 2005; Barnes et al., 2007). Herbal products have been employed for both types of diseases acute as well as chronic including depression, inflammation, and cardiovascular disorders, and also as immune system boosters (Moshi, 2005; Russo et al., 2007). Approximately 25,000–30,000 herbal formulations are traditionally used in India, for that 7800 manufacturing units are active which utilized about 2000 tonnes of herbs per year (Bandaranayake, 2010). All the data available suggests that herbal products have a growing market.

3.4 NUTRACEUTICALS

Stephen Defelice, MD, the founder chairperson and of the foundation for innovation in medicine, New Jersey (1989), united the two words nutrition + pharmaceutical as nutraceuticals (Keservani et al., 2010a; Keservani et al., 2020). Further, he defined nutraceuticals as the food or food parts that also possess health benefits by preventing or treating diseases (Trottier et al., 2010). A well-known, Green physician or father of medicines, Hippocrates described "Let food be your medicine" (Kalra, 2003).

Dietary Supplements, Health, and Education Act (DSHEA) mentioned that dietary supplements must contain more than one content such as vitamins, minerals, amino acids, etc. (FDA, 1994). Functional food must consist of components that can provide beneficial effect to the human body besides basic nutrition (Keservani et al., 2010b; Keservani et al., 2017).

TABLE 3.1 Some Components of Functional Food and their Examples (Faisal and Varma, 2009).

Functional food components	Examples
Dietary fibers	Fibers
Cereals and grains	Fibers
Antioxidants	Vitamin C
Digestive enzymes	Enzymes
Inorganic mineral supplements	Minerals
Vitamin supplements	Vitamins
Prebiotics	Digestive enzymes
Probiotics	*Lactobacillus acidophilus*
Health drinks	Fruit juice
Herbs as a functional food	Soya proteins
Phytochemicals	Carotenoids

TABLE 3.2 Vitamins, their Sources, and Health Benefits.

Vitamin name (Vit.)	Their sources	Health-related benefits
Vit. A (Retinol)	Liver, fish oil, spinach, papaya, pumpkin	Good for vision, and exert an antioxidant effect on skin and mucous membrane.
Vit. B1 (Thiamin)	Peas, nuts, wholegrain bread, fruits like bananas and oranges	Helps in the development of neurons.
Vit. B2 (Riboflavin)	Yogurt, cheese, eggs, salmon, dairy milk	Assists the production of energy inside the body.
Vit. B3 (Niacin)	Brown rice, fish, legumes, nuts, seeds, poultry	Assists energy production from the food and also helps in brain functions.
Vit. B5 (Nicotinic acid)	Avocado, nuts, seeds, mushrooms, chicken breast	Maintain proper neurological functions.
Vit. B6 (Pyridoxin)	Pork, poultry, oats, peanuts, fish, bananas	Production of protein inside the body and conversion of that protein into the energy.
Vit. B7 (Biotin)	Nuts, seeds, sweet potato, salmon, avocados, beef liver	Maintain metabolic balance which includes catabolism and anabolism.
Vit. B9 (Folic acid)	Brussels sprouts, leafy vegetables, liver, cereals	Blood cell formation especially RBCs and also essential during pregnancy.
Vit. C	Citrus fruits, peppers, strawberries, potatoes, broccoli	Antioxidant.
Vit. D	Egg yolk, liver oily fish, fortified foods such as cereals	Helps in the absorption and utilization of calcium by the body and is also essential for the formation of bones and teeth.
Vit. E	Sunflower oil, soybean, almonds, beet greens, pumpkin	Blood cell formation, an antioxidant and immunity booster.
Vit. K	Green leafy vegetables such as spinach, canola oil, soybean	Blood clotting factor.

- **Herbal nutraceuticals provides various health-related benefits:**

1. Reduction in adverse reactions.
2. Promote health.
3. Improve appetite.

4. Easily available as well as economical.
5. A part of natural dietary supplements, so they do not result in a serious type of adverse reactions (Zhao, 2007).

- **Various possible mechanism pathways for herbal nutraceuticals—They act as:**

    i. Cofactor in enzyme-assisted reaction.
    ii. Ligand and activate or antagonize various extracellular as well as an intracellular receptor.
    iii. Fermentation substrate for beneficial bacteria.
    iv. Act as a substrate in various enzyme reactions.
    v. Inhibit undesirable enzymatic reactions.
    vi. Increase the growth of beneficial bacteria.
    vii. Inhibit toxic chemicals and harmful bacteria.
    viii. Increase the absorption of beneficial nutrients (Dillard and German, 2000).

- **Classification of nutraceuticals (Kokate et al., 2002; Pandey et al., 2010):**

Nutraceuticals

1) Potential nutraceuticals
(Abundantly available in potent nutraceuticals, it becomes established)

2) Established nutraceuticals
(After finding the efficient clinical data of a nature)

According to their food sources, nutraceuticals may be classified into the following categories:

i. Dietary fibers
ii. Prebiotics
iii. Probiotics
iv. Polyphenols
v. Polyunsaturated fatty acids
vi. Spices
vii. Antioxidant vitamins (Kalia, 2005).

3.5 THE GLOBAL MARKET OF HERBAL NUTRACEUTICALS

There are three types of nutraceuticals industries according to their production:

a. Functional food industries.
b. Natural/herbal product industries.
c. Dietary supplements industries (Pandey et al., 2010).

In 2007, the global sales of nutraceuticals were $74.6 billion at an average annual growth rate (AAGR) of 9.9% but the present global market of nutraceuticals is estimated at around $117 billion. In India, the surveys revealed that the consumption of herbal nutraceuticals is increasing by 21% per year which has valued at INR 44 billion. Nutraceuticals are still seemed to be new to the Indian market, but at the same time, they are very quickly growing in comparison to the global market; it has been estimated that the compound annual growth rate (CAGR) is 18% especially for beverages and functional food, in previous three years (FICCI, 2006).

3.6 REGULATORY AUTHORITIES

If a food supplement claims that it would also provide health or medicinal benefit or would be used as a nutraceutical, then the product would require to comply with the regulatory guidelines in respect of quality, safety, efficacy, and also for marketing permission (Pandey et al., 2010). The Food and Drug Administration (FDA) ensures the safety and efficacy of dietary supplements along with accurate and truthful labeling of these products.

The Indian government introduced the Food Safety & Standard Act (FSSA) in 2006, with the main objective to ensure the quality and safety of food products. The Food Safety & Standard Authorities was appointed to run this activity.

- **Food Safety & Standard Act (FSSA)**

The main benefits of FSSA implementation are:

➢ To ensure the safety and efficacy of food supplements, ultimately provides good health to human beings.
➢ To minimize the corruption.
➢ Correspond to international regulations.
➢ To create the scientific standardization for food supplements.
➢ A harmonization step concerning global authorizations.

In the United States, Dietary Supplement Health and Education Act (DSHEA), 1994, regulates the importing, manufacturing, packaging, and marketing of food or dietary supplements and other nutraceuticals (FICCI, 2006). In India, the Indian Food Safety Standard Act (IFSSA), 2006, ensures the same, for which an Indian Food Safety Standard bill was passed in 2005 by the government of India. There are two main objectives for IFSSA:

i. To initiate a unique standard pattern for all the food supplements for their safety, quality, and efficacy as well.
ii. To ensure that food supplements are manufactured, imported, and marketed as per the scientific procedure.

- **Nutrition Labeling and Education Act (NLEA), 1990**

The NLEA, 1990, ensures the labeling of food products, dietary supplements, and nutraceuticals fulfills the standards, established and enforced by the FDA. There are some benefits of NLEA:

i. To stop the false or misleading claims for food products or nutraceuticals.
ii. To assist the innovation of food products and beverages, good for health as well.
iii. To provide legal security towards functional food or nutraceuticals.
iv. At the same time, regulations could not force anyone to make a false claim regarding products, in cases, where positive claims could not happen (Dureja et al., 2003).

3.7 FUTURE PROSPECTS

For decades, herbal medicines have been used and various methods of medicine preparation for herbs are also there, as they are passing from one generation to others. But there is no written document or recipes are in record for the preparation of these herbal medicines, therefore WHO took an initiative by publishing the method of drug preparation from herbs, so that these herbal medicines can be utilized worldwide (Modak et al., 2007). As a result, many developing countries also started to maintain the ethnopharmacological records on herbal medicines. These records help people to easily understand how safely and effectively herbal medicines can be utilized (Kamboj, 2000; Gunjan et al., 2015). Phytochemicals are abundantly present in plants; these phytochemicals are mainly responsible

TABLE 3.3 Some Herbal Nutraceuticals, their Main Chemical Constituents, and Health Benefits.

Biological name	Part of the plant used	Chemical constituent	Health benefits	References
Allium sativum (Liliaceae)	Dried bulb of garlic	Allicin, alliin	Antigout, antiinflammatory, brain tonic, and anti-bacterial	Thomson and Ali, 2003.
Cassia angustifolia (Leguminosae)	Dried leaves of senna	Sennosides A, B, C, D	Carminative and purgative	Tripathi, 1999.
Zingiberofficinalis (Zingiberaceae)	Rhizomes of ginger	Zingiberene and gingerols	Anti-diabetic, stimulant, anti-asthmatic, and throat pain	Terry et al., 2011.
Glycyrrhizaglabra (Leguminosae)	The dried root of liquorice	Liquirtin and glycyrrhizin	Anti-allergic, anti-inflammatory, and expectorant	Thakur and Raj, 2017.
Allium cepa (Liliaceae)	Dried bulb of onion	Alliin and allicin	Anti-diabetic, anti-atherosclerosis, and antibiotic	*Upadhyay, 2016.*
Aloe barbadensis (Liliaceae)	Dried juice of aloe-vera leaves	Aloesin and aloins	Wound healing property, emollient, anti-inflammatory, and capillaries dilator	Surjushe et al., 2008.
Panax ginseng (Araliaceae)	The dried root of ginseng	Panaxosides and ginsenosides	CNS stimulant and immune system stimulant	Kiefer and Pantuso, 2003.
Hypericumperforatum (Hypericaceae)	The dried aerial part of St. John's Wort	Hyperforin and hypericin	Antidepressants, anti-viral especially anti-HIV and anti-hepatitis C	Qin et al., 2017.
Ginkgo biloba (Ginkgoaceae)	Leaves of Maidenhair tree	Bilobalide and ginkgolide	Antioxidant, cognition enhancer	Mckenna et al., 2001.
Echinacaepurpurea (Asteraceae)	Dried herb of Echinacea	Echinacoside and Alkylamide	Antiviral, immunomodulatory, and anti-inflammatory	Kumar and Ramaiah, 2011.
Curcuma longa (Zingiberacae)	Rhizomes	Curcumin	Anti-cancer, antiseptic, antiarthritic, and antiinflammatory	Omosa et al., 2017.
Hydrastis Canadensis (Ranunculaceae)	The dried root of Goldenseal	Berberine and hydrastine	Astringent, coagulant, and antimicrobial	Scazzocchio et al., 2001.

TABLE 3.3 *(Continued)*

Biological name	**Part of the plant used**	**Chemical constituent**	**Health benefits**	**References**
Centellaasiatica (Umbelliferae)	Herbs of Brahmi	Madecassoside and asiaticoside	Anti-anxiety, brain tonic, and spasmolytic	Arora et al, 2002.
Ferula assafoetida (Umbelliferae)	Oleo-gum-resin of asafoetida	Umbellic acid and ferulic acid	Expectorant, carminative and stimulant	Fatehi et al., 2004.
Aegle marmelos (Rutaceae)	Unripe fruit of Bael	Marmelosin	Anti-diarrheal, digestive, and appetizer	Dhankhar et al., 2011.
Valerianaofficinalis (Valerianaceae)	The dried root of Valeriana	Valerate and valerenic acid	Tranquillizer, migraine, bronchial spasm	Nandhini et al., 2018

TABLE 3.4 Some Marketed Herbal Nutraceuticals Along with their Contents.

Functions	Brand name	Contents	Manufacture
Calcium supplement	Coral calcium	Calcium and mineral traces	Nature's answer
Nutritional supplement	GRD	Carbohydrates, vitamins, minerals, and proteins	ZydusCadila Ltd.
Calcium supplement	Calcirol D-3	Calcium and vitamins	Cadilla healthcare Ltd.
Protein supplement	Proteinex	Predigested proteins, minerals, carbohydrates, and vitamins	Pfizer, Ltd.
Immunomodulator	Amiripash (gold)	Swarnabhasma, RasSindur and Chyawanprash	Uap-Pharam Pvt. Ltd.
Immunity-booster	Celestial Healthtone	Dry fruit extract	Celestial Biolabs Ltd.
Immune supplement	Omega Women	Phytochemicald, for example, Lycopene, resveratrol, vitamins, and antioxidants	Wassen, Surrey, U.K.
Boost immune system	Chyawanprash	Pippali, amal, and ashwagandha	Dabur India Ltd.

for the therapeutic effect of plants. Also, these phytochemicals also are utilized in cosmetics and food supplements (Shakya and Shukla, 2011). However, these plant-derived products became popular for their good impact on health and they are accepted by almost every community across the world. Many cosmetics, perfumery, and food supplement industries are based on herbal sources as their raw materials. Therefore, herbal sources are playing two major roles:

a. Providing health benefits.
b. Economic growth of countries (Kalia, 2017).

In passing years, the demand for herbal materials have been increased and therefore herbal-based industries are growing business (Barnes and Bloom, 2007). Many countries are based on developing countries like India, China, and others to fulfill the demand for herbal raw materials or finished products. Because these developing countries have abundant sources of plants and therefore, these countries stand on top exporting countries of herbal materials across the world. The future of herbal products (e.g., food supplements, nutraceuticals, therapeutics, etc.) is bright because they are cheap, easily available, cause minimum side effects, etc.

3.8 CONCLUSION

Herbal nutraceuticals have been proven to provide health benefits to the human being by preventing diseases or promoting health. Regulatory toxicologists, nutritionists, and health professionals should work along with the utilization of herbal sources, manufacturing herbal nutraceuticals, and ultimately providing good health as well as therapeutic advantages to the human being. Therefore, regulatory bodies are required to ensure standardization in nutraceutical industries. A review of this topic related to nutraceuticals is a must due to the rapid growth of the nutraceutical industries. Herbal nutraceuticals have been proven effective against different acute or chronic nutritional diseases, therefore, herbal nutraceuticals promote better health, longevity, and quality of life as well.

CONFLICTS OF INTEREST

The authors declare that there are no conflicts of interest.

ACKNOWLEDGMENTS

Authors are thankful to Faculty of Pharmacy, DIT University, Dehradun, India, for constant encouragement.

KEYWORDS

- **dietary supplement**
- **current concepts**
- **NLEA**
- **vitamin supplements**
- **Food Safety and Standard Act**

REFERENCES

Ampofo, A. J.; Andoh, A.; Tetteh, W.; Bello, M. Microbiological Profile of Some Ghanaian Herbal Preparation-Safety Issue and Implications for the Health Professions. *Open J. Med. Microbiol.* **2012,** *2* (3), 121–130.

Arora, D.; Kumar, M.; Dubey, S. D. Centella Asiatica - A Review of its Medicinal uses and Pharmacological Effects. J. Natl. Remedies **2002,** *2* (2), 143–149.

Bandaranayake, W. M. Quality Control, Screening, Toxicity, and Regulation of Herbal Drugs. In *Modern Phytomedicine Turning Medicinal Plants into Drugs*; Wiley-VCH GmbH and Co. KGaA: Weinheim, 2006; pp 25–57.

Barnes, P. M.; Bloom, B. Complementary and Alternative Medicine Use Among Adults and Children: United States. *Natl. Health Stat. Rep.* **2007,** *12*, 1–24.

Barnes, P. M.; Bloom, B.; Nahin, R. L. Complementary and Alternative Medicine use Among Adults and Children: United States. *Natl. Health Stat. Rep.* **2008,** *12*, 1–23.

Bottcher, H. Miracle Drugs; Zagreb: Zora, 1965; pp 23–139.

Conboy, L.; Kaptchuk, T. J.; Eisenberg, D. M.; Gottlieb, B.; AcevedoGarcia, D. The Relationship Between Social Factors and Attitudes Toward Conventional and CAM Practitioners. *Complement. Ther. Clin. Pract.* **2007,** *13*, 146–157.

Das, L.; Bhaumik, E.; Raychaudhuri, U.; Chakraborty, R. Role of Nutraceuticals in Human Health. *J. Food Sci. Technol.* 2011.

Dhankhar, S.; Ruhil, S.; Balhara, M.; Dhankhar, S.; Chhillar, A. K. Aegle Marmelos (Linn.) Correa: A Potential Source of Phytomedicine. *J. Med. Plants Res.* 2011, *5* (9), 1497–1507.

Dillard, C. J.; German, J. B. Phytochemicals: Nutraceuticals and Human Health. *J. Sci. Food Agric.* **2000,** *80*, 1744–1756.

Dureja, H.; Kaushik, D.; Kumar, V. Developments in Nutraceuticals. *Indian J. Pharmacol.* **2003,** *35*, 363–372.

Ernst, E.; Schmidt, K.; Wider, B. CAM Research in Britain: The Last 10 Years. Complement. *Ther. Clin. Pract.* **2005,** *11*, 17–20.

Evans, M. A Guide to Herbal Remedies; Orient Paperbacks: New Delhi, India, 1994.

Faisal, N.; Varma, K. S. Nutraceuticals and its Impact on Health Care, 2009. http://farmacists.blogspot.in/2009/05/ nutraceuticals-and-itsimpact-on-health.html.

Farnsworth, N. R.; Bingel, A. S. Problems and Prospects of Discovery New Drugs from Higher Plants by Pharmacological Screening. In *New Natural Products and Plant Drugs with Pharmacological, Biological and Therapeutical Activity*; Wagner, H., Wolff, P., Eds.; Springer: Verlag, Berlin, 1977; pp 1–22.

Fatehi, M.; Farifteh, F.; Fatehi-Hassanabad, Z. Antispasmodic and Hypotensive Effects of Ferula Asafoetida Gum Extract. *J. Ethnopharmacol.* **2004,** *91* (2-3), 321–324.

FDA/CFSAN Resources Page. Food and Drug Administration Website. Dietary Supplement Health and Education Act, 1994. http://vm.cfsan.fda.gov/~dms/dietsupp.html

FICCI Study on Implementation of Food Safety and Standard Act. An Industry Perspective, 2006. http://www. Indiaenvironmentportal.org.in/Files/food_safety_study.pdf.

Gunjan, M.; Naing, T. W.; Saini, R. S.; Ahmad, A.; Naidu, J. R.; Kumar, I. Marketing Trends and Future Prospects of Herbal Medicine in the Treatment of Various Disease. *World J. Pharm. Res.* **2015,** *4* (9), 132–155.

Hasan, S. Z.; Misra, V.; Singh, S.; Arora, G.; Sharma, S. Current Status of Herbal Drugs and their Future Perspectives. *Biol. Forum Int. J.* **2009,** *1* (1), 12–17.

Israili, A. H. Humoral Theory of Unani Tibb. *Indian J. Hist. Sci.* **1981,** *16*, 95–99.

Kalia, A. N. In *Textbook of Industrial Pharmacognocy*; CBS Publisher and Distributor: New Delhi, 2005; pp 204–208.

Kalia, A. N. In *Text Book of Industrial Pharmacognosy*; Oscar Publication: New Delhi, India, 2017.

Kalra, E. K. Nutraceutical - Definition and Introduction. *AAPS Pharm. Sci.* **2003,** *5*, 25.

Kamboj, V. P. Herbal Medicine. *Curr. Sci.* **2000,** *78*, 35–39.

Katic, R. Medicine en Serbie au Moyen Age; Scientific Work: Beograd, 1958; pp 7–36.

Kelly, K. History of Medicine; Facts on File: New York, 2009; pp 29–50.
Keservani, R. K.; Kesharwani, R. K.; Sharma, A. K.; Vyas, N.; Chadoker, A. Nutritional Supplements: An Overview. *Int. J. Curr. Pharm. Rev. Res.* **2010b,** *1* (1), 59–75.
Keservani, R. K.; Kesharwani, R. K.; Vyas, N.; Jain, S.; Raghuvanshi, R.; Sharma, A. K. Nutraceutical and Functional Food as Future Food: A Review. *Der Pharmacia Lettre* **2010a,** *2* (1), 106–116.
Keservani, R. K.; Sharma, A. K.; Kesharwani, R. K. An Overview and Therapeutic Applications of Nutraceutical and Functional Foods. In *Recent Advances in Drug Delivery Technology*, 2017; pp 160–201. ISBN:9781522507543.
Keservani, R. K.; Sharma, A. K.; Kesharwani, R. K. Eds.; Nutraceuticals and Dietary Supplements: Applications in Health Improvement and Disease Management; CRC Press, 2020. ISBN: 9781771888738.
Kiefer, D.; Pantuso, T. Panax Ginseng. *Am. Fam. Physician* **2003,** *68* (8), 1539–1542.
Kokate, C. K.; Purohit, A. P.; Gokhale, S. B. Nutraceutical and Cosmaceutical. In *Pharmacognosy*; Nirali Prakashan: Pune, India, 2002; vol *21*, pp 542–549.
Kumar, K. M.; Ramaiah, S. Pharmacological Importance of Echinacea Purpurea. *Int. J. Pharma Bio Sci.* **2011,** *2* (4), 303–314.
Leslie, T. N. D. Plant Based Drugs and Medicines; Raintree Nutrition, 2000.
McKenna, D. J.; Jones, K.; Hughes, K. Efficacy, Safety, and Use of Ginkgo biloba in Clinical and Preclinical Applications. *Altern. Ther. Health Med.* **2001,** *7* (5), 70–86.
Modak, M.; Dixit, P.; Londhe, J.; Ghaskadbi, S.; Paul, T.; Devasagayam, A. Indian Herbs and Herbal Drugs used for the Treatment of Diabetes. *J. Clin. Biochem. Nut.* **2007,** *40* (3), 163–173.
Moshi, M. J. Current and Future Prospects of Integrating Traditional and Alternative Medicine in the Management of Diseases in Tanzania. *Tanzan Health Res. Bull.* **2005,** *7*, 159–167.
Nandhini, S.; Narayanan, K. B.; Ilango, K. Valeriana Officinalis: A Review of its Traditional Uses, Phytochemistry and Pharmacology. *Asian J. Pharm. Clin. Res.* **2018,** *11* (1), 36–41.
Omosa, L. K.; Midiwo, J. O.; Kuete, V. Curcuma Longa. Medicinal Spices and Vegetables from Africa, 2017; pp 425–435.
Pal, S. K.; Shukla, Y. Herbal Medicine: Current Status and the Future. *Asian Pac. J. Cancer Prev.* **2003,** *4*, 281–288.
Pandey, M.; Verma, R. K.; Saraf, S. A. Nutraceuticals: New Era of Medicine and Health. *Asian J. Pharm. Clin. Res.* **2010,** *3*, 11–15.
Pelagic, V. Pelagic Folk Teacher; Freedom: Beograd, 1970; pp 500–502.
Qin, X. N.; Nandini, V.; Collin, Y. Clinical use of Hypericum Perforatum (St John's Wort) in Depression: A Meta-Analysis. *J. Affect. Disord.* **2017,** *210*, 211–221.
Raghavendra, H. L.; Yogesh, H. S.; Gopalakrishna, B.; Chandrashekhar, V. M.; Sathishkumar, B. P.; Kumar, V. An Overview of Herbal Medicine. *Int. J. Pharm. Sci.* **2009,** *1*, 1–20.
Rane, B. R.; Bharath, M. S.; Patil, R. R.; Keservani, R. K.; Jain, A. S. Novel Approaches in Nutraceuticals, Chapter 11. In *Enhancing the Therapeutic Efficacy of Herbal Formulations through Novel Drug Delivery Systems*; Kesharwani, R. K.; Keservani, R. K.; Sharma, A. K., Ed.; IGI Global International Publisher: Pennsylvania, USA, 2020b; pp 241–266. ISBN13: 9781799844532.
Rane, B. R.; Patil, A. S.; Keservani, R. K.; Jain, A. S. Novel Approaches in Herbal Formulation, Chapter 2. In *Enhancing the Therapeutic Efficacy of Herbal Formulations through Novel Drug Delivery Systems*; Kesharwani, R. K.; Keservani, R. K.; Sharma, A. K.; IGI Global International Publisher: Pennsylvania, USA, 2020a; pp 43–68. ISBN13: 9781799844532.

Rane, B. R.; Tadavi, S. A.; Keservani, R. K. Naturopathy, Chapter 12. In *Herbal Product Development*; Sharma, A. K., Keservani, R. K., Gautam, S. P., Ed.; Apple Academic Press, CRC Press, Taylor & Francis Group, 2020c; pp 321–347. ISBN: 9781771888776.

Rogers, K. In *The 100 Most Influential Scientists of All Time*; Britannica Educational Publishing in Association with Rosen Educational Services: New York, 2010; vol 1.

Russo, M.; Spagnuolo, C.; Tedesco, I.; Russo, G. L. Phytochemicals in Cancer Prevention and Therapy: Truth or Dare? *Toxins* (Basel) **2010,** *2*, 517–551.

Scazzocchio, F.; Cometa, M. F.; Tomassini, L.; Palmery, M. Antibacterial Activity of Hydrastis Canadensis Extract and its Major Isolated Alkaloids. *Planta Med.* **2001,** *67* (6), 561–564.

Shakya, A. K.; Shukla, S. Evaluation of Hepatoprotective Efficacy of Majoon-eDabeed-ul-ward Against Acetaminophen Induced Liver Damage: A Unani Herbal Formulation. *Drug Develop. Res.* **2011,** *72* (4), 346–352.

Sharma, A. K.; Keservani, R. K.; Gautam Surya, P. Herbal Product Development; Apple Academic Press, CRC Press, Taylor & Francis Group, 2020. ISBN: 9781771888776.

Surjushe, A.; Vasani, R.; Saple, D. G. Aloe Vera: A Short Review. Indian J. Dermatol. **2008,** *53* (4), 163–166.

Terry, R.; Posadzki, P.; Watson, L. K.; Ernst, E. The Use of Ginger (Zingiber officinale) for the Treatment of Pain: A Systematic Review of Clinical. *Pain Med.* **2011,** *12*, 1808–1818.

Thakur, A. K.; Raj, P. Pharmacological Perspective of Glycyrrhiza glabra Linn. A Mini-Review. *J. Anal. Pharm. Res.* **2017,** *5* (5), 156.

Thomson, M.; Ali, M. Garlic [Allium sativum]: A Review of its Potential Use as an Anti-Cancer Agent. *Curr. Cancer Drug Targets* **2003,** *3*, 67–81.

Tilburt, J. C.; Kaptchuk, T. J. Herbal Medicine Research and Global Health: An Ethical Analysis. *Bull. World Health Organ.* **2008,** *86*, 594–599.

Tripathi, Y. C. Cassia angustifolia, A Versatile Medicinal Crop. *Int. Tree Crops J.* **1999,** *10* (2), 121–129.

Trottier, G.; Boström, P. J.; Lawrentschuk, N.; Fleshner, N. E. Nutraceuticals and Prostate Cancer Prevention: A Current Review. *Nat. Rev. Urol.* **2010,** *7*, 21–30.

Upadhyay, R. K. Nutraceuticals, Pharmaceutical and Therapeutic Uses of Allium cepa: A Review. *Int. J. Green Pharm.* **2016,** *10* (1), 46.

Wiart, C. In *Etnopharmacology of Medicinal Plants*; Humana Press: New Jersey, 2006; pp 1–50.

Wink, M. Introduction: Biochemistry, Physiology and Ecological Functions of Secondary Metabolites. *Ann. Plant Rev.* **2010,** *40* (2), 1–19.

World Health Organization. National Policy on Traditional Medicine and Regulation of Herbal Medicines, Report of WHO Global Survey, Geneva, 2005.

World Health Organization. Traditional Medicine Strategy: 2014–2023; SAR, China, World Health Organization, Hong Kong, 2013.

Zhao J. Nutraceuticals, Nutritional Therapy, Phytonutrients, and Phytotherapy for Improvement of Human Health: A Perspective on Plant Biotechnology Application, Bentham Science Publishers, 2007. http://www.benthamscience.com/biot/ samples/biot1-1/Zhao.

PART II

ROLE OF HERBAL NUTRACEUTICALS IN COMBATING DISEASE

CHAPTER 4

NUTRACEUTICALS: A POTENTIAL SOURCE IN TREATING HUMAN AILMENTS

HIMANSHU JOSHI[1], ARUN JOSHI[2], ANANT BHANDARKAR[2], and PRANAYA CHODANKAR[2]

[1]*Department of Pharmacology, College of Pharmacy, Graphic Era Hill University, Bhimtal, Uttarakhand, India*

[2]*Department of Pharmacognosy, Goa College of Pharmacy, Goa University, Goa, India*

ABSTRACT

The term "nutraceuticals" was forged in 1989 by the Foundation for Innovation in Medicine (New York, US) to provide a name for this expeditiously growing area of biomedical research. It may be defined as any substance appraised as food or part of a food and contributes toward medical or health benefits, including preventing and treating diseases. Nutraceuticals may extend from isolated nutrients, dietary supplements, and diets to genetically engineered "designer" foods, herbal products, cereals, soups, and beverages. Undoubtedly, many products possess proper physiological functions and valuable biological activities. The ongoing research will lead to a new generation of foods, which will cause the interface between food and drug to become increasingly apparent. The accumulated knowledge about nutraceuticals is unquestionably an excellent challenge for nutritionists, physicians, food technologists, and chemists. Public health authorities contemplate prevention and treatment with nutraceuticals, a powerful instrument in maintaining health and acting against nutritionally influenced acute and

Herbals As Nutraceuticals: Their Role in Healthcare.
Raj K. Keservani, Rajeshwar Kamal Kant Arya, and Rajesh K. Kesharwani (Eds.)

chronic diseases, encouraging longevity, optimal health, and quality of life. Current lifestyle changes are comparable with a substantially escalated risk of chronic diseases, thereby creating a global medical care and commercial burden to society. Therefore, investigators aim to provide adequate treatment for ailing diseases with negligible adverse effects. Recently, many examinations on functional foods have been instigated to obtain foods with fewer adverse effects and increased remedial activity in the past few years. Consequently, efforts were made to isolate constituents from therapeutically active food products. The purpose and spectrum of the manuscript are to raise the readers' cognizance about the use of nutraceuticals in managing various ailments through readily available ingredients and tend to show protective action against multiple disorders. The present study highlights the potential role of nutraceuticals in neurodegeneration, psychotic, cancer, obesity, and cardiovascular diseases.

4.1 INTRODUCTION

In today's lifestyle, people have changed their essential food habits. Consumption of processed food has increased to a greater extent, leading to many diseases like obesity, heart diseases, cancer, osteoporosis, arthritis, and many more. There is a need for complementary or alternative beneficial products, as consumers are frustrated with modern medicine, being expensive and causing various side effects. The condition of the hour is quoted by Hippocrates "Let food be thy medicine and medicine be thy food," this holds good for today's scenario. Nutraceuticals are the soaring class of natural products that fade the line between food and drugs (Adelaja and Schilling, 1999). Nutraceuticals have a long history; only recently, with scientific nutritional and medicinal evidence, nutraceuticals have emerged as a potential tool for treating various ailments (Keservani et al., 2010a; Keservani et al., 2010b; Keservani et al., 2017; Dillard and German, 2000). Since time immemorial, people have relied on natural products to cure various ailments, exhibiting conspicuous results (Bungau and Popa, 2015). Nonetheless, advances in science and technology have led to the discovery and dissemination of numerous phytoconstituents with significant therapeutic activity from plant and non-plant sources, thereby resulting in resurgence in the study of nutrition and human health, paving the way for developing ideal dietary substances and a new term known as nutraceuticals.

The term nutraceuticals (Keservani et al., 2020) have been foraged from the words "nutrition" and "pharmaceutical" by Stephan De Felice, the founder and chairman of the Foundation for Innovation in Medicine (FIM), Cranford, NJ, in 1989 (Maddi et al., 2007). Canada defines Nutraceuticals as a product prepared from foods but traded in pills or powder or any medicinal form, not related foods (Wildman 2001 and Bull, 2000). The European dietary supplements market in 2015 was ca. €14 billion. Italy's biggest market is €1424 million, followed by Russia with €888 million and Germany with €967 million (Statista The statistics Portal, 2016). On the other hand, according to various sources, the US DS market is ca. three to eight times bigger and reached US$40–121 billion.

According to the World Health Organization, the United States spent 17% of its GDP on health care, while India spent 4%. The global nutraceuticals market was worth US$379.061 billion in 2017 and is anticipated to be worth US$734.601 billion by 2026. (Globe News Wire, 2019). Nutraceuticals and functional foods would be worth US$8.5 billion by 2022, according to the Association Chamber of Commerce and Industry of India (ASSOCHAM) (Indian Nutraceutical Market Outlook). From US$1.8 billion to US$5.2 billion, the Indian dietary supplement industry is predicted to develop at a CAGR of 16 percent (Indian Nutraceuticals Market Outlook: Vision 2022).

Today, as Nutraceuticals are at their peak, the scientific community should emphasize the standardization of Nutraceuticals compounds and products. It also includes the development and execution of clinical studies to justify health claims and to produce an impact on consumers and nutraceutical companies making strategic investments (Das et al., 2014). The current state of Nutraceuticals and assimilated information implies that physicians, nutritionists, chemists, and food technologists face a huge problem (Peterson et al., 2017). People have relied on spices and natural ingredients for curing various ailments since the dawn of time, and the results have been astounding (Bungau and Popa, 2015). In accession, science and technology have permitted the analysis and application of a variety of therapeutic ingredients originating from both herbs and nonherbal sources, resulting in a revival in nourishment and human health research, as well as the production of novel dietary components.

Indians, Chinese, Egyptians, and Sumerians are among ancient civilizations that furnished proof of the efficient use of dietary products in medicine and poor health chaos (Orlando et al., 2018), despite being backed by Ayurveda for 5000 years. In a nutshell, a nutraceutical is a functional food that has been shown to have health advantages and nutritional features.

According to evidence, nutraceuticals are developing as a potential technique in caring for various chronic diseases, including nervine disorders.

Herbs were used as a therapy source for numerous diseases all across the planet during the prehistoric age. Herbalism, apothecary, ethnopharmacology, and phytotherapy are examples of alternative medicine that have historically been linked to nutraceuticals (Georgiou et al., 2011; Williams, 2015). Vegetables, animals, mineral-sourced substances, and medicinal plants created the therapy with instinctual and magical elements. Scientists examined the importance of nutrition for both public and individual health before the emergence of dietary supplements. From the time of Hippocrates (460–377 BC) to the rise of contemporary medicine, it was predicted that the variation in illness depends on the diet consumed by society.

Investigating molecules separated from conventional therapies and how they can assist in incapacitating and degenerative ailments is the focus of ongoing research in nutraceuticals (Abdel-Daim et al., 2019). Despite their potential benefits, nutraceuticals have several drawbacks, such as poor bioavailability, brain permeability, metabolism, and so on, which restrict their therapeutic effects (Brown et al., 2009). Regardless, when used in conjunction with certain pharmaceuticals, nutraceuticals can boost the therapeutic effects of certain medications by enhancing multiple pathways, such as enriched re-uptake of blocked monoamines, resulting in remarkable neurological effects.

The New York Institution for Innovation in Medicine, an educational foundation, forged the word "nutraceuticals" in 1989 to encourage rapid biomedical research (Kuhnau, 2013). Europeans profited the most from traditional knowledge gathered from Asian countries (Chanda et al., 2019; Menon et al., 2010). With the concurrent development of recent drugs and clinical trials, the sensible use of drugs enabled the role of pharmacists in findings the principles of drug activity. The Ayurveda (including Sushruta, and Charaka), Ashtavaidya, Unani, and Siddha systems of medicine are learned to have the art of healing techniques.

Clinical testing on animals and people is required in the pharmaceutical development process, and the data gained verify the drug's therapeutic benefits (Granato et al., 2020; Bogue et al., 2017; Champagne et al., 2018). While there were no defined procedures for verifying the medicinal effects of nutritional foods in the past, recently, food molecules have been scientifically demonstrated to prevent lifestyle-related diseases. Nutraceuticals provide various benefits, including increasing the importance of a balanced diet and assisting in living a longer life. Apart from its medicinal benefits, it also has demonstrated psychological benefits; as a result, it is the most popular in preventing neurological illness conditions (Casey et al., 2010; Nicastro et al., 2015). As a result, more populations, particularly the elderly, gravitate toward nutrient-rich meals for lifestyle-related illnesses due to minor perceived adverse effects.

The aim and the scope of this chapter are to boost the understanding of the scholars about the use of nutraceuticals in the governance of various human ailments by the use of components that are readily accessible and manage to exhibit intense pharmacological effects.

4.2 CLASSIFICATION OF NUTRACEUTICALS

Nutraceuticals are categorized as follows.

4.2.1 CONVENTIONAL NUTRACEUTICALS

It consists of food-based commodities acquired from surroundings without varying their chemical nature. These include grains, fish, meat, eggs, vegetables, and milk products which supply many gains ahead of fundamental nutrition (Bhaskarachary, 2016; Bhat and Bhat, 2011).

4.2.1.1 NUTRIENTS

Vitamins, fatty acids, minerals, and amino acids, being primary metabolites, possess nutritional properties in metabolic pathways. The various disorders can be cured by combining these nutrients with plant or animal products. The prevention of fragile bones, inspiriting hemoglobin, boosting muscle power and neuronal transmission can be achieved by nutrients. In addition, brain functioning is enhanced by fatty acids and the reduction in cholesterol available in arteries, consequently proving its hypolipidemic action (Ramalingum and Mahomoodally, 2014; Prakash and Boekel, 2010).

4.2.1.2 HERBS OR EXTRACTS AND CONCENTRATES OF BOTANICAL PRODUCTS

Including mental health, herbals and nutrients provide an incredible impact on life-related disorders (Dohrmann, 2019; Barba, 2020). Lavender containing tannins helps in releasing stress and reduces blood pressure. Peppermint and Menthol contain terpenoids and are used to treat respiratory conditions. Aloe vera illustrates anti-inflammatory and dilating effects, accordingly used in wound healing. Ma-hung retains widening of bronchi and constriction of blood vessels and therefore used in bronchospasms (Williams et al., 2016). Ginger and garlic possess anti-inflammatory, chemotherapeutic, anti-hypertension, and immunity booster properties (Putnik et al., 2019; Putnik et al., 2019; Poojary et al., 2017). The constituents as Nutraceuticals can be obtained from herbal products. For example, vegetables contain carotenoids and thus boost immunity and exhibit anticarcinogenic properties (Montesano et al., 2018). Chickpeas and soya beans help in dissolving cholesterol. Curcumin contains phenolic acids and exhibits maximum antioxidant properties, thus acting as an anti-inflammatory agent. Antioxidant-rich foods, namely ginger, green tea, cumin etc., have demonstrated potential effects on weight loss (Pillitteri et al., 2008).

4.2.1.3 PROBIOTIC MICRO-ORGANISMS

These are a culture of live microorganisms (either an individual microorganism or a blended culture) that possess pharmaceutical benefits to the recipient when administered in a suitable dose (FAO and WHO, 2006). The history of probiotics is as old as 2000 years. Metchnikoff, a Russian Immunologist, coined the word. He disclosed why Bulgarian peasants are healthy and long-lived by consuming fermented milk containing *Lactobacillus* bacteria.

The following list of bacteria are incorporated as probiotics:

i. Lactobacilli such as *L. acidophilus, L. casei, L. delbrueckii, L. Brevis, L.cellobiosus.*
ii. Gram-positive cocci such as *Lactococcus lactis, Streptococcus salivaeius*, and *Enterococcus faecium.*
iii. Bifidobacteria such as *B.bifidun, B.adolescentis, B. infantis, B. longum, B. thermophilum.*

Commercially probiotics are available in powder form, liquid form, gel or paste or granule form, capsule form, etc. They exhibit properties of non-pathogenic, nontoxic resistance to gut epithelial tissues, producing antibacterial substances (Suvarna and Boby, 2005). Evidence indicates that probiotics administration decreases the risk of allergy, asthma, and cancer, including ear and urinary tract infections (Lenoir-wijnkoop et al., 2007).

4.2.1.4 ENZYMES AS NUTRACEUTICALS

Enzymes are biocatalysts made up of proteins and are synthesized in the cells. They provoke metabolic functions to transpire more rapidly and are generally valuable in health issues associated with gastroesophageal reflux disease, constipation, diarrhea, etc. In addition, the use of enzyme supplements can cure recent rare conditions like Hunter syndrome and Gaucher disease of enzyme supplements. Moreover, they are admiringly cost-effective as they are acquired from plants and animal origins.

Food-based nutraceuticals offer a large number of advantages. Nutraceuticals obtained from foods, such as turmeric, ginger, garlic, carotenoids, and dairy products, are much healthier and can provide all the vital nutrients needed by our body. They are readily available in grocery stores and prevent the aggravation of diabetes including cancers. Good mental health along with good diet can serve as a better option for neuroprotection. However, there is a critical need to examine functional foods for their protection prior to release. All substances are poison unless consumed in a small amount. Apart from the anticarcinogen effect, highly functional foods can concurrently act as cardiotoxic. Thus, management of the preferred dose is essential (Putnik et al., 2019; Poojary et al., 2017; Tapal and Kaul, 2019).

4.2.2 NON-TRADITIONAL NUTRACEUTICALS

They consist of foods derived from agricultural products and nutrients, viz., calcium-fortified orange juice, vitamins, and minerals in cereals. The nutritional content of the crops is being effectively altered by the scientists adopting special procedures (Singh and Sinha, 2012; Sapkale et al., 2012).

4.2.2.1 FORTIFIED NUTRACEUTICALS

These are nutraceuticals created through agricultural breeding to improve the content of nutrients, namely iron in flour, folic acid, minerals in cereals, calcium, and milk strengthened with cholecalciferol to treat vitamin D insufficiency, among other things (Ottaway, 2008; Street, 2015).

4.2.2.2 RECOMBINANT NUTRACEUTICALS

These are nutraceuticals derived from applying biotechnology to food items. Dairy products, cheese, and bread are utilized to obtain therapeutically beneficial enzymes (Williams et al., 2015; Andlauer and Fürst, 2002). Nontraditional nutraceuticals have arisen to improve the number of nutrients extracted in current food supplements to contribute more significant benefits for the same amount of food consumed. It has several advantages and is

TABLE 4.1 Different Sources of Nutraceutical Enzymes (Singh and Sinha, 2012).

Plants enzymes	Source	Animals enzymes	Source	Microbes enzymes	Source
Hemicellulase	Plant cell walls	Oxbile	Ox	Hemicellulase	Mushrooms
Pectinase	Cell wall	Pancreolipase	Pancreatic juice	Catalase	
αGalactosidase	Beans, cabbage, Brussels Sprouts, Broccoli,	Trypsin	Pancreatic juice	Amyloglucosidase	Ascomycetes
	Asparagus, other vegetables, and whole grains				
β-Amylase	Higher plants	Chymotrypsin	Allclasses of vertebrates	Glucoamylase	*A. niger,* *Saccharomyco psisfibuligera*
Bromelain	Pineapple	Pepsin	Animals tracheal secretions	Cellulase	All living cells
Biodiastase	Soybean	Lysozyme	Saliva, tears, egg white, and many animal fluids	Invertase Sucrase	Yeast
Glucoamylase	Callus and suspension cultures of sugar beets (*Beta vulgaris* L.) as well as in mature roots	A-Amylase	Saliva	Lactase	β-Galactosidase (bacteria)

a boon in this age of seated behavior. The regulations of the FDA in this are yet to be finalized development of nutraceuticals (Dietary Supplement Health and Education Act of 1994). Many corporations create low-quality nontraditional nutraceuticals to increase sales and margins. Furthermore, increasing nutritional bioavailability is not measured and is often inadequate in many settings. Nutraceutical testing is not as regulated as pharmaceutical testing, implying no set guidelines.

4.3 ON THE BASIS OF MECHANISM OF ACTION

Nutraceuticals have been categorized into antiobesity, antioxidants, antifungal, anti-inflammatory, and antibacterials based on their therapeutic abilities to define their task and analyze their applicability. Food poisoning is responsible for a number of infection-related mortalities. Many bioactive chemicals, such as carbonic acid (terpenoids), quercetin (polyphenols), and others, have been employed as effective antibacterial therapies in treating infectious diseases. They can be found in various fruits and vegetables (Gutiérrez-Del-Río et al., 2018). For many years, traditional foods have been consumed for their anticancer, antidiabetic, and other therapeutic properties. According to various research, medicinal mushrooms have emerged as a strong contender for antifungal medication with the least side effects. This type of mushroom is prevalent. Because of their anti-inflammatory properties, nutraceuticals have been utilized to treat various inflammatory illnesses, including rheumatoid arthritis (Giavasis, 2014; Al-Okbi, 2014). Anti-inflammatory action is higher in foods high in antioxidants, carotenoids, and tocopherols. Nutraceuticals are known for their antioxidant and free radical scavenging effects. As a result of its antioxidant activity, it has a role in treating other illnesses, such as obesity (Cornelli, 2009). As previously said, consuming too much of anything is poisonous. Nutraceutical consumption depending on the mechanism of action is quite beneficial. Consumption of any substance for therapeutic purposes reduces the likelihood of visible toxic features and benefits the individual for the preferred ill condition. In addition, the administration of nutraceuticals in a controlled manner avoids any dangerous side effects.

4.4 ON THE BASIS OF CHEMICAL NATURE OF THE PRODUCTS

The presence of phytoconstituents such as amino acids, carbohydrates, and fatty acids-based substances helps organize nutraceuticals, as each nutraceutical's source is distinctive (Chintale, 2013). The activity a substance is connected with is determined by its chemical composition. Therefore, the consumption of nutraceuticals based on their chemical composition can improve therapeutic activity while diminishing the danger of adverse effects. The outcomes, however, are not as expected. Every person's functioning is unique, and some active chemicals have a proclivity for reacting with host molecules and causing harmful effects.

4.5 NUTRACEUTICALS IN THE TREATMENT OF DISEASES

4.5.1 NUTRACEUTICALS IN THE TREATMENT OF ALZHEIMER'S DISEASE (AD)

Alzheimer's disease (AD) is also recognized as dementia of the Alzheimer's type (SDAT) or the primary degenerative dementia of the Alzheimer's type (PDDAT). It is the most typical form of memory loss (Linseman, 2009). Management of Alzheimer's disease can be achieved by Nutraceuticals, which are helpful in the control of the disease. Vital antioxidants can treat chronic ailments due to oxidative stress, critical in neurological disorders, including AD (Frisardi, 2010). Oxidative stress is accelerated by a lack of intake of dietary antioxidants and ageing, leading to disease progression and stimulation (Nunomura, 2001). Further, investigators suggest that AD prohibition is not as tricky as supposed.

The intake of food commodities rich in polyunsaturated fatty acids and saturated and trans fatty acids tends to repress neurodegeneration. Conversely, foods rich in unsaturated fatty acids can improve neurodegeneration (Dohrmann et al., 2019).

Following is the list of compounds beneficial in treating Alzheimer's disease.

4.5.1.1 FLAVONOIDS

Catechin, epicatechin, epigallocatechin, and epigallocatechin gallate are the flavonoids (Keservani and Sharma, 2014; Keservani et al., 2016a; Keservani et al., 2016b) used in neurodegenerative disorders, mainly Alzheimer's disease. These are a group of polyphenolic constituents found in the human diet. Vegetables, fruits, and drinks, such as coffee, tea and wine, are the resources for flavonoids. They maintain nervine-modulating activities and are being regarded to connect with the neuronal-glial signaling route, associated with the existence and functioning of neurons (Glevitzky et al., 2019). In addition, the flow of blood through the cerebrum is controlled by the upregulated action of antioxidant enzymes and proteins, which induces synaptic plasticity and restores neuronal functions by hindering neuropathological processes in the brain associated with AD (Mecocci et al., 2014).

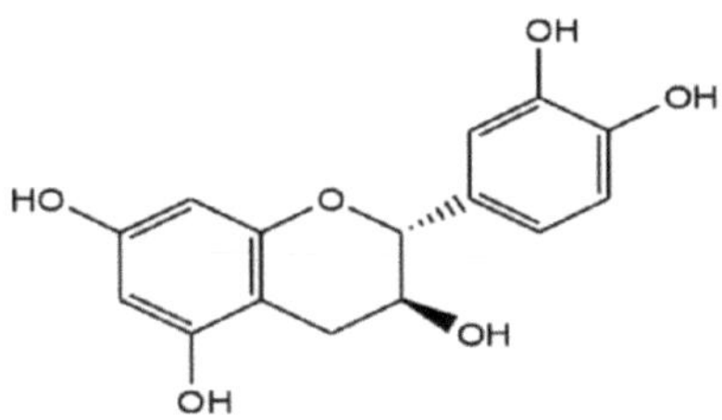

Catechin

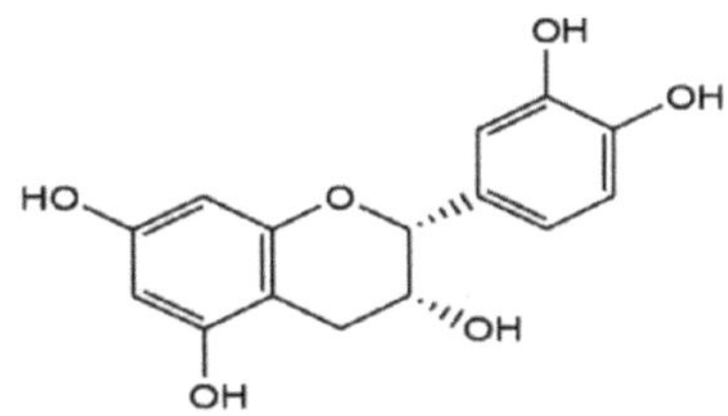

EpiCatechin

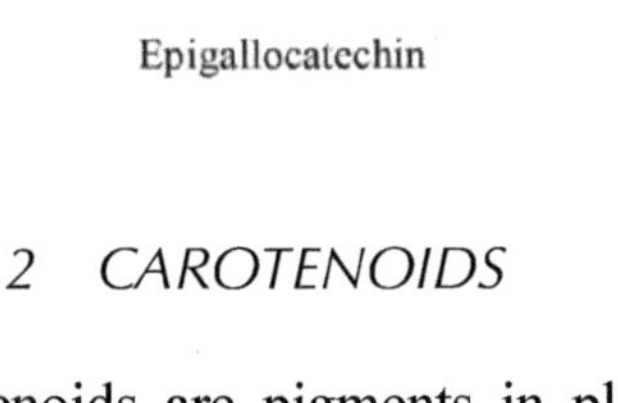

Epigallocatechin

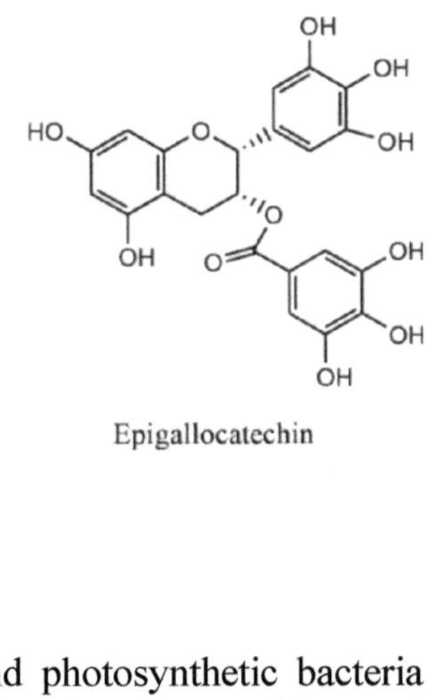

Epigallocatechin

4.5.1.2 CAROTENOIDS

Carotenoids are pigments in plants and algae and photosynthetic bacteria producing bright yellow, red, and orange colors in plants, vegetables, and fruits. More than 600 different carotenoids are identified, out of which 40 are found in human tissues and blood. α and β carotene, β-cryptoxanthin, lutein, zeaxanthin, and lycopene are the major carotenoids in humans. Carotenoids are fat-soluble compounds, unlike some protein-rich foods and vegetables. Therefore, cooking and chopping carotenoid-rich food increase the strength of nutrients when they enter the bloodstream. Astaxanthin belongs to the class of carotenoids and appears as red pigment. It is found naturally in freshwater microalgae, *Haematococcus pluviallis* and the yeast fungus *Xanthophylomyces dendrorhou*. It has been analyzed for its antioxidant properties in both *in vitro* and *in vivo* instances. In addition, it is identified as a potent neuroprotective compound by studying microcirculatory and mitochondrial protective functions. Patients suffering from moderate or severe AD lack major carotenoid constituents namely beta carotene and lutein, compared with those suffering from mild AD (Bungau et al., 2019; Castellani et al., 2001).

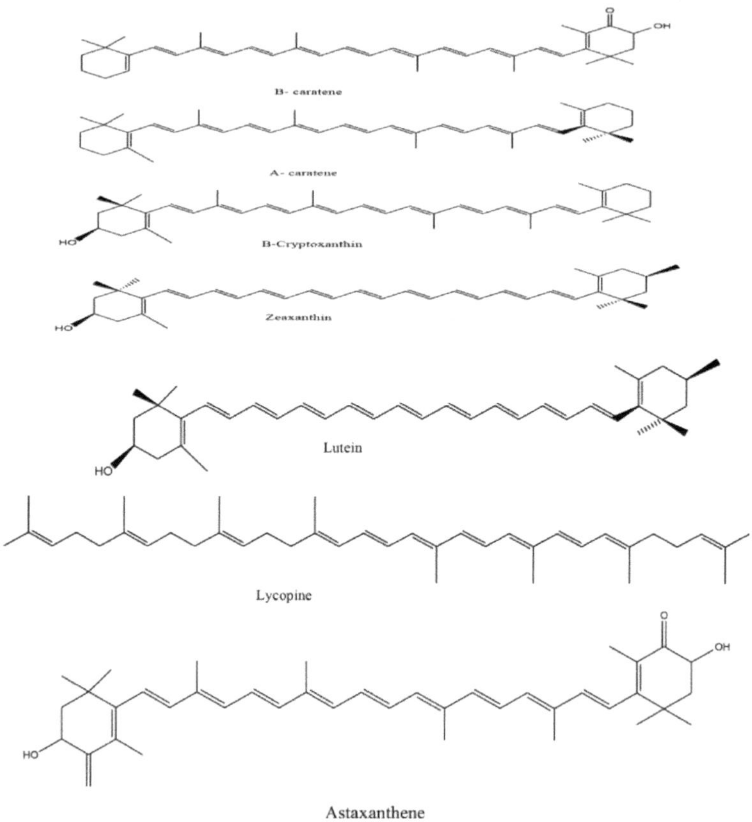

4.5.1.3 CROCIN

Crocin is a major phytoconstituent obtained from the dried stigma of *Crocus sativus* (Iridaceae). They belong to a class of mono and di-glucosyl esters of polyene dicarboxylic acid crocetin (Lozano et al, 1999). It appears a deep red color and forms crystals with a melting point of 186°C. It possesses carminative, gingival sedative, neurine sedative, expectorant, stimulant, and antispasmodic effects for ages. Crocin is believed to improve learning and enhance memory and is utilized in neurodegenerative disorders. Crocin enhances cognition by CDR-SD and ADAS-Cog-mediated enzymes in mild to medium AD patients. A literature survey revealed that crocin could drastically tweak the degree of oxidative markers in the hippocampus territory and terminate the adverse consequences on learning and memory due to chronic stress (Boccardi et al., 2017).

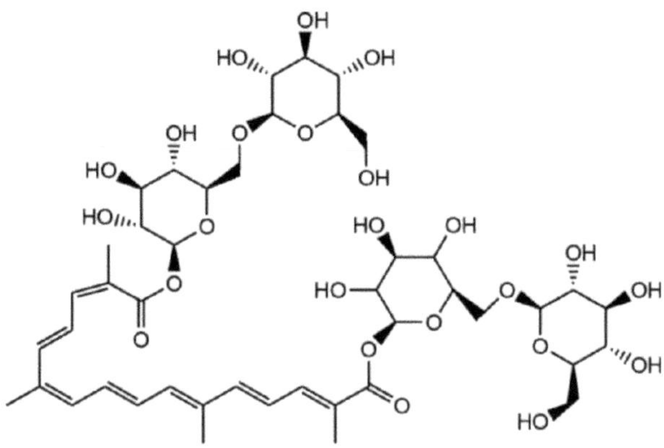

Crocin

4.5.1.4 CYANIDIN

It is a natural organic compound under the class of anthocyanidins. It is a pigment found in many red-berried, including grapes, blueberry, bilberry, blackberry, cherry, chokeberries, cranberries etc. It exerts a potent anti-inflammatory and neuroprotective effect by hindering the activation of proinflammatory cytokines and eventually harming the brain cell. The activity is mainly due to the blockage of phospholipase A2, which is primarily concerned with the signaling of proinflammatory cytokines and oxidative stress spectrum, the discretion of which offers outstanding neuroprotection.

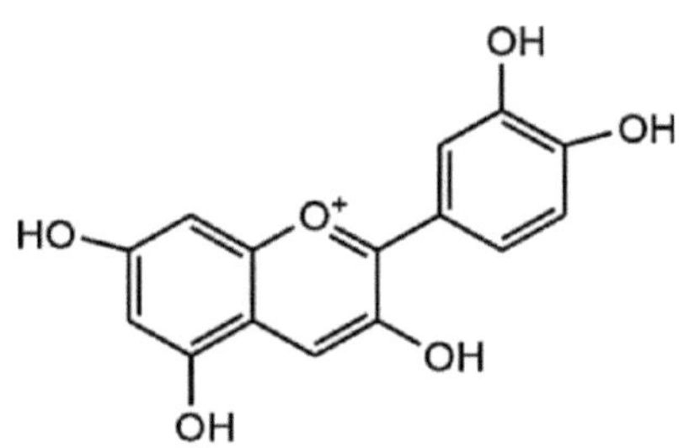

Cyanidin

4.5.1.5 LUTEOLIN

Flavons such as Luteolin and apigenin possess remarkable neuroprotective activity. Rosemary, parsley, and celery are the principal sources of these compounds. These phytoconstituents possess the capacity to guard DNA against hydrogen peroxide-mediated toxicity and help in discouraging inflammation and cell damage in Alzheimer's disease (Sawmiller et al., 2014; Wang et al., 2016).

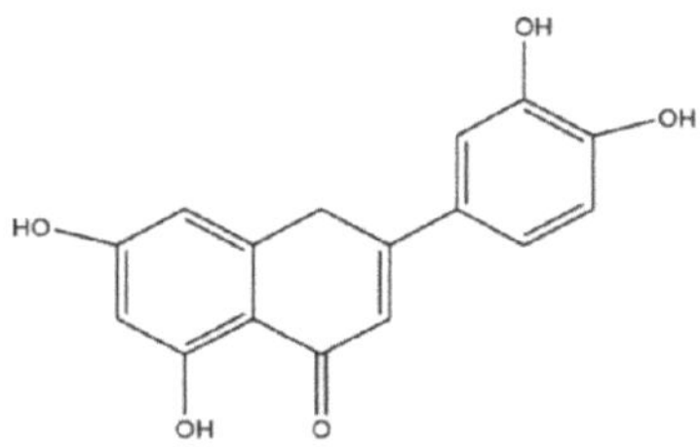

Luteolin

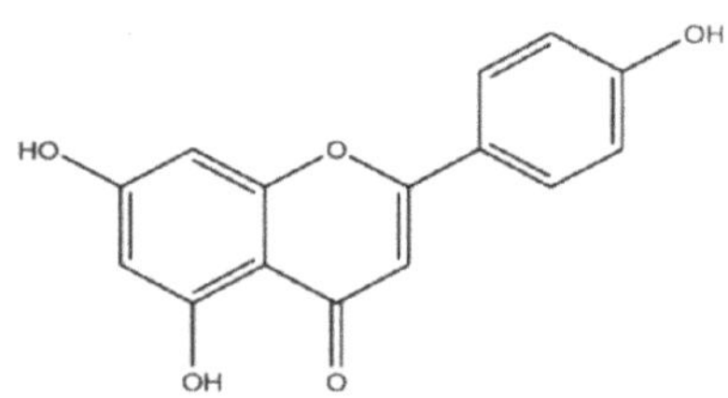

Apigenin

4.5.2 NUTRACEUTICALS IN THE TREATMENT OF DEPRESSION

Depression is a psychic disorder with glum or despairing mood symptoms merged with a diminished interest in civic involvement, directing to a disrupted schedule. Its occurrence is about 15%, with a yearly incidence of 7%. Omega-3 fatty acids and vitamin B9 are usually adequate against major depressive disorder, specifically as ancillary therapy, expanding proof of its effectiveness as a therapy (Ceskova and Silhan, 2018). In addition, zinc, vitamin B9, omega-3 fatty acids, micro and macronutrients can initiate the functioning of the brain and manage the depression condition (Van der Burg, 2019).

St. John's Wort (*Hypericum perforatum*) is extensively used as an antidepressant. Its major components are flavanol glycosides, including hypericin, quercetin, rutin, and hyperforin. The antidepressant activity of the plant is due to the inhibition of the enzyme monoamine oxidase (MAO). Carbon dioxide (CO_2) extracts fortified with adhyperforin and hyperforin subdued the reuptake of norepinephrine, dopamine, and serotonin and exhibited antidepressant effects (Butterweck and Schmidt, 2007).

TABLE 4.2 Commonly Employed Nutraceuticals in The Management of Depression (Lama et al., 2020; Ceskova and Silhan, 2018).

Constituents	Mechanism of action
Omega-3fatty acid molecules	These benefit neurotransmission by enhancing the fluidity in cell membranes by reducing monoamine re-uptake during neurological transmission. These molecules also reduce the production of inflammatory mediators, prevent depression, and boost neurogenesis.
N-acetyl cysteine	It primarily consists of antioxidant and anti-inflammatory properties. They promote glutathione rejuvenation and neurogenesis. It prevents mitochondrial damage and inhibits depression via modulating the glutamate system.
S-adenosyl methionine	It is involved in the synthesis and bioconversion of neurotransmitters since it is a major methyl donor. It reduces prolactin release while increasing phosphatidylcholine conversion.
L-Tryptophan /5-HTP	Tryptophan is essential to convert serotonin in the presence of pyridoxin and magnesium to form 5-HTP. The enhancement of (2S)-2-amino-3-(1H-indol-3-yl) propanoic acid with a variety of antidepressants effectively increases the effect. It is utilized with various antidepressants, patients with unregulated serotonergic pathways or protein deficiency.
Cholecalciferol	It is a "neurosteroid" that serves as a polypeptide for receptors found in the prefrontal cortex, substantia nigra, hypothalamus, and brain. It is primarily responsible for regulating the genetic expression that leads to summarizing the protein tyrosine hydroxylase.
Zinc	Zinc is the most abundant microelement in the brain's neocortex, amygdale, and hippocampus territory. By boosting BNDF, it primarily causes amplification of neurogenesis in hippocampal areas. In addition, the glutamate and NMDA receptors' activity is also altered.

4.5.3 NUTRACEUTICALS IN THE TREATMENT OF PSYCHOTIC DISORDERS

Nutraceuticals also play a vital role in managing schizophrenia and bipolar disorder (Cloutier et al., 2016). Omega-3 fatty acids and vitamins are commonly used nutraceuticals in psychosis.

The human body contains two varieties of polyunsaturated fatty acids. Arachidonic acid (AA) belongs to omega-6 series and is obtained from linoleic acid, whereas alpha-linolenic acid belongs to the omega-3 series. All these elements are necessary for the survival of the human body as they are not synthesized in the body. Instead, they are supplied through diet and regulate the function of mitochondria, a primary source of oxidative stress (Brown and Roffman, 2016; Sarris et al., 2015). In addition, they safeguard against toxicity due to cell death and normalize gene expression of brain-derived neurotrophic factors (BDNF). As the human body cannot synthesize vitamins in adequate amounts, they need to be supplied through the diet.

TABLE 4.3 Commonly Used Nutraceuticals in Various Neurological Aliments (Ceskova and Silhan, 2018; Lama et al., 2020; Preethi and Sampath, 2018).

Disorders	Mechanism of action	Examples of frequently used nutraceuticals
Neuro-degenerative disorders	Protein misfolding is a major cause of neurodegenerative disorders. Nutraceuticals prevent such disorders by interfering with the activation and production of proinflammatory cytokines and adjuvant pathways, which prevents misfolding of proteins.	Brahmine, Bacoside A, bacoside B, withanine, kaempferol, somniferine, Asiatic acid, bhilavanol A and B.
Alzheimer's disease	It is chiefly associated with an upsurge in oxidative stress and free radicals. Hence in managing Alzheimer's disease, nutraceuticals with antioxidant potential are utilized.	Flavonoids (fruits, vegetables, tea, wine coffee); carotenoids (lutein, zeaxanthine, lycopene, β-cryptoxanthine including α and β carotenes); anthocyanidins (cyanidin); flavones (luteolin, apigenin)
Parkinson's disease	Factors such as uninhibited oxidative stress and free radicals, abnormally misfolded proteins, neuroinflammation, and dysfunctional mitochondria compromise cellular metabolism and energy. This affects the normal functioning of the brain and leads to neurogenerative disorders including PD.	Vitamin A, Omega-3 fatty acids, lycopene, vincamine, gallic acid, curcumin, Mito Q.
Depression	Depression can be managed with nutraceuticals that inhibit the re-uptake of monoamines, having antioxidant and anti-inflammatory properties.	Omega-3 fatty acids, folic acid, S- adenosyl methionine, zinc, N-acetyl cysteine, L-Tryptophan/5-HTP, Vitamin D.
Psychosis	Nutraceuticals that can enhance neurotransmission in dopaminergic serotoninergic neurons can be engaged to manage psychosis.	Omega 3 fatty acids and various vitamins.

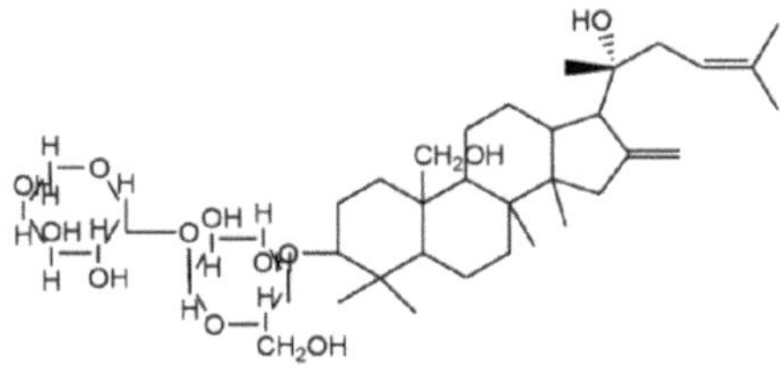

Bacoside A

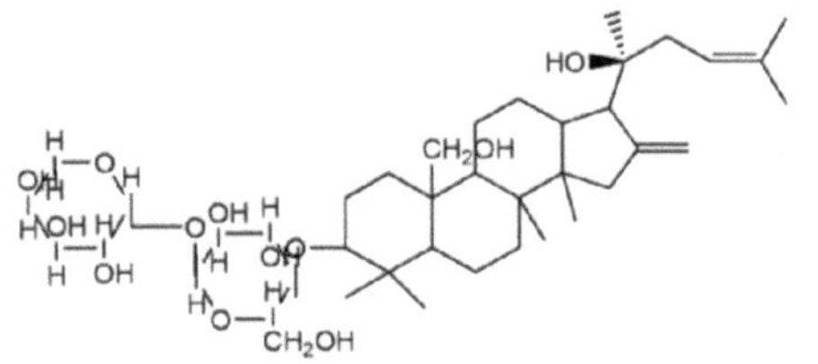

Bacoside B

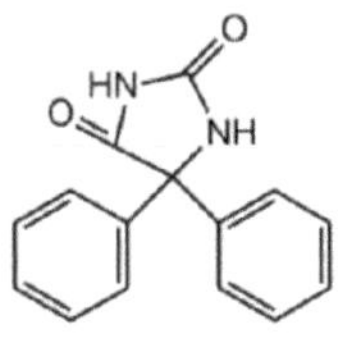

Brahmine

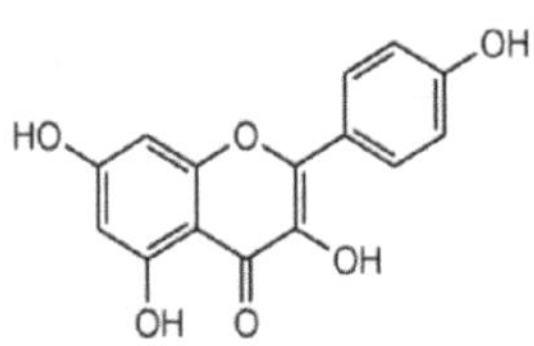

Kaempferol

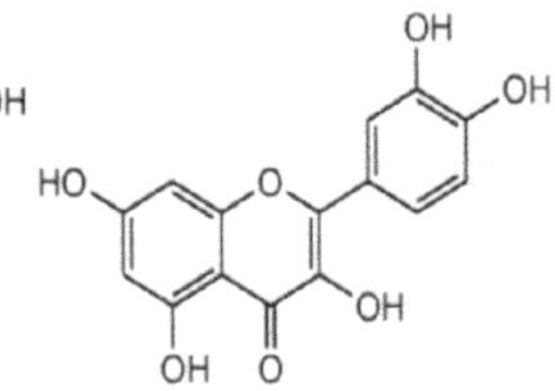

Quercetin

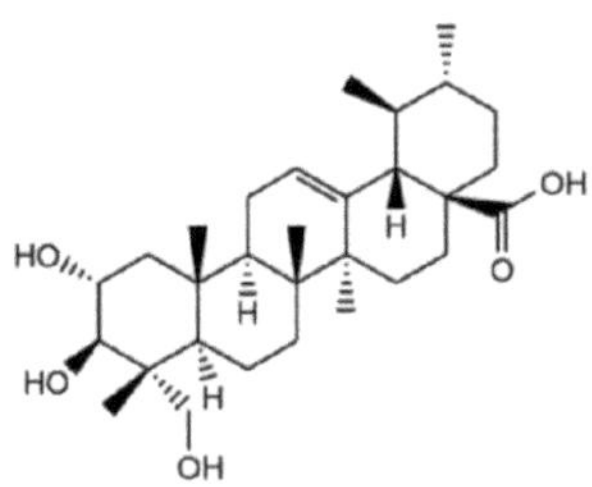

Asiatic acid

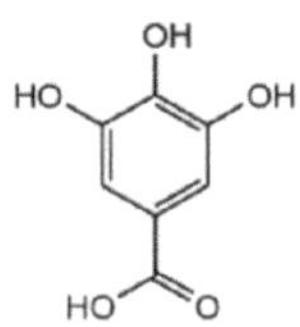

Gallic acid

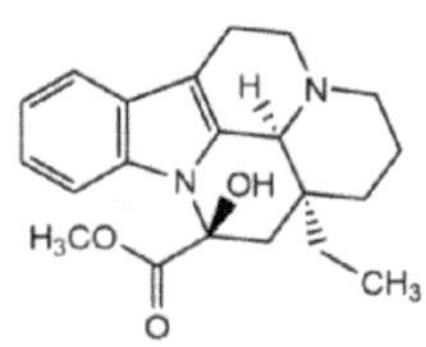

Vincamine

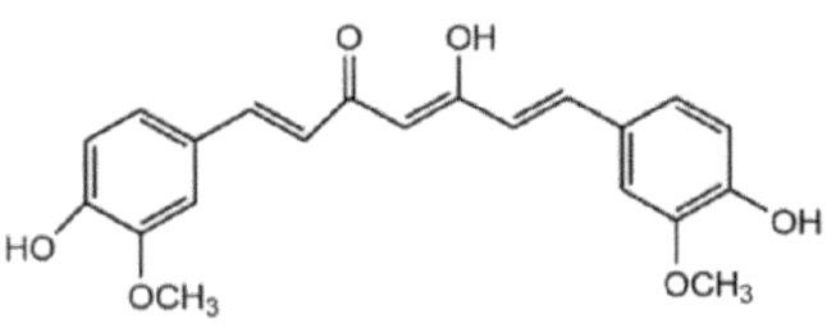

Curcumin

4.5.4 NUTRACEUTICALS IN THE TREATMENT OF PARKINSON'S DISEASE

Parkinson's disease (PD) is a neurological disease with deficient dopaminergic neurons in the substantia nigra par compacta part of the brain, directing to an extreme dopamine deficit (DA). Depletion of antioxidants, oxidative stress, impairment of mitochondria, etc., are the factors that contribute to neurodegeneration leading to PD (Mythri and Kumar, 2015). However, vitamins and their supplements displayed mixed consequences in managing PD symptoms. Hence, there is a necessity to establish evidence of their action against PD. Folates, cholecalciferol, alpha-tocopherol, and ascorbic acid can be utilized in Parkinson's disease (Tatton et al., 2003). However, drugs currently utilized prevent disease progression by furnishing symptomatic alleviation only. However, products obtained from natural and herbal (Sharma et al., 2020; Rane et al., 2020a; Rane et al., 2020b; Rane et al., 2020c) sources have been investigated clinically for PD treatment (Jenner and Olanow, 1996). Natural products exercise favorable results on PD by inhibiting the various pathologic pathways influencing it, viz. apoptosis dysfunction of mitochondria, oxidative stress, and neuroinflammation.

The presence of eicosapentaenoic and docosahexaenoic omega-3 fatty acids in fish oil exhibits neuroprotective effects by numerous routes. Epigallocatechin-3 gallate (EGCG) is a dominant polyphenol acquired from *Camellia sinensis* and has proved to exhibit neuroprotective actions by surpassing the blood–brain barrier (BBB). The catechol-like structure of EGCG is responsible for the iron chelation and radical scavenging activity. It considerably elevates the DA levels in the striatal territory of the brain in diseased patients and thus decreases neurotoxicity. Ginseng (*Panax ginseng* belonging to the family Araliaceae) and its by-products, ginsenosides, displayed neuroprotective activity in several investigations on PD. The antioxidant activity of ginsenoside is related to its ability to manage glutathione levels and the reactive oxygen species-mediated NF-kB pathway and governance of the transport of iron and affiliated proteins, thereby causing depleted stores of iron in the nigral region of the brain. A monoterpenoid indole alkaloid, Vincamine, obtained from *Catharanthus roseus*, has demonstrated anti-PD activity via various mechanisms of action. It maintains vasodilation activity and induces muscle relaxation of the capillaries in neurons, generating an increased flow of nutrients and glucose to the brain with a parallel growth in ATP generation through the Krebs cycle.

Oxidative stress and iron are also targeted by vincamine to enhance the generation of DA and lessen neuronal impairment. By reducing the synthesis of reactive oxygen species (ROS) and iron-chelating molecules namely mito Q, the role of vincamine and its derivatives, vinpocetine, can be summarized in the management of PD. Triphenylphosphine, a lipophilic cation of mito Q, is accountable for antioxidant activity and regulates the performance of the respiratory chain. In addition, apocynin, a natural antioxidant, is scrutinized for its PD-protective action (Navarro and Boveris, 2009; Lama et al., 2020).

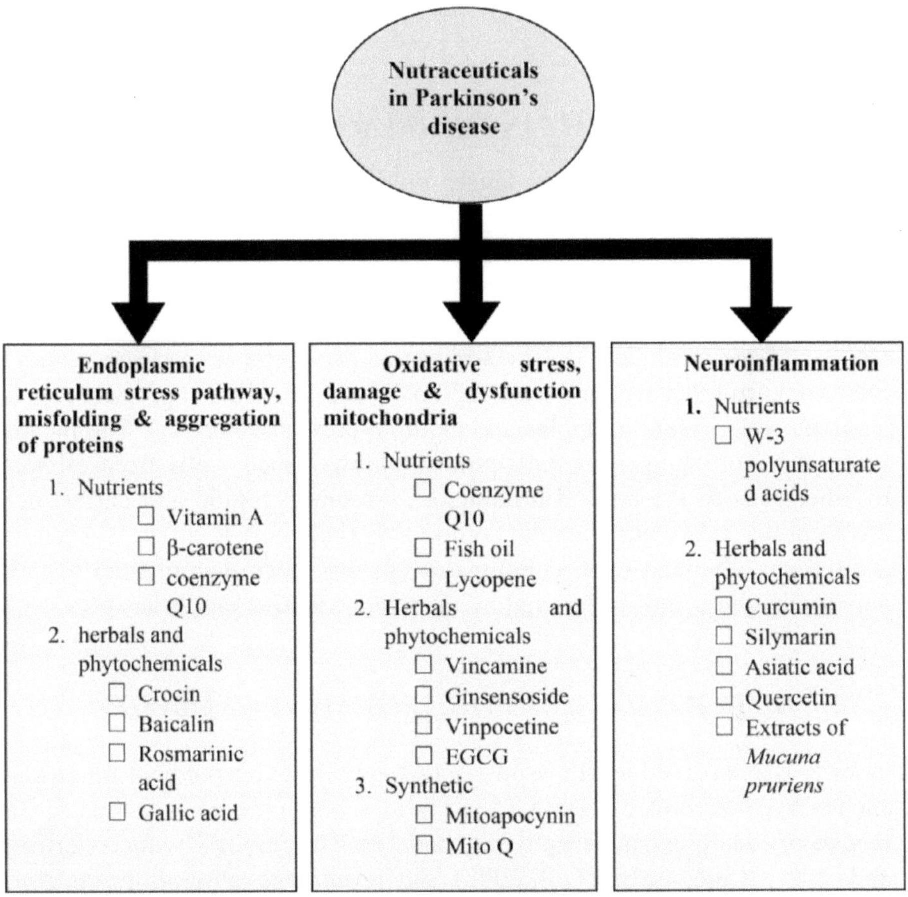

FIGURE 4.1 Mechanism of action of nutraceuticals in Parkinson's disease.

4.5.5 NUTRACEUTICALS IN THE TREATMENT OF CARDIOVASCULAR DISEASES (CVD)

CVD influences the heart or blood vessels. Intake of less vegetables and fruits will increase CVD risk (Rafieian-Kopaei, 2012). However, reports suggest that a fruit and vegetable-rich diet prevents CVD risk. Therefore, nutraceuticals like vitamins, minerals, dietary fibers, omega-3 polyunsaturated fatty acids and antioxidants, generally with a physical workout, are recommended to avoid and treat CVD. Polyphenols reduce the chance of arterial diseases by modifying cellular metabolism and cell signaling (Asgary et al., 2014). Flavonoids play a vital role in the interruption of CVD. Platelet aggregation is prevented by blocking angiotensin-converting and cyclooxygenase enzymes (Nasri et al., 2014).

4.5.6 NUTRACEUTICALS IN THE TREATMENT OF CANCER

Cancer is considered to be a severe public health problem. Carotenoids, a secondary metabolite, exhibit antioxidant activity and are helpful in the prevention of cancer. Lycopene, a Carotenoids, is beneficial in preventing cancer disease (Willis and Wians, 2003). Vegetables and fruits rich in Lycopene possess an anticancer effect by reducing oxidative stress and injury to DNA (Shirzad et al., 2013). The sources of Lycopene are tomatoes, guava, pink grapefruit, papaya, and watermelon. Pectin can prevent prostate cancer, a soluble fiber found in apples and exhibits preventive effects. The mechanism of action against prostate cancer inhibits cancer cells from clinging to other cells in the body. Gallic acids, curcumin, ferulic and caffeic acid are naturally occurring phenolic compounds (Nasri et al., 2014). Curcumin, a diterpene obtained from Curcuma longa, has been documented to have anticarcinogenic, anti-inflammatory, and antioxidative properties.

4.5.7 NUTRACEUTICALS IN THE TREATMENT OF OBESITY

Obesity is a universal health issue associated with severe medical conditions, such as hypertension, congestive heart failure, angina pectoris, osteoarthritis, respiratory disorders, hyperlipidemia, renal vein thrombosis, reduced fertility, and cancer (Caterson and Gill, 2002). The prominent cause of obesity is the consumption of a high-fat diet. Therefore, nutraceuticals are currently being considered potential for managing obesity. Nutraceuticals such as capsaicin conjugated linoleic acid, *Citrus aurantium*, *Momordica charantia*, and Psyllium fiber possess antiobesity properties (Nasri et al., 2014, De Freitas et al., 2017). In addition, natural stimulants, such as caffeine, green tea, ephedrine, and chitosan, effectively facilitate bodyweight loss.

4.6 DOSAGE OF NUTRACEUTICALS

The protective role of nutraceuticals has been proved by conducting both in vitro and in vivo studies. However, in many cases, it is devoid of clinical proof. At dietary doses, Nutraceuticals are considered safe (Gonzalez-Suarez et al., 2016). However, higher doses of nutraceuticals intake may cause toxic genomic changes in target tissues. For instance, Natural biologics produced from plants and Vitamins demonstrate beneficial effects on human health, but purified vitamins can also be toxic. High input of polyphenols results in harmful consequences. Herbs and spices are harmless but, in higher doses, produce toxicity. The effect is due to their interactions with other pharmaceuticals. For example, the consumption of uncooked garlic juice (4 mL/kg) or garlic (100 mg/kg) may lead to a loss in weight, anemia, kidney, liver, and heart toxicity (Banerjee et al., 2003). Garlic (5–6 cloves) and onion are suitable for lowering cholesterol (Das et al., 2012). However, onion at a dose of 500 mg/kg causes damage to the tissues and lungs (Ali et al. 2000). Diabetes can be prevented by the consumption of onion (50 g), garlic (5–6 cloves), fenugreek seeds (25–50 g), and turmeric powder (1 pinch) daily. A reduction in blood lipid level (33%) is observed upon regular intake of curcuminoids (0.5 g). Hence, taking an ideal dose of any food item is essential to benefit health. Consumption of Nutraceuticals within the optimum recommended dietary input will control diseases and helps in maintaining good health.

4.7 FUTURE CHALLENGES

- The concept of nutraceuticals should be redefined and also be able to differentiate from food supplements, botanicals, and pharmaceuticals.
- A shared regulation is essential for food supplements and nutraceuticals, as they possess different approaches and principles in various countries worldwide.
- The researchers and scientists should address the potential use and mode of action in facing numerous pathological health conditions.
- Nutraceuticals must include clinical evidence substantiated by clinical data in assessing pathological target identification, safety, and efficacy.
- Micro and nano-Nutraceuticals are emerging intending to target pathological conditions for treatment. The gap between prevention of diseases and management of chronic diseases can be shortened by using natural, sustainable, and environmentally friendly substances.
- The issue of bio-availability of Nutraceuticals to be addressed due to limited bioaccessibility, poor absorption, and chemical transformation within the gastro-intestinal tract.
- Need for pertinent description and specific classification (Daliu et al., 2019).

4.8 CONCLUSIONS

The use of nutritional supplements is endorsed by many people and is growing fast because of their all-natural origin. The demand for fewer synthetic pharmaceuticals is triggering this interest and stimulating the industry to develop and put new products that claim beneficial health effects. However, there is a firm limit on the lack of a shared regulation for nutraceuticals, often confused with other food-derived preparations, e.g. food supplements. This gap should permeate by conceding the unicity of nutraceuticals as pharma foods, categorically accepting an added value in prevention and therapy. Nutraceuticals cannot replace pharmaceuticals but can be a robust high, value tool for prevention and aid in therapy of some pathological conditions. Therefore, regulation is essential to differentiate them sharply from all the other food-derived products available (e.g. functional foods, pre and probiotics, food supplements, botanicals) and pharmaceuticals.

Nutraceuticals can serve as alternative natural remedies that prevent and cure many serious ailments. They are more valued than pharmaceutical drugs in the commercial market as they are more economical and readily available. Long-term research is essential to understand the benefits and mechanisms of curing diseases. The study of the effects of Nutraceuticals on prognosis and activity against diseases is the need of the hour. The impact of secondary metabolites on other food ingredients is essential. The effect of this interaction on the usefulness of secondary components to function as a drug is needed. The microbial instability of these active constituents needs to be optimized upon their incorporation into the food. More rapid, precise, and standardized clinical trials are needed to boost their acceptability in the commercial market globally. Nutraceuticals are proven to possess health benefits, and their intake will keep diseases at bay and helps in maintaining good health.

ACKNOWLEDGMENTS

The authors thank the officials of the Government of Goa and the Principal of Goa College of Pharmacy for their encouragement and kind support. Special thanks to Dr. Rajashree Gude and Dr. Shailendra Gurav for their help in preparing the manuscript.

KEYWORDS

- **nutraceuticals**
- **phytoconstituents**
- **neurodegeneration**
- **psychotic**
- **cancer**
- **obesity**
- **cardiovascular diseases**

REFERENCES

Abdel-Daim, M. M.; El-Tawil, O. S.; Bungau, S. G.; Atanasov, A. G. Applications of Antioxidants in Metabolic Disorders and Degenerative Diseases: Mechanistic Approach. *Oxidative Med. Cell. Longev*. **2019**.

Adelaja, A. O.; Schilling, B. J. Nutraceuticals: Blurring the Line Between Food and Drugs in the Twenty-first Century. *Choices*. **1999,** *14* (4), 35–39.

Ali, M.; Thomson, M.; Afzal, M. Garlic and Onions: Their Effect on Eicosanoid Metabolism and its Clinical Relevance. *Prostaglandins Leukot. Essent. Fatty Acids*. **2000,** *62* (2), 55–73.

Al-Okbi, S. Y. Nutraceuticals of Anti-Inflammatory Activity as Complementary Therapy for Rheumatoid Arthritis. *Toxicol. Ind. Health.* **2014,** *30*, 738–749.

Andlauer, W.; Fürst, P. Nutraceuticals: A Piece of History, Present Status and Outlook. *Food Res. Int.* **2002,** *35,* 171–176.

Asgary, S.; Sahebkar, A.; Afshani, M. R.; Keshvari, M.; Haghjooyjavanmard, S.; Rafieian-Kopaei, M. Clinical Evaluation of Blood Pressure Lowering, Endothelial Function Improving, Hypolipidemic and Anti-inflammatory Effects of Pomegranate Juice in Hypertensive Subjects. *Phytother. Res.* **2014,** *28* (2), 193–199.

Banerjee, S. K.; Mukherjee, P. K.; Maulik, S. K. Garlic as an Antioxidant: The Good, the Bad and the Ugly. *Phytother. Res*. **2003,** *17* (2), 97–106.

Barba, F. J.; Putnik, P.; Kovacevic, D. B. *Agri-food Industry Strategies for Healthy Diets and Sustainability: New Challenges in Nutrition and Public Health*; Press, A., Ed.; MPS Limited Tamil Nadu, India, 2020.

Bhaskarachary, K. Traditional Foods, Functional Foods and Nutraceuticals. *Proc. Indian Natl. Sci. Acad.* **2016,** *82,* 1565–1577.

Bhat, Z. F.; Bhat, H. Milk and Dairy Products as Functional Foods: A Review. *Int. J. Dairy Sci*. **2011**, *6,* 1–12.

Boccardi, V.; Tinarelli, C.; Mecocci, P. Nutraceuticals and Cognitive Dysfunction. Neuroprotective E_. Phytochem. *Neurol. Disord.* **2017,** 561–579.

Bogue, J.; Collins, O.; Troy, A. J. Market Analysis and Concept Development of Functional Foods. In Developing New Functional Food and Nutraceutical Products; DebasisBagchi, S. N., Ed.; Academic Press: Cambridge, MA, USA. **2017,** 29–45.

Brown, H. E.; Roffman, J. L. Emerging Treatments in Schizophrenia: Highlights from Recent Supplementation and Prevention Trials. *Harv. Rev. Psychiatry*. **2016,** *24,* e1–e7.

Brown, L. A.; Riby, L. M.; Reay, J. L. Supplementing Cognitive Aging: A Selective Review of the Effects of Ginkgo Biloba and a Number of Everyday Nutritional Substances. *Exp. Aging Res*. **2009,** *36,* 105–122.

Bull E. What is Nutraceutical? *Pharm J.* **2000**, *265,* 57–58.

Bungau, S.; Abdel-Daim, M. M.; Tit, D. M.; Ghanem, E.; Sato, S.; Maruyama-Inoue, M.; Yamane, S.; Kadonosono, K. Health Benefits of Polyphenols and Carotenoids in Age-Related Eye Diseases. *Oxidative Med. Cell. Longev*. **2019**.

Bungau, S. G.; Popa, V. C. Between Religion and Science Some Aspects Concerning Illness and Healing inAntiquity. *Transylv. Rev*. **2015,** *24,* 3–18.

Butterweck, V.; Schmidt, M. St. John'sWort: Role of Active Compounds for Its Mechanism of Action and Effcacy. *Wien. Med. Wochenschr*. **2007,** *157,* 356–361.

Casey, C.; Slawson, D. C.; Neal, L. R. Vitamin D supplementation in Infants, Children, and Adolescents. *Am. Fam. Physician*. **2010,** *81,* 745–748.

Castellani, R. J.; Harris, P. L.; Sayre, L. M.; Fujii, J.; Taniguchi, N.; Vitek, M. P.; Founds, H.; Atwood, C. S.; Perry, G.; Smith, M. A. Active Glycation in Neurofibrillary Pathology of Alzheimer Disease: N?-(carboxymethyl) Lysineand Hexitol-lysine. *Free Radic. Biol. Med.* **2001,** *31,* 175–180.

Caterson, I. D.; Gill T. P. Obesity: Epidemiology and Possible Prevention. *Best Pract. Res. Clin. Endocrinol. Metab*. **2002,** *16* (4), 595–610.

Ceskova, E.; Silhan, P. Novel Treatment Options in Depression and Psychosis. *Neuropsychiatr. Dis. Treat*. **2018,** *14,* 741.

Champagne, C. P.; Gomes da Cruz, A.; Daga, M. Strategies to Improve the Functionality of Probiotics in Supplements and Foods. *Curr. Opin. Food Sci*. **2018,** *22,* 160–166.

Chanda, S.; Tiwari, R. K.; Kumar, A.; Singh, K. Nutraceuticals Inspiring the Current Therapy for Lifestyle Diseases. *Adv. Pharmacol. Pharm. Sci*. **2019,** *2019,* 6908716.

Chintale Ashwini, G.; Kadam Vaishali, S.; Sakhare Ram, S.; Birajdar Ganesh, O.; Nalwad Digambar, N. Role of nutraceuticals Nn various Diseases: A Comprehensive Review. *Int. J. Res. Pharm. Chem*. **2013,** *3,* 290–299.

Cloutier, M.; Aigbogun, M. S.; Guerin, A.; Nitulescu, R.; Ramanakumar, A. V.; Kamat, S. A.; DeLucia, M.; Du_y, R.; Legacy, S. N.; Henderson, C. The Economic Burden of Schizophrenia in the United States in 2013. *J. Clin. Psychiatry*. **2016,** *77,* 764–771.

Cornelli, U. Antioxidant Use in Nutraceuticals. *Clin. Dermatol*. **2009,** *27,* 175–194.

Das, L.; Bhaumik, E.; Raychaudhuri, U.; Chakraborty, R. Role of Nutraceuticals in Human Health. *J. Food Sci. Technol.* **2012,** *49* (2), 173–183.

De Freitas Junior, L. M.; De Almeida, E. B. Medicinal Plants for the Treatment of Obesity: Ethnopharmacological Approach Andchemical and Biological Studies. *Am. J. Transl. Res.* **2017,** *9* (5), 2050–2064.

Dietary Supplement Health and Education Act of 1994 [Online]. https://ods.od.nih.gov/About/DSHEA_Wording.aspx (Accessed March 1, 2022).

Dillard, C. J.; German, J. B., Phytochemicals: Nutraceuticals and Human Health. *J. Sci. Food Agri*. **2000,** *80* (12), 1744–1756.

Dohrmann, D. D.; Putnik, P.; Bursa´cKovaˇcevi´c, D.; Simal-Gandara, J.; Lorenzo, J. M.; Barba, F. J. Japanese, Mediterranean and Argentinean Diets and Their Potential Roles in Neurodegenerative Diseases. *Food Res. Int*. **2019,** *120,* 464–477.

Frisardi, V.; Panza, F.; Solfrizzi, V.; Seripa, D.; Pilotto, A. Plasma Lipid Disturbances and Cognitive Decline. *J. Am. Geriatr. Soc.* **2010,** *58,* 2429–2430.

Georgiou, N. A.; Garssen, J., Witkamp, R. F. Pharma-nutrition Interface: The Gap is Narrowing. *Eur. J. Pharmacol.* **2011,** *651,* 1–8.

Giavasis, I. Bioactive Fungal Polysaccharides as Potential Functional Ingredients in Food and Nutraceuticals. *Curr. Opin. Biotechnol.* **2014,** *26,* 162–173.

Glevitzky, I.; Dumitrel, G. A.; Glevitzky, M.; Pasca, B.; Otrisal, P.; Bungau, S.; Cioca, G.; Pantis, C.; Popa, M. Statistical Analysis of the Relationship Between Antioxidant Activity and the Structure of FlavonoidCompounds. *Rev. Chim.* **2019,** *70,* 3103–3107.

Globe News Wire; [Online] 2019. https://www.globenewswire.com/newsrelease/2019/01/29/1706747/0/en/Global-Nutraceuticals-MarketReport-2018-Market-Accounted-for-379-06-Billion-in-2017-andExpected-to-Reach-734-60-Billion-by-2026.html (Accessed March, 4, 2022).

Gonzalez-Suarez, I.; Martin, F.; Hoeng, J.; Peitsch, M. C. Mechanistic Network Models in Safety and Toxicity Evaluation of Nutraceuticals. In *Nutraceuticals*; Academic Press, 2016, pp 287–304.

Granato, D.; Barba, F. J.; Bursa´cKovaˇcevi´c, D.; Lorenzo, J. M.; Cruz, A. G.; Putnik, P. Functional Foods: Product Development, Technological Trends, Efficacy Testing, and Safety. *Annu. Rev. Food Sci. Technol.* **2020,** *11,* 93–118.

Gutiérrez-Del-Río, I.; Fernández, J.; Lombó, F. Plant Nutraceuticals as Antimicrobial Agents in Food Preservation: Terpenoids, Polyphenols and Thiols. *Int. J. Antimicrob. Agents.* **2018,** *52,* 309–315.

Indian Nutraceuticals Market Outlook: Vision [Online] 2022. http://publication.assocham.tv/data/product-file/88 Indian Nutraceutical Market Opportunities_Report_LR.pdf (Accessed March 4, 2022).

Jenner, P.; Olanow, C. W. Oxidative Stress and the Pathogenesis of Parkinson's Disease. *Neurology.* **1996,** *47,* 161–170.

Keservani, R. K.; Sharma, A. K. Flavonoids: Emerging Trends and Potential Health Benefits. *J. Chinese Pharma. Sci.* **2014,** *23* (12), 815.

Keservani, R. K.; Kesharwani, R. K.; Vyas, N.; Jain, S.; Raghuvanshi, R.; Sharma, A. K. Nutraceutical and Functional Food as Future Food: A Review. *Der Pharmacia Lettre.* **2010a,** *2* (1), 106–116.

Keservani, R. K.; Sharma, A. K.; Kesharwani, R. K. Nutraceutical and Functional Foods for Cardiovascular Health. In *Food Process Engineering*; Apple Academic Press; CRC Press, 2016a; pp 291–312, pp 257–278. ISBN: 9781771884020

Keservani, R. K.; Sharma, A. K.; Kesharwani, R. K. Medicinal Effect of Nutraceutical Fruits for the Cognition and Brain Health. *Scientifica.* **2016b,** *3109254,* 10. doi:10.1155/2016/3109254

Keservani, R. K.; Sharma, A. K.; Kesharwani, R. K. An Overview and Therapeutic Applications of Nutraceutical and Functional Foods. *Recent Adv. Drug Delivery Technol.* **2017,** 160–201. ISBN:9781522507543.

Keservani, R. K.; Sharma, A. K.; Kesharwani, R. K. (Eds.). *Nutraceuticals and Dietary Supplements: Applications in Health Improvement and Disease Management*; CRC Press, 2020; ISBN: 9781771888738.

Keservani, R. K.; Kesharwani, R. K.; Sharma, A. K.; Vyas, N.; Chadoker, A. Nutritional Supplements: An Overview. *Int. J. Curr. Pharm. Rev. Res.* **2010b,** *1* (1), 59–75.

Kuhnau, J. Flavonoids. A Class of Semi-essential Food Components: Their Role in Human Nutrition. *World Rev. Nutr. Diet.* **2013,** *24,* 117–191.

Lama, A.; Pirozzi, C.; Avagliano, C.; Annunziata, C.; Mollica, M. P.; Calignano, A.; Meli, R.; MattaceRaso, G. Nutraceuticals: An Integrative Approach to Starve Parkinson's Disease. *Brain Behav. Immun. Health.* **2020,** *2,* 100037.

Lenoir-Wijnkoop, I.; Sanders, M. E.; Cabana, M. D.; Caglar, E.; Corthier, G.; Rayes, N.; Sherman, P. M.; Timmerman, H. M.; Vaneechoutte, M.; Van Loo, J.; Wolvers, D. A. Probiotic and Prebiotic Influence Beyond the Intestinal Tract. *Nutr. Rev.* **2007,** *65* (11), 469–489.

Linseman, D. A. Targeting Oxidative Stress for Neuroprotection. *Antioxid Red. Signal.* **2009,** *11* (3), 421–424.

Lozano, P.; M. R. Castellar; M. J. Simancas.; J. L. Iborra. Quantitative High Performance Chromatographic Method to Analyse Commercial Saffron (crocus sativus L.) Products. *J. Chromatogr. A.* **1999,** *830,* 477–483.

Maddi, V. S.; Aragade P. D.; Digge V. G.; Nitaliker M. N. Importance of Nutraceuticals in Health Management. *Phcog Rev.* **2007,** *1,* 377–379.

Mecocci, P.; Tinarelli, C.; Schulz, R. J.; Polidori, M. C. Nutraceuticals in Cognitive Impairment and Alzheimer's Disease. *Front. Pharmacol.* **2014,** *5,* 147.

Menon, I.; Spudich, A. The Ashtavaidya Physicians of Kerala: A Tradition in Transition. *J. Ayurveda Integr. Med.* **2010,** *1,* 245–250.

Montesano, D.; Rocchetti, G.; Putnik, P.; Lucini, L. Bioactive Profile of Pumpkin: An Overview on Terpenoids and Their Health-promoting Properties. *Curr. Opin. Food Sci.* **2018,** *22,* 81–87.

Mythri, R.; Kumar, A.; Mms, B. Nutraceuticals and Other Natural Products in Parkinson's Disease Therapy: Focus on Clinical Applications. In *Bioactive Nutraceuticals and Dietary Supplements in Neurological and Brain Disease*; Academic Press: Cambridge, MA, USA, 2015, 421–431.

Nasri, H.; Baradaran, A.; Shirzad, H.; Rafieian-Kopaei, M. New Concepts in Nutraceuticals as Alternative for Pharmaceuticals. *Int. J. Prev. Med.* **2014,** *5* (12), 1487–1499.

Navarro, A.; Boveris, A. Brain Mitochondrial Dysfunction and Oxidative Damage in Parkinson's Disease. *J. Bioenerg. Biomembr.* **2009,** *41,* 517–521.

Nicastro, H. L.; Ross, S. A.; Milner, J. A. Garlic and Onions: Their Cancer Prevention Properties. *Cancer Prev. Res.* **2015,** *8,* 181–189.

Nunomura, A.; Perry, G.; Aliev, G.; Hirai, K.; Takeda, A.; Balraj, E. K.; Jones, P. K.; Ghanbari, H.; Wataya, T.; Shimohama, S., et al. Oxidative Damage is the Earliest Event in Alzheimer Disease. *J. Neuropathol. Exp. Neurol.* **2001,** *60,* 759–767.

Orlando, J. M. Behavioral Nutraceuticals and Diets. *Vet. Clin. Small Anim. Pract.* **2018,** *48,* 473–495.

Ottaway, P. B. *Food Fortification and Supplementation: Technological, Safety and Regulatory Aspects*; Woodhead Publishing: Cambridge, UK, 2008.

Pallavi, M. C. P.; Kumar, H. M. S. Nutraceuticals in Prophylaxis and Therapy of Neurodegenerative Diseases. In *Discovery and Development of Neuroprotective Agents from Natural Products*; 2018; pp 359–376.

Peterson, C. T.; Denniston, K.; Chopra, D. Therapeutic Uses of Triphala in Ayurvedic Medicine. *J. Altern. Complement. Med.* **2017,** *23,* 607–614.

Pillitteri, J. L.; Shiffman, S.; Rohay, J. M.; Harkins, A. M.; Burton, S. L.; Wadden, T. A. Use of Dietary Supplements for Weight Loss in the United States: Results of a National Survey. *Obesity.* **2008,** *16,* 790–796.

Poojary, M. M.; Putnik, P.; Bursa´cKova˘cevi´c, D.; Barba, F. J.; Lorenzo, J. M.; Dias, D. A.; Shpigelman, A. Stability and Extraction of Bioactive Sulfur Compounds from Allium Genus Processed by Traditional and Innovative Technologies. *J. Food Compost. Anal.* **2017,** *61,* 28–39.

Prakash, V.; Boekel. Nutraceuticals: Possible Future Ingredients and Food Safety aspects. In *Ensuring Global Food Safety*; Academic Press: Cambridge, MA, USA, 2010; pp 333–338.

Putnik, P.; Gabrić, D.; Roohinejad, S.; Barba, F. J.; Granato, D.; Lorenzo, J. M.; Bursać Kovačević, D. Bioavailability and Food Production of Organosulfur Compounds from Edible Allium species. In *Innovative Thermal and Non-Thermal Processing, Bioaccessibility and Bioavailability of Nutrients and Bioactive Compounds*; Francisco, J., Barba, J. M. A. S., Giancarlo Cravotto, J., Lorenzo, M., Eds.; Woodhead Publishing: Cambridge, UK, 2019, pp 293–308.

Putnik, P.; Gabrić, D.; Roohinejad, S.; Barba, F. J.; Granato, D.; Mallikarjunan, K.; Lorenzo, J. M.; Bursać Kovačević, D. An Overview of Organosulfur Compounds from Allium Spp.: From Processing and Preservation to Evaluation of Their Bioavailability, Antimicrobial, and Anti-Inflammatory Properties. *Food Chem.* **2019,** *276,* 680–691.

Rafieian-Kopaei, M. Medicinal Plants and the Human Needs. *J. Herbmed. Pharmacol.* **2012,** *1* (1), 1–2.

Ramalingum, N.; Mahomoodally, M. F. The Therapeutic Potential of Medicinal Foods. *Adv. Pharmacol. Sci.* **2014,** *2014,* 354264.

Rane, B. R.; Bharath, M. S.; Patil, R. R.; Keservani, R. K.; Jain, A. S . Novel Approaches in Nutraceuticals. In *Enhancing the Therapeutic Efficacy of Herbal Formulations through Novel Drug Delivery Systems*. Kesharwani, R. K., Keservani, R. K., Sharma, A. K., Eds.; IGI Global International Publisher: Pennsylvania, USA, 2020b; Chapter 11, pp 241–266. ISBN13: 9781799844532.

Rane, B. R.; Patil, A. S.; Keservani, R. K.; Jain, A. S. Novel Approaches in Herbal Formulation. In *Enhancing the Therapeutic Efficacy of Herbal Formulations through Novel Drug Delivery Systems*; Kesharwani, R. K., Keservani, Raj K., Sharma, A. K., Eds.; IGI Global International Publisher: Pennsylvania, USA, 2020a; Chapter 2, pp 43–68. ISBN13: 9781799844532.

Rane, Bhushan R.; Tadavi, Sandip A.; Keservani, Raj K. Naturopathy. In *Herbal Product Development*; Sharma, A. K., Keservani, R. K., Gautam, S. P., Eds.; Apple Academic Press, CRC Press, Taylor & Francis Group, 2020c; Chapter 12, pp 321–347. ISBN: 9781771888776.

Sapkale, A. P.; Thorat, M. S.; Vir, P. R.; Singh, M. C. Nutraceuticals—Global Status and Applications: A Review. *Int. J. Chem. Pharm.* **2012,** *1,* 1166–1181.

Sarris, J.; Logan, A. C.; Akbaraly, T. N.; Amminger, G. P.; Balanzá-Martínez, V.; Freeman, M. P.; Hibbeln, J.; Matsuoka, Y.; Mischoulon, D.; Mizoue, T., et al. Nutritional Medicine as Mainstream in Psychiatry. *Lancet Psychiatry*. **2015,** *2,* 271–274.

Sawmiller, D.; Li, S.; Shahaduzzaman, M.; Smith, A. J.; Obregon, D.; Giunta, B.; Borlongan, C. V.; Sanberg, P. R.; Tan, J. Luteolin Reduces Alzheimer's Disease Pathologies Induced by Traumatic Brain Injury. *Int. J. Mol. Sci.* **2014,** *15,* 895–904.

Sharma, Anil K.; Keservani, Raj K.; Gautam Surya P. *Herbal Product Development, Apple Academic Press*; CRC Press, Taylor & Francis Group, 2020. ISBN: 9781771888776.

Shirzad, M.; Kordyazdi, R.; Shahinfard, N.; Nikokar, M. Does Royal Jelly Affect Tumor Cells? *J. Herbmed. Pharmacol.* **2013,** *2* (2), 45–48.

Singh, J.; Sinha, S. Classification, Regulatory Acts and Applications of Nutraceuticals for Health. *Int. J. Pharma Bio. Sci.* **2012,** *2* (1), 177–187.

Statista, The statistics Portal, 2016. Accessed on 4 march [Online] 2022. https://www.statista.com/statistics/589452/value-dietary-supplementsmarkets-europe-by-country.

Street, A. Food as Pharma: Marketing Nutraceuticals to India's Rural Poor. *Crit. Public Health.* **2015,** *25,* 361–372.

Suvarna, V. C.; Boby V. U. Probiotics in Human Health: A Current Assessment. *Curr Sci.* **2005,** *88,* 1744–1748.

Tapal, A.; Kaul Tiku, P. Nutritional and Nutraceutical Improvement by Enzymatic Modification of Food Proteins. In Enzymes in Food Biotechnology; Kuddus, M., Ed.; Academic Press: Cambridge, MA, USA, **2019,** 471–481.

Tatton, W.; Chalmers-Redman, R.; Brown, D.; Tatton, N. Apoptosis in Parkinson's Disease: Signals for Neuronal Degradation. *Ann. Neurol.* **2003,** *53* (3), 61–70.

Van der Burg, K. P.; Cribb, L.; Firth, J.; Karmacoska, D.; Sarris, J. Nutrient and Genetic Biomarkers of Nutraceutical Treatment Response in Mood and Psychotic Disorders: A systematic review. *Nutr. Neurosci.* **2019,** 1–17.

Wang, H.; Wang, H.; Cheng, H.; Che, Z. Ameliorating Effect of Luteolin on Memory Impairment in an Alzheimer's Disease Model. *Mol. Med. Rep.* **2016,** *13,* 4215–4220.

Wildman, R. E. C.; ed. *Handbook of Nutraceuticals and Functional Foods*. CRC Press; Boca Raton, 2001, pp 13–30.

Williams, R. J.; Mohanakumar, K. P.; Beart, P. M. Neuro-nutraceuticals: Further insights into Their Promise for Brain Health. *Neurochem. Int.* **2016,** *95,* 1–3.

Williams, R. J.; Mohanakumar, K. P.; Beart, P. M. Neuro-nutraceuticals: The Path to Brain Health via Nourishment is not so Distant. *Neurochem. Int.* **2015**, *89,* 1–6.

Willis, M. S.; Wians, F. H. Jr. The Role of Nutrition in Preventing Prostate Cancer: A Review of the Proposed Mechanism of Action of Various Dietary Substances. *Clin. Chim. Acta.* **2003,** *330* (1–2), 57–83.

CHAPTER 5

ANTIMICROBIAL PROPERTIES OF HERBAL NUTRACEUTICALS

KOMAL CHANDRA[1], SURESH KUMAR[2], VIRENDER KAUR[1], and TIRATH KUMAR[1]

[1]*Department of Pharmaceutical Sciences, Sir J.C. Bose Technical Campus Bhimtal, Kumaun University, Nainital, India*

[2]*Department of Biotechnology, Department of Pharmaceutical Sciences, Sir J.C. Bose Technical Campus Bhimtal, Kumaun University, Nainital, India*

ABSTRACT

Opportunistic human pathogens are threats of great concern to the world, and millions of people die globally per year, as the pathogens have the ability to resistance against antibiotics leading to a menace to humans. Concern regarding the development of alternative therapy has aroused some opportunity to achieve this goal by using herbal nutraceuticals and can be explored worldwide. Nutraceuticals are similar to dietary supplements like the medicinal product and are traditionally sold over-the-counter (OTC). Herbal nutraceuticals are well defined as food/part of food having medicinal properties. Nutraceuticals are key to the scientific concept of a personalized diet that has become a part of healthy life. Several herbal nutraceuticals like clove (*Syzygium aromaticum*), turmeric (*Curcuma longa*), garlic (*Allium sativum*), onion (*Allium cepa*), cinnamon (*Cinnamomum verum*), and fennel (*Foeniculum vulgare*) owed to the presence of secondary metabolites such as alkaloids, phenols, carbohydrates, saponins and flavonoids have potent antimicrobial properties which can target one or more steps of pathogen life cycle and kill/inhibit the growth of microbes.

Herbals As Nutraceuticals: Their Role in Healthcare.
Raj K. Keservani, Rajeshwar Kamal Kant Arya, and Rajesh K. Kesharwani (Eds.)

5.1 INTRODUCTION

In the 20th century antimicrobials have the greatest contribution to therapeutics in terms of curing the disease, but in current scenario resistance development among microbes toward antibiotics has made them less effective and potent. Microbes are living together with us in our environment and they are causing millions of deaths every year globally which has become a threat to humans (Gupta et al., 2019; Tripathi, 2014). The strategy to fight against these resistant microbes is to develop novel molecules or to look for an alternative that has the potency to encounter these microbes and safeguard health. In this context, researchers are directing toward one of the widely used ancient systems of medicine of India "Ayurveda." In the ayurvedic system of medicine for treatment/diagnosis/prevention herbs were used which are a hub of numerous phytochemicals with potent antioxidant, antimicrobial, analgesics, and anti-inflammatory activities. Traditionally, herbs were not used only as a part of food but were also used as therapeutic agents in illness, so this traditional ideology of therapy is again gaining the attention of researchers in the hope of discovering novel molecules to target pathogens. "Let food be thy medicine and medicine be thy food" is a famous saying of Hippocrates is the base of innovative complementary medicine, i.e., nutraceuticals which have attracted scientists worldwide and this era is of nutraceuticals (Ubaoji and Nwosu, 2020; Khameneh et al., 2019). In 1989, veteran researcher Dr. Stephen DeFelice coined the term "nutraceuticals" meaning a cluster of products that has therapeutic properties such as herbal products, food or segment of food, dietary supplements, isolated nutrients, functional foods, minerals, and vitamins (Keservani et al., 2010a, 2010b, 2017, 2020). It has been derived from two words "nutrient" and "pharmaceutical" (Ganesh et al., 2015). The word nutrient comes from the Latin word "*nutrire*" and the word pharmaceutical comes from the Late Latin word "*parmaceuticus*" which together comprise the term nutraceuticals. According to Dr. DeFelice, the nutraceuticals is a substance of food origin that offers health/medical benefits and prevention/treatment of a disorder or any disease but it has no regulatory definition. They are not considered similar everywhere and vary with different jurisdiction like in Canada they are marketed as drug/food and falls under Food and Drugs Act and Regulations provisions, in the United States they are regulated as drug/dietary supplement/food ingredient/food and in India, nutraceuticals are regarded as food by Food Safety and Standard Act, 2006. Apart from seeing nutraceuticals distinct, they are termed differently in different countries and regions and numerous laws govern nutraceuticals. Canadians denote them as Natural Health Product, Australians as complementary medicines, Indians as Foods for special dietary usage, European Union and the USA as Food supplements. Another definition of nutraceuticals (NCs) is given by European Nutraceutical Association (ENA); they describe NCs as a substance that is different from pharmaceuticals which are synthetic/chemical products formulated for specific indications. Different scholars also came forward to define nutraceuticals, but the core concept of all definitions remained identical, i.e., "Food as medicine" and the philosophy behind NCs is prevention which was the idea of the Father of medicine and Greek physician HIPPOCRATES (Thakkar et al., 2020).

5.2 REGULATION OF NUTRACEUTICALS IN DIFFERENT COUNTRIES

- **Canada:** Nutraceuticals in Canada for safety and efficacy are regulated by Health Canada (HC) and fall under Natural Health Products Regulations (NHPR), 2004. Natural Health Product (NHP) framed some products under natural sources or categories, like minerals, herbals, vitamins (Singh et al., 2017b), and homeopathic medicines. These are deliberate to human health benefits, while quality of the product is regulated by the Canadian Food and Drugs Act.
- **United States of America:** Nutraceuticals play a pivotal role as a dietary supplement in the United States. Dietary supplements are substances that add nutrients to a diet regularly, help to overcome the risk concerning health complications worldwide. Dietary supplements are the products that are used in a diet as a supplement to increase dietary intake. The US-FDA—Food Drug and Cosmetic Act (Zeisel, 1999) regulated dietary supplements. According to FD & C Act, dietary supplements do not claim as to the therapeutic/preventive/diagnostic agents of disease. New dietary supplements safety is ensured through New Dietary Ingredient Notification (NDIN) process before marketing by a manufacturer under the dietary supplement, health and education act (DSHEA). They are marketed in different dosage forms like tablets, powders, liquids, extracts, soft gels, and capsules (Denham, 2011).
- **Australia:** In Australia, nutraceuticals are regarded as complementary medicine (TGA, 2013) and are regulated under Therapeutic Goods Act (TGA), 1989. Complementary medicine includes all

products/practices other than conventional like herbs, vitamins, minerals, nutritional supplements, homeopathy, microorganisms etc.

5.3 NUTRACEUTICALS ARE CATEGORIZED INTO DIFFERENT CATEGORIES

1. **Nutrient nutraceuticals:** Substances like minerals and vitamins which have established nutritional functions.
2. **Herbal nutraceuticals:** Substances of herbal origin (phytochemicals, botanical product, part of the herb, extract) that have therapeutic properties.
3. **Dietary supplements:** Substances that supplement diet like diet to lose/gain weight, health maintenance diet, and sport diet (Chauhan et al., 2013).

5.4 HERBAL NUTRACEUTICALS

Herbs are used since antiquity as therapeutic agents by Indians, Chinese, and Egyptians. Human civilization and herbals have a long history together. Humans have used herbs exorbitantly for the treatment of acute and chronic ailments and according to World Health Organization, 70–80% of the world population depends upon the herbal source of medicine. In the current scenario, people are more conscious about their health, food, nutrition, diet, and medicines. They believe in herbal (Sharma et al., 2020; Rane et al., 2020a, 2020b, 2020c) products more than prescribed drugs and are turning toward them when standard treatments are failing to treat diseases. The present and forthcoming era are of herbs as peoples are truly frustrated, tired, and disappointed with resistant-developing drugs; standard specific ailment treatment failure; expensive medicines, diagnosis tests, medical-hospital care; populated clinics and hospitals and doctor's appointments (Ward et al., 2002). They are looking forward and whole heartedly accepting a complementary system of medicine. Herbals are a source of alternative medicine which is economically affordable, has natural supplements of diet, and adds health benefit effects. Herbals apart from being less costly are much safer, efficacious, and improve health; therefore, herbs are esteemed equivalent to jewels/gold nowadays. COVID-19 pandemic which was terrifying, threatening, destructive, stressing and with no selective inhibitor to inhibit the growth of severe acute respiratory syndrome coronavirus 2 (SARS-CoV-2) positive single-stranded RNA virus leads to dependence upon herbal nutraceuticals like *Pippali (Piper longum), piperine(Piper nigrum), turmeric (Curcuma longa), licorice (Glycyrrhiza glabra), golden milk, Giloy (Tinospora cordifolia), black seed (Nigella sativa),* Fishwort *(Houttuynia cordata), African geranium (Pelargonium sidoides),* Kosterm *(Lindera aggregate), Tongue Fern (Pyrrosia lingua), red spider lily (Lycoris radiata), and sweet sagewort (Artemisia annua)* for prevention, personal protection, and supportive care (Velavan and Meyer, 2020); (Ciotti et al., 2020; https://www.ayush.gov.in; Tahir et al., 2020). This pandemic gave herbals a massive degree of elevation and in return, they gave a ray of hope to fight against CoVID-19 to the public when the world was facing crises due to lack of medicines, medical facilities, and proper treatment.

Herbs owning a wide variety of medicinal properties from antimicrobial to anticancer formulate a great combination for novel lead compounds with nutraceuticals for treating numerous diseases and can be an alternative to pharmaceutics which are toxic and less efficacious. Interest in herbal nutraceuticals has increased among researchers because many studies have indicated that they are the source of a varied number of lead molecules.

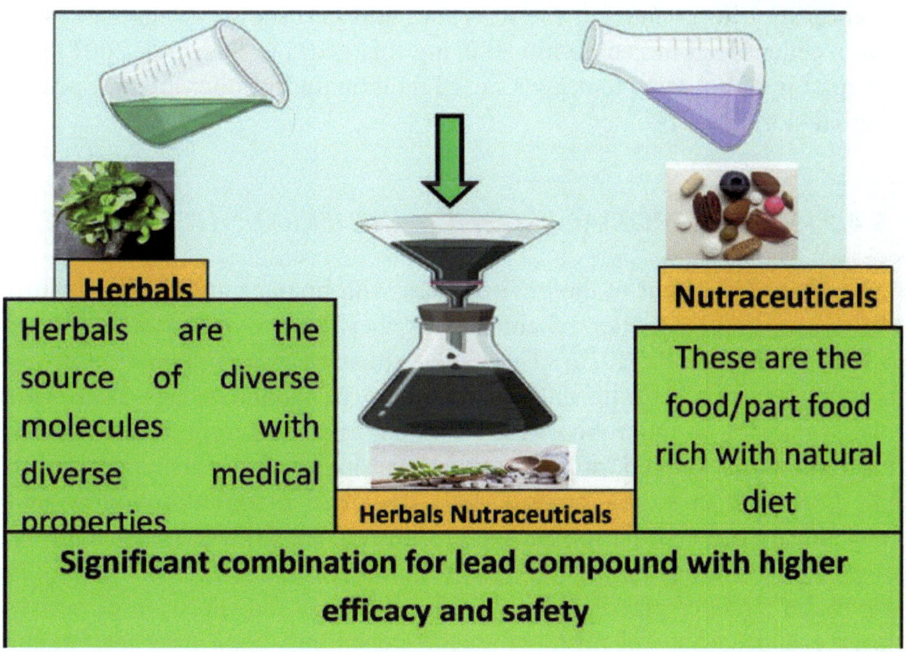

FIGURE 5.1 Concept of herbal nutraceuticals

5.4.1 PROPERTIES OF HERBAL NUTRACEUTICALS

Herbs are used as medicine and food from ancient times due to the presence of a wide number of bioactive which have significant medicinal properties. Herbals are one of the most important categories of nutraceuticals and commonly used complementary medicine. Parts of herb/whole herbs are rich in phytochemicals like polyphenols, tannins, essential oils, and saponins which offer a synergistic effect in treating/preventing several disorders and providing defense to the body from several microbes (Jhansi and Manjula, 2013). They are valued in recent times because concern regarding hygiene and health has been increased among the public. Herbal nutraceuticals previously were mainly known for nutritional properties, but they also provide plenty of other properties like antioxidant, antimicrobial, antiaging, anti-inflammatory, anticancer, and antiviral (Parham et al., 2020). Presently we have focused on a highly active field of research in recent scenarios, i.e., antimicrobial therapy/antimicrobial agents. Our environment is surrounded by microbes that are affecting health conditions directly/indirectly and they are surviving in extremely adverse conditions and due to mutation and development of resistance against synthetic antimicrobial agents, their growth is uncontrollable. Therapy for antimicrobial generally requires a synergistic combination to counteract the microbes and herbals nutraceuticals can be an excellent lead molecule in achieving this goal (Williamson, 2001). This chapter is an initiative to draw a ray of light on the antimicrobial property of herbal nutraceuticals.

5.4.2 ANTIMICROBIAL PROPERTY OF HERBAL NUTRACEUTICALS

Microbes are present in the environment with humans and their uncontrollable growth is a matter of concern as they lead to hazardous problems (Scheepmaker et al., 2020). The Covid situation makes people realize the importance of the health and sustainability of human beings after which the search for antimicrobial agents has increased. Microorganism includes bacteria, viruses, fungi, protozoa, and the property of an agent to kill/inhibit the growth of these microorganisms is regarded as antimicrobial property and those agents which possess this property are known as antimicrobial agents. Antimicrobials can be classified into various categories depending upon the mode of classification, chemical nature and spectrum of activity, source and activity against pathogen. Depending upon microbes against which antimicrobial agent is primarily active, antimicrobial agents are classified as antibacterial, antivirals, antifungal, anti-helminthic, and anti-protozoal. Herbs are nature's gift to humankind to treat several illnesses and due to the presence of alkaloids, carbohydrates, flavonoids (Keservani and Sharma, 2014; Keservani et al., 2016a; Keservani et al., 2016b), protein (Singh et al., 2017), phenols, saponins as secondary metabolites they have the potent antimicrobial property (Hussain et al., 2013; Murugan T et al., 2013) which can target various growth cycle steps and structural parts of pathogens as mentioned in Table 5.1

TABLE 5.1 Secondary Metabolites and their Mode of Action.

Secondary metabolite	Mechanism of action
Quinones	Complexing irreversibly to exposed adhesions, cell wall polypeptides, and membrane bound enzymes in microbial cell and leading to inactivation/ loss of function of target protein.
Flavones, flavonoids, and flavonols	Lipophilic flavonoids disrupt microbial membranes and they also form complex with soluble and extracellular proteins and inactivate them.
Tannins	Inhibit viral transcriptase
Terpenoids	Membrane disruption
Alkaloids	Intercalate with DNA

There are several other potent antimicrobial phytochemicals that are present in herbal nutraceuticals, but their mechanism of action is not clear and required more research. Herbal nutraceuticals which we are using in our day-to-day life have a number of medicinal properties and we are using them for centuries. Some of the herbal nutraceuticals with antimicrobial properties have been mentioned in Table 5.2 which can be used in effective natural remedy development against resistant microbes (Parham S et al., 2020).

5.5 CONCLUSION AND FUTURE DIRECTION

Herbal nutraceuticals are an emerging field of therapy and are essential for maintaining the quality of life. Herbal nutraceuticals are a repository source of phytochemicals with potent antimicrobial activity. The current research approaches to finding the active sources of antimicrobial agents, nutraceuticals that open out the research area and attract the interest of researchers toward this field. The nutraceuticals-based substances performed outstanding activities against the opportunistic human pathogen. The studies revealed that these compounds, such as phytochemicals/secondary metabolites, showed

TABLE 5.2 Herbal Nutraceuticals with Antimicrobial Property.

Common name	Biological source	Family	Chemical constituents	Properties
Garlic	*Allium sativum*	Amaryllidaceae	Alliin and allicin	Antibacterial, antimicrobial, and antioxidant
Echinacea	*Echinacea purpurea*	Asteraceae	Alkylamide and echinacoside	Antiviral
St. John's wort	*Hypericum perforatum*	Hypericaceae	Hypericin and hyperforin	Used against HIV and hepatitis-c virus
Turmeric	*Curcuma Longa*	Zingiberacae	Curcumin	Antiseptic
Onion	*Allium cepa* Linn.	Liliaceae	Allicin and alliin	Antibiotic
Goldenseal	*Hydrastis Canadensis*	Ranunculaceae	Hydrastine and berberine	Antimicrobial
Clove	*Syzygium aromaticum*	Myrtaceae	Eugenol and eugenyl acetate	Antimicrobial, antioxidant, anti-inflammatory, Antimutagenic, anticancer, and antiallergic
Portulaca	*Portulaca oleracea*	Portulacaceae	Ascorbic acid, a-tocopherols, quercetin, omega-3-fatty, apigenin, kaempferol, and gallotannins	Antimicrobial and antioxidant activity
Tribulus	*Tribulus terrestris*	*Zygophyllaceae*	Flavonoid, phenolic acids, and tannin	Antimicrobial, antioxidant, antihypertensive, anticancer, and analgesic
Eryngium		Apicaceae	Flavonoids, phenolic acid, and coumarins.	Antimicrobial, antidiuretic, antitus-sive, and antioxidant
Cinnamon	*Cinnamomum verum* and *Cinnamomum zeylanicum*	Lauraceae	Cinnamaldehyde and eugenol	Antimicrobial and antioxidant
Ginger	*Zingiber officinale roscoe*	Zingiberaceae	Phenolic acid, gingerols, paradols, and shogaols	Antimicrobial, Antihypertensive, analgesic, and antiulcer
Thyme	*Thymus vulgaris*	Lamiaceae	Carvacrol, thymol, and phenol	Antimicrobial, anticancer, antitus-sive, and antibacterial

TABLE 5.2 *(Continued)*

Common name	Biological source	Family	Chemical constituents	Properties
Pennyroyal	*Mentha pulegium*	Lamiaceae	Neo-menthol, pulegone, and menthone	Antimicrobial, antioxidant, and antibacterial
Fennel	*Foeniculum vulgare*	Apiaceae	Phenolic compounds	Antimicrobial
Chamomile	*Matricaria chamomilla*	Asteraceae	Flavonoids, terpenoids, phenolic compounds, apigenin, and matricin	Antimicrobial, antioxidant, and anti-inflammatory
Mint	*Mentha spicata*	Lamiaceae	Phenolic compounds	Antimicrobial
Burdock	*Arctium lappa*		Caffeic acid, rutin, o-hydrobenzoic acid, chromogenic acid, p-coumaricacid, phenolic acid, and flavonoids	Antimicrobial
Eucalyptus	*Eucalyptus globulus*	Myrtaceae	Flavonols, hydroxybenzoic acids, and hydrolyzable tannins	Antimicrobial
Primrose			Phenolic acids, flavonoids, sterols, hydrocarbons, and tocopherols.	Antimicrobial
Lemon balm	*Melissa officinalis*	Lamiaceae	Phenolic compounds such as thymol and carvacrol	Antimicrobial
Mallows	*Malva sylvestris*	Malvaceae	Flavonoids, vitamin E, polyphenols, vitamin C, and β-cartene.	Antimicrobial, antiviral, antibacterial, and antioxidant

tremendous medicinal properties to inhibit the growth of these pathogens and provided the direction to develop novel and effective agents to encounter microbes growing at an uncontrollable rate in our environment and causing health hazards (Gonelimali et al., 2018). The herbal nutraceuticals have shown excellent properties against pathogens. Hence, it is significant to explore more studies on these compounds and their possible therapeutical applications and regulation of these products into the market, appropriate monitoring and formulation of these herbal nutraceuticals are also required to avoid health-related problems and toxic effects.

KEYWORDS

- **nutraceuticals**
- **herbal nutraceuticals**
- **antimicrobial agents**
- **antimicrobial property**
- **microbes**

REFERENCES

Chauhan, B.; Kumar, G.; Kalam; N.; Ansari, S. H. Current Concepts and Prospects of Herbal Nutraceutical: A Review. *J. Adv. Pharm. Technol.* **2013,** *4* (1), 4–8.

Ciotti, M.; Ciccozzi, M.; Terrinoni, A.; Jiang, W. C.; Wang, C. B. ; Bernardini, S. The COVID-19 Pandemic. *Crit. Rev. Clin. Lab. Sci.* **2020,** *57* (6), 365–388.

Denham, B. E. J. J. Dietary Supplements-Regulatory Issues and Implications for Public Health. *JAMA*. **2011,** *306* (4), 428–429.

Ganesh, G. N. K.; Ramchandran, A.; Suresh, K. R.; Senthil, V.; Baviya, P. R. Nutraceuticals—A Regulatory Teview. *Int. J. Drug Regul. Aff.* **2015,** *3* (2), 22–29.

Gonelimali, F. D.; Lin, J.; Miao, W.; Xuan, J.; Charles, F., Chen, M.; Hatab, S. R. Antimicrobial Properties and Mechanism of Action of Some Plants Extracts Against Food Pathogens and Spoilage Microorganisms. *Front. Microbiol.* **2018,** 9, 1639.https://www.ayush.gov.in

Hussain, S. A.; Panjagari, N. R.; Singh, R. R. B.; Patil, G. R. Potential Herbs and Herbal Nutraceuticals: Good Applications and Their Interactions with Food Components. *Crit. Rev. Food Sci. Nutr.* **2015,** *55* (1), 94–122.

Jhansi, D.; Manjula, K. Functional and Nutraceutical Properties of Herbals and Its Applications in Food. *Int. J. Sci. Res.* **2016,** 5 (7).

Keservani, R. K.; Kesharwani, R. K.; Sharma, A. K.; Vyas, N.; Chadoker, A. Nutritional Supplements: An Overview. *Int. J. Cur. Pharm. Rev. Res.* **2010b,** *1* (1), 59–75.

Keservani, R. K.; Kesharwani, R. K.; Vyas, N.; Jain, S.; Raghuvanshi, R.; Sharma, A. K. Nutraceutical and Functional Food as Future Food: A Review. *Der. Pharm. Let.* **2010a,** *2* (1), 106–116.

Keservani, R. K.; Sharma, A. K. Flavonoids: Emerging Trends and Potential Health Benefits. *J. Chine. Pharm. Sci.* **2014,** *23* (12), 815.

Keservani, R. K.; Sharma, A. K.; Kesharwani, R. K. Nutraceutical and Functional Foods for Cardiovascular Health. In *Food Process Engineering*; Apple Academic Press, CRC Press, 2016a; pp 291–312, pp 257–278. ISBN: 9781771884020.

Keservani, R. K.; Sharma, A. K.; Kesharwani, R. K. Medicinal Effect of Nutraceutical Fruits for the Cognition and Brain Health. *Scientifica* **2016b,** *3109254,* 10. doi: 10.1155/2016/3109254.

Keservani, R. K.; Sharma, A. K.; Kesharwani, R. K. An Overview and Therapeutic Applications of Nutraceutical and Functional Foods. *Recent Adv. Drug Del. Technol.* **2017,** 160–201. ISBN:9781522507543.

Keservani, R. K.; Sharma, A. K.; Kesharwani, R. K.; Eds. *Nutraceuticals and Dietary Supplements: Applications in Health Improvement and Disease Management.* CRC Press, 2020. ISBN: 9781771888738.

Khameneh, B.; Iranshahy, M.; Soheili, V.; Bazzaz, B. S. F. Review on Plant Antimicrobials: A Mechanistic Viewpoint. *Antimicrob. Resist. Infect. Control.* **2019,** *8,* 118.

Murugan, T.; Wins, A. J.; Murgan, M. Antimicrobial Activity and Phytochemical Constituents of Leaf Extract of *Cassia Auriculata. Ind. J. Pharm. Sci.* **2013,** *75* (1), 122–125.

Nwosu, O.; Ubaoji, K. I. Nutraceuticals: History, Classification and Market Femand. *J. Funct. Foods.* **2020,** 13–22.

Parham, S.; Kharazi, A. Z.; Bakhsheshi-Rad, H. R.; Nur, H.; Ismail, A. F.; Sharif, S., RamaKrishna, S.; Berto, F. Antioxidant, Antimicrobial and Antiviral Properties of Herbal Materials. *Antioxidants.* **2020,** *9,* 1309.

Rane, B. R.; Bharath, M. S.; Patil, R. R.; Keservani, R. K.; Jain, A. S. Novel Approaches in Nutraceuticals. In *Enhancing the Therapeutic Efficacy of Herbal Formulations through Novel Drug Delivery Systems*; Kesharwani, R. K., Keservani, R. K., Sharma, A. K., Eds.; IGI Global International Publisher: Pennsylvania, USA, 2020b; Chapter 11, pp 241–266. ISBN13: 9781799844532.

Rane, B. R.; Patil, A. S.; Keservani, R. K.; Jain, A. S. Novel Approaches in Herbal Formulation, In *Enhancing the Therapeutic Efficacy of Herbal Formulations through Novel Drug Delivery Systems.* Kesharwani; R. K., Keservani, R. K., Sharma, A. K., (Eds.). IGI Global International Publisher: Pennsylvania, USA, 2020a; Chapter 2, pp 43–68. ISBN13: 9781799844532.

Rane, B. R.; Tadavi, S. A.; Keservani, R. K. Naturopathy In *Herbal Product Development.* Sharma, A. K., Keservani, R. K.. Gautam, S. P., Eds., Apple Academic Press, CRC Press, Taylor & Francis Group, 2020c; Chapter 12, pp 321–347. ISBN: 9781771888776.

Sharma, A. K.; Keservani, R. K.; Gautam, S. P. *Herbal Product Development.* Apple Academic Press, CRC Press, Taylor & Francis Group, 2020. ISBN: 9781771888776.

Singh, P.; Kesharwani; R. K.; Keservani, R. K. Antioxidants and Vitamins: Roles in Cellular Function and Metabolism. In *Sustained Energy for Enhanced Human Functions and Activity*; Academic Press, 2017a; pp 385–407. ISBN:9780128054130.

Singh, P.; Kesharwani; R. K.; Keservani, R. K. Protein, Carbohydrates, and Fats: Energy Metabolism. In *Sustained Energy for Enhanced Human Functions and Activity.* Academic Press, 2017b; pp 103–115. ISBN:9780128054130.

Tahir, A.; Javed, M.; Hussain, Z. Nutraceuticals and Herbal Extracts: A Ray of Hope for COVID-19 and Related Infections (Review). *Int. J. Funct. Nutr.* **2020,** *1,* 6.

Thaisa, M. A. M.; T P S Clerci, M. Burdock (Articalappa L) Roots as a Source of Insuin-type Fructans and Other Bioactive Compounds. Current Knowledge and Future Perspectives for Food and Non-food Applications. *Food Res. Int.* **2020**, *141,* 109889.

Thakkar, S.; Anklam, E.; Xu, A.; Ulberth, F.; Li, J.; Hugas, M.; Sarma, N.; Crerar, S.; Swift, S.; Hakamatsuka, T.; Curtui, V.; Yan, W.; Geng, X.; Slikker, W.; Tong, W. Regulatory Landscape of Dietary Supplements and Herbal Medicines from a Global Perspective. *Regul. Toxicol. Pharmacol.* **2020,** *114,* 104647.

Ward, P. M. L.; Fasitsas, S.; Katz, S. E. Inhibition, Resistance Development, and Increased Antibiotic and Antimicrobial Resistance Caused by Nutraceuticals. *J. Food Prot.* **2002,** *65* (3), 528–533.

Williamson, E. M. Synergy and Other Interactions in Phytomedicines. *Phytomedicine*. **2001,** *8* (5), 401–409.

Zeisel, S. H. Regulation of Nutraceuticals. *Science*. **1999,** *285* (5435), 1853–1855.

CHAPTER 6

NUTRACEUTICAL HERBS IN CARDIOVASCULAR DISEASES

MOHIT AGRAWAL[1], MANMOHAN SINGHAL[2], RUPA GUPTA[3], SAMIR BHARGAVA[2], DHEERAJ BISHT[4], RAJESHWAR KAMAL KANT ARYA[4], VIJAY SINGH RANA[2], and NEERAJ KUMAR SETHIYA[2]

[1]*Department of Pharmacology, Rajiv Academy for Pharmacy, Mathura, Uttar Pradesh, India*

[2]*Faculty of Pharmacy, DIT University Dehradun, Uttarakhand, India*

[3]*Amity Institute of Pharmacy, Amity University, Gurugram, Haryana, India*

[4]*Department of Pharmaceutical Sciences, Sir J.C. Bose Technical Campus Bhimtal, Kumaun University, Nainital, India*

ABSTRACT

Nutraceuticals is a term commonly implied for medicinal and nutritional food products having an important role in overall health including total well-being maintenance, immunity modulation, and prevention from specific diseases. Despite many other diseases, cardiovascular disease (CVD) is associated with many risk factors and can be prevented through the involvement of intervention of nutraceutical products. Many nutraceutical products are recognized through various studies that have positively controlled and prevented the influencing risk factors associated with cardiovascular diseases. Further, some of the associated conditions such as hypertension, atherosclerosis, *diabetes mellitus*, and cardiac heart failure are well established

Herbals As Nutraceuticals: Their Role in Healthcare.
Raj K. Keservani, Rajeshwar Kamal Kant Arya, and Rajesh K. Kesharwani (Eds.)

to be directly associated with CVD. The approaches to utilized medicinal plants and their isolated components as referred in traditional medicine are a possible alternative method to prevent onsets of CVD in earlier stages. In recent past much research on bioactive plant-based molecules demonstrated their significance as nutraceutical food for the prevention of CVD has been customized. Several studies pertaining to *in vitro* and/or *in vivo* activity of medicinal plants and their isolated components have been explored for possibilities to prevent CVD through various pharmacological screening models. Therefore, the authors tried to customize some of the selected recent plant-based nutraceutical research findings to be useful against CVD prevention in the present chapter. In this context, databases obtained from online search engine from various sources consulted and relevant studies were selected to summarize the present chapter. Further, patents involving technologies related to treatment and prevention of CVD are reviewed and presented in the current book chapter.

6.1 INTRODUCTION

The term "nutraceutical (meaning nutritions having pharmaceuticals benefits)" was coined for the first time in 1989 by Stephen DeFelice (founder and chairman of the Foundation of Innovation Medicine (FIM), Crawford, New Jersey) as a food (or food parts) having health or medicinal benefits for the prevention and/or treatment of several diseases (Das et al., 2012). In the past about 2500 years back Hippocrates already quoted that "Let food be your medicine and medicine be your food" (Jeyasekaran et al., 2015). In summary from many published research on nutraceuticals, it was found to be helpful to improve overall health performance, improve energy levels, relieve from anxiety, improve mental fitness, enhance sleep, prevent from chronic illness or diseases, reduce drug cravings, delay the aging process, increase total life expectancy, and improve various body functions (Nasri et al., 2014). The natural food products used as nutraceuticals (Keservani et al., 2010a, 2010b, 2017, 2020) can be broadly categorized as fibers from dietary sources, prebiotics and probiotics products, antioxidant vitamins (Singh et al., 2017a), polyphenols, polyunsaturated fatty acids (Singh et al., 2017b), and spices (Das et al., 2012). Additionally, these are further classified into three basic categories including nutrients, herbals, and dietary supplements (Chauhan et al., 2013). Various roles of nutraceutical products for the promotion of good health and the prevention of disease are described in Figure 6.1 (Lokhande, 2018). As per the current estimate the market size value for nutraceuticals in 2020 is USD 412.7 billion, which was estimated to be USD 722.5 billion in 2027 with an annual growth of 8.3% from 2020 to 2027 expanded majorly on USA, Canada, Mexico, Germany, France, UK, Russia, Italy, Spain, China, Japan, Australia, New Zealand, Brazil, Argentina, South Africa, and India (https://www.grandviewresearch.com/industry-analysis/nutraceuticals-market; retrieved on 06 October 2021). The incidence of cardiovascular diseases has been increasing since decades throughout the world predominantly in India as per number of reported cases (Huffman et al., 2011).

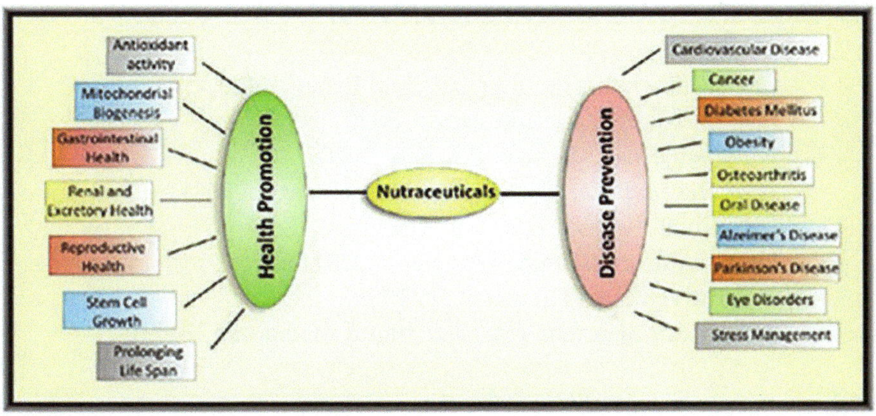

FIGURE 6.1 Role of nutraceutical in health promotion and disease prevention.

Source: Reprinted with permission from Dutta et al., 2018. Open Access

6.2 ABOUT CARDIOVASCULAR DISEASES (CVD)

Cardiovascular diseases (CVD), mainly ischemic heart disease and stroke, are the most common current health problem faced by the populations across the world and may become fatal at later stages, if not properly cared (American Heart Association, 2021). CVD are not associated with any isolated single risk factor but associated with consequences of several merged complex processes that commence during early stages in life. In this connection several studies suggested the involvement of behavioral, environmental, socioeconomic, and genetic risk factors (Dzau et al., 2006; Kumar and Sinha, 2020). In summary obesity, atherosclerosis, dyslipidemia, diabetes, peripheral vascular disease, cerebrovascular disease (stroke), coronary heart disease (heart attack), heart failure, hypertension (even beginning with younger generations), and cardiac heart failure are some of the associated predominantly risk factors of CVD (Petrie et al., 2018; Powell-Wiley et al., 2021). According to WHO, globally CVD (heart attacks and strokes constitute 85%) are the number one cause of death and constitute an estimated 31% of all deaths worldwide with reports of 17.9 million deaths each year of which 75% cases are mainly from low- and middle-income countries (American Heart Association, 2021). As per WHO, one-fifth of all these deaths trend higher especially in young generation with death rate of 272 per 100,000 populations (global average is 235) caused in India (Kumar and Sinha, 2020).

6.2.1 ASSOCIATED DISEASE WITH CVD

- Congestive heart failure (Latic and Erben, 2020)
- Hypertension (Latic and Erben, 2020)
- Atherosclerosis (Latic and Erben, 2020)
- *Diabetes mellitus (*Deedwania and Srikanth, 2008)
- Fatty liver disease (Lee et al., 2021).
- Hypercholesterolemia (Wong et al., 2016).
- Ulcerative colitis (Iizuka et al., 1999).
- Pulmonary disorders (Mizoguchiand Matsubara, 2009).

6.2.2 VARIOUS DIAGNOSTIC TOOLS FOR CVD

Cardiovascular diseases are diagnosed by adopting various tools and technology including many laboratory tests and imaging studies. The primary and most important part of diagnosis is patient medical and family histories, associated risk factors, physical examination, and finally coordination of all these findings with the results from tests and procedures (Garami and Alexandrov, 2009; Hoeper et al., 2013; Windecker et al., 2014; Afonso et al., 2017; Gopalan et al., 2017; DeFilippis et al., 2019; Buono et al., 2019; Saxena et al., 2019). Some of the common tests used for the diagnosis of cardiovascular diseases include:

- Blood tests
- EKG/ECG (electrocardiogram)
- Stress testing
- Echocardiography
- Coronary angiography and cardiac catheterization
- Chest X ray

- Electron-beam computed tomography or EBCT
- Cardiac MRI

6.3 NUTRACEUTICAL HERBS AS A TREATMENT AND PREVENTION OF CVD

Nutraceuticals obtained from various sources have been claimed to have several physiological benefits or prevented from many of associated risk factors of CVDs as mentioned in Figures 6.2 and 6.3. In the last decade, the food industry has focused on the characterization and identification of bioactive molecules contained in traditional foods or in natural matrices having nutritional value with health benefits in the treatment of chronic diseases (Hasler, 2002; José et al., 2021). This process started mainly because these natural compounds are considered to be consumer friendly because of myth to have fewer side effects in comparison with other pharmacological therapies (Ekor, 2013). However, when compared with other active pharmaceutical ingredients (API), these plant-based nutraceuticals do not have IPR protection norms, unless novelty is proved through designing some formulation or process (Chong et al., 2019). Further, the definition of nutraceuticals in many parts overlaps with functional foods and often both are considered to be synonymous to each other (Aronson, 2017). In brief, functional foods are defined to any foods article fortified with bioactive compounds aimed to enhance nutritional and health requirements without obligation of any dosage forms (Hasler, 2002). However, nutraceuticals products are obliging to present in the market with any forms of dosage form such as tablets, capsules, or powder (Granato et al., 2020; Roberfroid, 2000). Additionally, CVD is considered dietary and lifestyle disease and can be stopped or reversed with modification (Casas et al., 2018). Herbs and spices are mainly obtained from plant sources, where fresh part of a plant mostly referred as herbs and dried parts comes under spices to add flavor with multiple health benefits (Embuscado, 2015). Since ancient at the beginning of civilization, herbs and some derivatives (e.g., aspirin, reserpine, and digitalis) are only the reliable sources for medical treatments and mainstays of human pharmacotherapy (Mashour et al., 1998). As per research conducted from past to present many of herbs and its derived product have been validated for treatment of CVD conditions such as congestive heart failure, systolic hypertension, angina pectoris, atherosclerosis, cerebral insufficiency, venous insufficiency, and arrhythmia (Sosnowska et al., 2017). Functional foods are the term which was commonly employed to those substances that brought physiological benefits and/or reduced the risk of chronic disease beyond their basic nutritional functions (Alissa and Ferns, 2012).

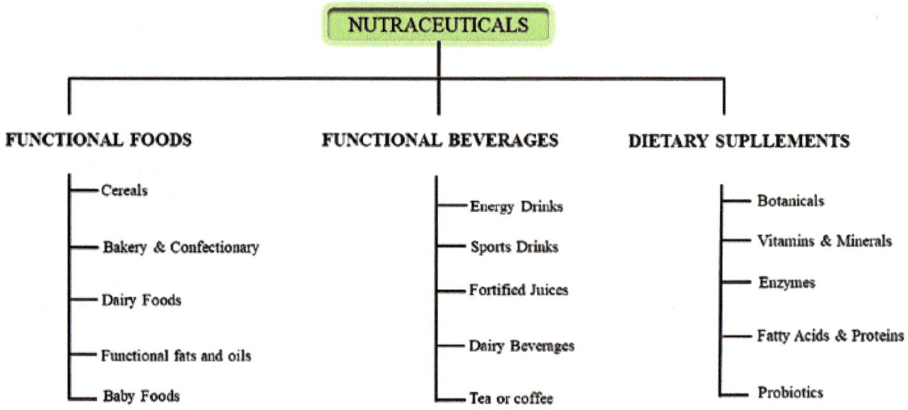

FIGURE 6.2 Sources of nutraceutical products for health promotion and disease prevention.
Source: Adapted from Damián et al., 2022

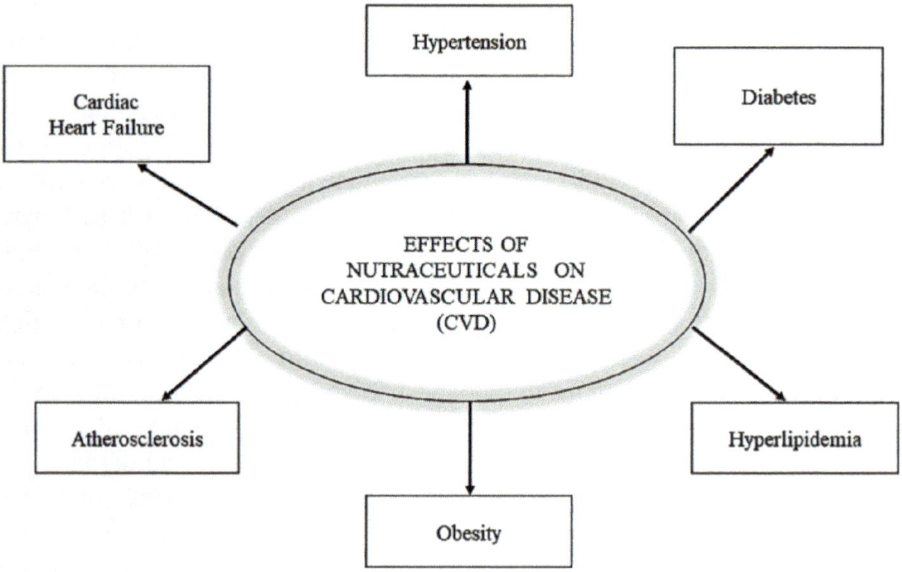

FIGURE 6.3 Effects of nutraceuticals on cardiovascular disease (CVD)
Source: Adapted from Chen et al., 2023

6.4 BIOACTIVE HERBS AS A NUTRACEUTICAL FOOD FOR CVD

Several herbs have been studied in past to investigate for cardiovascular diseases through series of *in vitro* and *in vivo* studies (Table 6.1). The outcomes of some medicinal herbs as potential sources of nutraceutical for the prevention and treatment of CVD are depicted herewith:

6.4.1 *COCOA (SEEDS OF THEOBROMA COCOA)*

Cocoa is also known by cocao seed, cocoa bean, and chocolate tree has been known for nutraceutical sources for the management of CVD as evidence from various research conducted through both preclinical and clinical studies (Galleano et al., 2009; Corti et al., 2010; Barthelmes et al, 2017; Tanghe et al., 2021). Both cocoa and its compounds were found to exhibit activation of nitric oxide, antioxidant, and anti-inflammatory effects (Corti et al., 2009). A number of reviews on human dietary intervention trials using flavanol-containing cocoa products have demonstrated improvements in endothelial and platelet function, as well as blood pressure (Erdman et al., 2008). Further, flavan-3-ol (flavanol) a flavonoid found in cocoa has been extensively studied for cardiovascular health benefits (Mehrinfar and Frishman, 2008). Many mechanisms have been postulated for the beneficiary effect of Cocoa and its derivative as a nutraceutical for associated conditions for cardiovascular health is well studied (Hertog et al., 1993; Kurosawa et al., 2005; Grassi et al., 2005; Paramo, 2008; Grassi et al., 2008; Yasuda et al., 2008; Rostami et al., 2015; Rabadan-Chavez et al., 2016; Guan et al., 2016; Pelaez-Soto et al., 2017; Ramirez-Sanchez et al., 2013; Larson et al., 2012; Cordero-Herrera et al., 2013, 2014).

6.4.2 *BRASSICA: PLANTS CABBAGE AND MUSTARD FAMILY (BRASSICACEAE)*

Brassica is species of genus mainly including *Brassica balearica, Brassica carinata, Brassica elongata, Brassica fruticulosa, Brassica hilarionis, Brassica juncea, Brassica napus, Brassica narinosa, Brassica nigra, Brassica oleracea, Brassica perviridis, Brassica rapa (syn. B. campestris), Brassica rupestris, Brassica spinescens*, and *Brassica tournefortii* having high nutraceuticals potential for cardiovascular health. It mainly acts as an antioxidant, induces ACE and renin inhibition, as evidence from past study (Han et al., 2015; Wu et al., 2004). One-month administration of Brassica improves AMPK/SIRT1/PGC1α/PPARα/UCP2 axis and produces strong renal and

TABLE 6.1 Medicinal Plants as a Nutraceutical Against Cardiovascular Diseases.

Plant name	Extract/derived compound	Model	Doses	Outcome	Reference
Chamaemelum Nobile	Aqueous extract	Male rats	140 mg/kg	Hypotensive activity	Zeggwagh et al., 2009
Chamaemelum Nobile	Aqueous extract	Smooth muscle cells Normal Wistar rats	10 and 20 mg/mL	Reduced pD2 value of arotic ring, showing dose-dependent vasorelaxant effect.	Zeggwagh et al., 2013
Adonis aestivalis	Hydroalcohol extract from aerial parts	Sheep	75 mg/kg/iv	Extract induced bradycardia and several ECG abnormalities. These changes suggest that isolates have potential to produce cardiac effects	Maham and Sarrafzadeh-Rezaei, 2014
Stevia rebaudiana	Aqueous extract of leaves	Albino rats	400 mg/kg/bw	Reduction in body appetite stimulation, weight gain, serum cholesterol, triglycerides, low density lipoprotein, very low-density lipoprotein levels, and LDL/HDL ratios.	Ahmad et al., 2018a
Portulaca oleracea	Dried powder infusion	Swiss albino mice	10% in hyper-lipidemic diet	Hypolipidemic effect demonstrated	El-Newary, 2016
Inula racemosa	Hexane extract of roots	Frogs (Rana tigrina)	10, 20, 30, 40 mcg/mL	Induces relaxation, dilation, and decrease in heart rate	Lokhande et al., 2006
Zingiber officinale	Fresh aqueous crude extract	Sprague Dawley rats	3–10 mg/kg	Dose-dependent hypotensive effect	Ghayur et al., 2005
Allium sativum (Pulp), *Androgaphis paniculata* (Leaves), *Boerhavia diffusa* (leaves), *Moringa oleifera* (bark), *Piper betle* (leaves) and *Piper longum* (seed).	Aqueous solution of polyherbal formulation	Adult male Wistar rats	250 mg/kg/bw	The Myocardial Ischemia treated group had higher concentration of high density lipoproteins, and prevent cardiomyocytes from damage.	Ishwarya et al., 2017

TABLE 6.1 *(Continued)*

Plant name	Extract/derived compound	Model	Doses	Outcome	Reference
Ocimum gratissimum	Essential oil from leaves	Male Wistar rats	1–20 mg/kg/i.v.	The induced hypotension is independent of effect by central autonomic nervous system. However, the actions have also been attributed due to Eugenol.	Lahlou et al., 2004
Croton nepetaefolius	Essential oil from bark and leaves	Male Wistar rats	1–20 mg/kg/i.v.	Administration of essential oil shows strong evidence of hypotensive action	Lahlou et al., 1999
Mentha x villosa	Essential oil from leaves	Male Wistar rats	1–20 mg/kg/i.v.	The administration of oil produced hypotensive action, reduced mean aortic pressure and heart rate.	Lahlou et al., 2002
Trachyspermum ammi	Seed powder	Albino rabbits	0.5–2 g/kg	T. ammi seed powder can be correlated with a decrease in LDL–cholesterol, an increase in HDL-cholesterol, and a decrease in plasma triglyceride T. ammi seed powder can be correlated with a decrease in LDL-cholesterol, an increase in HDL-cholesterol, and a decrease in plasma triglyceride *T. ammi* seed powder can be correlated with a decrease in LDL-cholesterol, an increase in HDL-cholesterol, and a decrease in plasma triglyceride Reduction in LDL-cholesterol and Plasma triglycerides was observed, with enhancement in HDL-cholesterol (up to 65%).	Javed et al., 2009
Vaccinium vitis-idaea	Polyphenols and dietary fiber	Male mice	Fat diet containing 44% lingonberries	Reduced triglyceridemia and decrease in size of atherosclerotic plaque.	Matziouridou et al., 2016
Allium sativum	Diallyldisulfide	Adult male Wistar rats	20 mg/kg/bw	Reduced low level lipoproteins, increased availability of NO, reduced protein carbonyls by preventing cardiovascular damage.	Kumar Sharma et al., 2012

TABLE 6.1 *(Continued)*

Plant name	Extract/derived compound	Model	Doses	Outcome	Reference
Solanum and Dioscorea species	Diosgenin	Male albino Wistar rats	40 mg/kg/day	Improves eNOS expression in aortas of diabetic rats, enabling vasorelaxation. It lowers ACE, preventing abnormal vasodilatations.	Roghani-Dehkordi et al., 2015
Solanum and Dioscorea species	Diosgenin	Male albino Wistar rats	40 mg/kg/day	Reduces ACE in Vasculature, causing renewal of level of endothelial NO synthase (eNOS) and avoiding age-related congealing within aorta.	Manivannan et al., 2013
Tacca chantrieri	Powder from rhizomes of plant	Male Sprague-Dawley rats	5 mg/kg/bw	Reduced blood pressure by 50% at mentioned dose. The activity might be due to presence of diarylheptanoid glucosides	Tiamjan et al., 2007
Asphodelus tenuifolius	Aqueous Ethanol extract	Sprague-Dawley rats	3–30 mg/kg/bw	Hypotensive, vasorelaxant, and diuretic activities, observed due to stoppage of vasorelaxant action and Ca2+ channel blocking activities.	Aslam et al., 2016
Ficus carica L.	Aqueous methanol extract of fruits	Sprague-Dawley rats	250–1000 mg/kg/bw	Dose-dependent decrease in different BP was observed with a prominent effect at 1000 mg/kg	Alamgeer et al., 2017
Lantana camara	Ethanol extract of leaves	Male Wistar albino rats	200 and 400 mg/kg/day	Hypertensive rats were protected against internal renal and vascular injuries. Concentration of potassium was found to be increased at 400 mg/kg dose.	Matta et al., 2015
Senecio serratuloides	Hydroethanol extract	Female Wistar rats	150 and 300 mg/kg	Reduced systolic BP by 15% and 25%, which was found to be comparable to hypertensive effect of Captopril at the end of Week 8. While diastolic reduced by 18 and 40%, compared to 20% reduction by Captopril.	Tata et al., 2019
Parastrephia quadrangularis	Hydro-ethanol extract	Male Sprague-Dawley rats	10–80 mg/kg	Dose-dependent reduction in BP by increase in vasodilation and decrease of heart rate.	Cifuentes et al., 2019
Berberis vulgaris	Aqueous extract of root bark	Sprague-Dawley rats	1–20 mg/kg/iv	Significant reduction in mean arterial BP was observed	Ahmed et al., 2018b

TABLE 6.1 *(Continued)*

Plant name	Extract/derived compound	Model	Doses	Outcome	Reference
Populus ciliata	Aqueous ethanol solution of coarse plant powder	Sprague Dawley albino rats. Rabbit and rat aorta (2–3 mm) from thoracic region	1–30 mg/kg/iv	Dose-dependent reduction in BP was observed which was comparable to Verapamil. Relaxed the induced contractions	Saqib et al., 2021
Eugenia dysenterica	Aqueous leaf extract	Male Wistar rats	15 mg/kg/iv	The hypotensive action has been attributed to L-type calcium channel blockage activity and my endothelial gap junction signaling.	Fidelis-de-Oliveira et al., 2020
Ajuga iva	Aqueous extracts of whole plants	Male SHR-SP and wistar rats	500 mg/kg/bw	The treatment reduced hypertension in stroke-prone hypertensive rats. The probable action has been postulated to be blockage of α1-receptor blockage.	El-Hilaly et al., 2021
Anagallis arvensis	Crude extract	Sprague-Dawley albino rats	1–30 mg/kg	There was prominent hypotensive activity with fall in BP	Saqib and Janbaz, 2021
Ganoderma lucidum	*Ganoderma lucidum* ethanol extract	Male Wistar rats	26, 44 and 70 mg/kg/bw	The antihypertensive activity of extract has been suggested due to trametenolic acid, at a dose of 70 mg/kg.	Adeyi et al., 2021
Evernia prunastri	Aqueous extract of plant material	Albino adult male rats and aorta isolated from male Wistar rats	200 mg/kg/bw and 50–1350 μg/mL	Systolic BP was reduced significantly after 6 h. Relaxation of aortic rings precontracted by Norepinephrine (NE) and KCl in a dose-dependent manner	Amssayef et al., 2021

cerebral protective effects in the stroke-prone spontaneously hypertensive rat model (Rubattu et al., 2015). However, no reduction of BP was observed, suggesting that renal and brain protection were independent of systemic BP regulation, being likely the consequence of molecular mechanisms at the vascular level. A clinical trial demonstrated on patients suggests that GSTT1 carrier is more responsive against BP improvement when compared to others (Han et al., 2015). In summary many mechanisms have been postulated for the beneficiary effect of various Brassica species and its derivative compounds as a nutraceutical for associated conditions for cardiovascular health are well studied and established through various preclinical and clinical studies (An et al., 2010; Palomaki et al., 2010; Bahadoran et al., 2011, 2013; He et al., 2013; Duran et al., 2016; Šamec et al., 2019; Roos et al., 2021; Raiola et al., 2017).

6.4.3 *SPIRULINA PLATENSIS (A BLUE-GREEN ALGAE)*

Spirulina platensis extracts were found to act against early atherosclerotic responses induced by histamine (Vo and Kim, 2013). It was found to reduce total cholesterol serum levels, TG and LDL-C levels, while improving levels of HDL-C in both diet-induced hypercholesteraemic rabbit and mice models (Cheong et al., 2010; Sengupta et al., 2018). In summary many mechanisms have been postulated for the beneficiary effect of various *Spirulina platensis* and its derivative compounds as a nutraceutical for associated conditions for cardiovascular health are well studied and established through various preclinical and clinical studies (Khan et al., 2006; Sato et al., 2008; Lu et al., 2010; Pan et al., 2015; Heo et al., 2017; Nasirian et al., 2017; Sadek et al., 2017; Szulinska et al., 2017; Sommella et al., 2018; Hamedifard et al., 2019; Lin et al., 2020).

6.5 BIOACTIVE NUTRACEUTICAL-BASED PHYTOCONSTITUENTS

Several phytochemicals have been studied in recent past to investigate for cardiovascular diseases through series of *in vitro* and *in vivo* studies that are reported in Table 6.2. Further, chemical structure of each compound exhibiting cardio-protective effects is mentioned in Figure 6.4. The outcomes of some bioactive phytochemicals as potential sources of nutraceutical for the prevention and treatment of CVD are depicted herewith:

TABLE 6.2 Bioactive Phyto-constituents as a Nutraceutical Against Cardiovascular Diseases.

Phytoconstituents/ derived compounds	Model	Outcome	References
3-epi-periplogenin (1)	Article in German	Reduced hypertension significantly	Peter and Max, 1980
Adonitoxin (2)	*Cats*	Found to be potent cardioprotective aglycone	Kovaříková and Chen, 1965
Terpinen-4-ol (3)	Male Wistar rats	Reduced mean Aortic pressure and the induced hypotension is independent of effect by central autonomic nervous system	Lahlou et al., 2002
Eucalyptol (4)	Male Wistar rats	Reduced contractile activity in rat cardiac muscle, the present results suggest that eucalyptol depresses force development, probably acting as a calcium channel blocker.	Soares et al., 2005
Methyleugenol (5)	Male Wistar rats	Relaxed potassium-induced contractions. The induced hypotension is independent of effect by central autonomic nervous system.	Lahlou et al., 2004
Piperitenone oxide (6)	Male Wistar rats	Hypotensive effect due to vasodilation effects.	Lahlou et al., 2001
14-deoxy-12-hydroxyandrographolide (7) and 14-deoxyandrographolide (8)	Sprague-Dawley rats	Showed evidence of reducing hypertension and found to be effective for treatment of angina. Vasodilation was due to release of nitric oxide and activation of the guanylate cyclase pathway.	Awang et al., 2012
Diallyldisulfide (9)	Adult male Wistar rats	Reduced low level lipoproteins, increased availability of NO, and reduced protein carbonyls by preventing cardiovascular damage.	Sharma et al., 2012
10-gingerol (10), 8-gingerol (11) and 6-gingerol (12)	Endothelium containing thoracic aorta, isolated from Sprague-Dawley rats and right aorta from Guinea-pigs	Exhibited a dose-dependent vasodilator effect on Phenylephrine hydrochloride	Ghayur et al., 2005
Catechin (13)	Sprague-Dawley albino rats of either sex	Exhibited a significant role in cardiovascular diseases	Saqib et al., 2021

TABLE 6.2 *(Continued)*

Phytoconstituents/ derived compounds	Model	Outcome	References
Diosgenin (14)	Male Wistar rats	Prevented diminishing of arterial endothelial function (in response to acetylcholine which is a vasodilatory signaling molecule). There is reduction of plasma glutathione levels and glutathione reductase activity	Szabó et al., 2018
Cyclovirobuxine D (15)	Neonatal Sprague-Dawley	Increased cardiomyocyte viability and pretreated with adriamycin.	Hu et al., 2007
Cyclovirobuxine D (15)	Isolated rat cardiomyocytes	Cyclovirobuxine D decreased L-type Ca2+ current in a concentration-dependent manner and increased intracellular Ca2+ mobilization induced by caffeine in rat ventricular cardiomyocytes.	Chen et al., 2004
Cyclovirobuxine D (15)	Rats	Significant cytoprotective outcome was observed in rats after the destruction of collecting oxygen species, followed by repairing of mitochondrial membrane potential after diabetic cardiomyopathy	Jiang et al., 2020
Epicatechin (16)	Rats	Animals showed enhanced cardiac (-38%), left-ventricular (-49%), and renal (-93%) weight indices. Epicatechin increased the transcription of nuclear factor-E2-related factor-2 (Nrf2) and Nrf2 target genes in aortas.	Gómez-Guzmán et al., 2012
Epicatechin (16)	Rats	In aorta, enhanced nitric oxide synthase (NOS) activity and elevated NO-dependent vasorelaxation. In the left heart ventricle, increased NOS activity without changing gene expressions of nNOS, iNOS, and eNOS.	Kluknavsky et al., 2016; Bernatova, 2018.
Scopoletin (17) and Rutin (18)	Male Wistar rats	Hypotensive action has been observed	Wigati et al., 2017
Trametenolic acid (19)	Male Wistar rats	Antihypertensive activity was observed	Adeyi et al., 2021

TABLE 6.2 *(Continued)*

Phytoconstituents/ derived compounds	Model	Outcome	References
Coumaric acid (20), vanillic acid (21), and quercetin (22)	Sprague Dawley Albino rats of either sex	Exhibited a significant role in cardiovascular diseases	Saqib et al., 2021
K-Strophanthoside	*Guinea pigs*	K-Strophanthoside concentration in heart was twice in 30 minutes, when compared to concentration of skeletal muscles.	Marzo et al., 1973
Cocoa	Zucker diabetic fatty rats.	Prevented vascular inflammation	Álvarez-Cilleros et al., 2020
Naringenin and apigenin 7-O-glucoside	Aorta isolated from male rats	Induced vascular relaxations in endothelium in a concentration-dependent mechanism.	El-Hilaly et al., 2021

6.5.1 RESVERATROL

Resveratrol (3,4,5-trihydroxy-trans-stilbene) is one of the scientifically proven powerful antioxidants mostly found in significant concentrations in skin of grape which are well researched and reviewed to maintain CVD through various actions (Delmas et al., 2005). Several studies have shown antihypertensive effects of resveratrol through antioxidant action, endothelial nitric oxide (NO) production inhibition, vascular inflammation inhibition, and the prevention of platelets aggregation (Smoliga et al., 2011). Interestingly, it also has protective effects in animals with dyslipidemia, causes insulin resistance, reduces cardiac hypertrophy and contractile dysfunction (Rivera et al., 2009; Chan et al., 2011; Rimbaud et al., 2011; Dolinsky et al., 2013; Zordoky et al., 2015). The major mechanisms behind are involvement of the AMPK, SIRT-1, and Nrf2 pathways, improved availability of NO (by increased eNOS activity), and reduction in vascular smooth cell contractility (via inhibition of Ang-II) (Hollenberg, 2003; Arunachalam et al., 2010; Cao et al., 2014; Zordoky et al., 2015). Clinically through double-blind, randomized, placebo-controlled it was shown to decrease expression of endothelial cell ICAM, VCAM, and IL-8 (Agarwal et al, 2013). In summary many mechanisms through several research have been postulated for the beneficiary effect of various Resveratrol as a nutraceutical component for CVD and its associated conditions are well studied and established through various preclinical and clinical studies (Wang et al., 2007; Cho et al., 2008; Gnoni and Paglialonga, 2009; Szkudelska and Szkudelski, 2010; Novaes et al., 2012; Xuan et al., 2012; Voloshyna et al.,

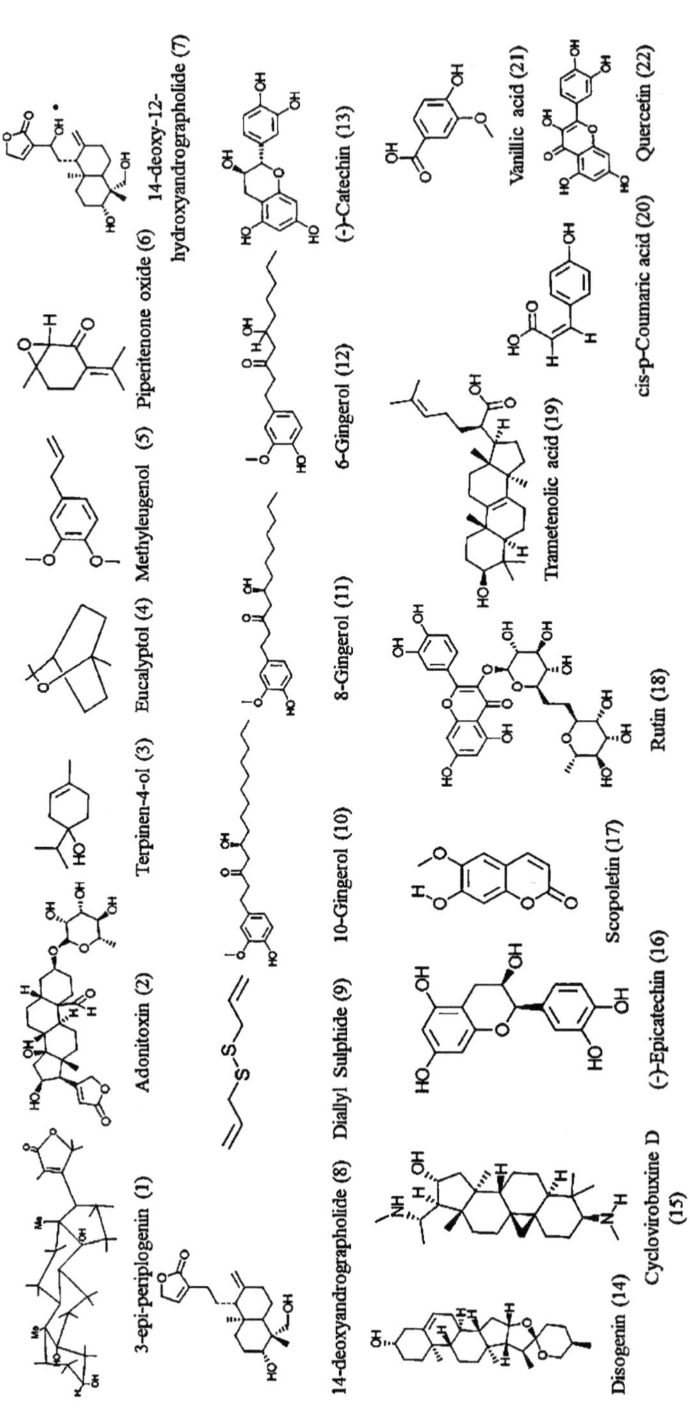

FIGURE 6.4 Chemical structures of bioactive nutraceuticals for managements of cardiovascular disease.

2012; Song et al., 2013; Vallianou et al., 2013; Shen et al., 2013; Ren et al., 2013; Gu et al., 2014; Gupta et al., 2014; Peredo-Escarcega et al., 2015; Zhang et al., 2016; Sadeghi et al., 2017; Fan et al., 2018; Annunziata et al., 2019).

6.5.2 QUERCETIN

Quercetin is a flavonoid (Keservani and Sharma, 2014; Keservani et al., 2016) that has been shown to be effective against CVD through interleukin, TNF-α pathways, antioxidative, cyclooxygenase/lipoxygenase pathways, as scavenger agent, maintaining BP levels by vascular compliance and elastic resistance and downregulate the autonomic nervous system (basal cause of hypertension) (Kim et al., 1998; Kempuraj et al., 2005; Bureau et al., 2008; Lee et al., 2010; Endale et al., 2013). Further, the ability of quercetin to regulate total blood volume by the reduction of ENaC gene expression and decreasing of body fluid volume have been investigated (Marunaka et al., 2018). In particular, it causes sodium reabsorption, less water retention, less body fluids, and blood volume with a consequent decrease of BP levels in experimental models was observed (Asano et al., 2009; Marunaka et al., 2017). In human trials, quercetin has been observed to possess anti-inflammatory properties by decreasing levels of CRP, thus acting as an important agent in the prevention of atherosclerosis (Schulz et al., 2005). Quercetin reduced blood LDL cholesterol levels and inhibited LDL oxidation by macrophages in overweight subjects with high cardiovascular risk (Mbikay et al., 2014; Xue et al., 2017). In summary many mechanisms through several research have been postulated for the beneficiary effect of various Quercetin as a nutraceutical component for CVD and its associated conditions are well studied and established through various preclinical and clinical studies (Coskun et al., 2005; Boots et al., 2008; Stewart et al., 2009; Kim et al., 2011; Kleemann et al., 2011; Askari et al., 2012; Kaur and Meena, 2012; Panchal et al., 2012; Bhaskar et al., 2013; Eid et al., 2015; Hung et al., 2015; Lu et al., 2017; Xiao et al., 2017).

6.5.3 CURCUMIN

Curcumin is a natural pigment (yellow/orange) having several beneficial effects on CVD through reduction of BP levels and prevention of vascular smooth cell proliferation by antioxidative, anti-inflammatory, and antiproliferative effects in association with calcium homeostatic properties (Satoskar et al., 1986; Aggarwal et al., 2007). During experimentation on rats, curcumin was found to prevent hypertension through reduction of AT1R expressions in arteries, cause GATA 4 acetylation, increase NO bioavailability, and improve Nrf2 expressions (Li et al., 2004; Tapia et al., 2012; Yao et al., 2016; Santos-Parker et al., 2017). Curcumin is one of the most effective anti-atherosclerotic nutraceuticals as it decreases in circulating LDL, increases in HDL, reduces AR expression and helps 12% reduction of cholesterol level (Pawar et al., 2012; Vareed et al., 2013; Cox et al., 2015). In summary many mechanisms through several research have been postulated for the beneficiary effect of various Curcumin as a nutraceutical component for CVD and its associated conditions are well studied and established through various preclinical and clinical studies (Alwi et al., 2008; Hatcher et al., 2008; Morimoto et al., 2008; Strimpakos and Sharma, 2008; Weisberg et al., 2008; Tang et al., 2009; Rungseesantivanon et al., 2010; Wickenberg et al., 2010; Wongcharoen et al., 2012; Meng et al., 2013; Zhang et al., 2013; Wongeakin et al., 2014; Ghosh et al., 2015; Nabavi et al., 2015; Hajavi et al., 2017; Oharomari et al., 2017; Azhdari et al., 2019).

6.5.4 BERBERINE

Berberine is an alkaloid having proven records for beneficial effects on CVD through both endothelium and vascular smooth muscle vasodilatation, NO release (by eNOS increased expression), reduction of ACE effect, balancing of K^+ channels activation, and intracellular Ca^{2+}channels release (Chiou et al., 1991; Olmez and Ilhan, 1992; Ko et al., 2000; Lan et al., 2015). Further, study also correlates long-term protective action toward hypertension via inhibitions of PDGF-induced vascular smooth cell growth and activation, reducing overall proliferation, leads to antagonizing structural and functional vascular alterations (Liang et al., 2008; Cheng et al., 2013). In summary many mechanisms through several research have been postulated for the beneficiary effect of various berberine as a nutraceutical component for CVD and its associated conditions are well studied and established through various preclinical and clinical studies (Marin-Neto et al., 1998; Zhou et al., 2001; Zeng et al., 2003; Khan et al., 2005; Wu and Wang, 2008; Zhang et al., 2008; Li et al., 2009; Xu et al., 2009; Affuso et al., 2010; Izzo et al., 2010; Wu et al., 2010; Wang et al., 2011; Chatuphonprasert et al., 2013; Li et al., 2014; Chang et al., 2015; Pang et al., 2015; Wu et al., 2015; Zhang et al., 2015; Feng et al., 2017).

6.6 PATENTS

Several technologies using various nutraceutical formulation and processes have been recently filed Intellectual Property Right and have been mentioned in Table 6.3.

TABLE 6.3 Important Patents Filled with Reference to Natural Product or Herbal Formulation for Management of Cardiovascular Diseases.

International patent	
Patent application number	**Patent title**
PCT/IN2009/000324	Ayurvedic formulation for treating coronary heart disease
PCT/IB2008/00301 11	Embllca officinalis plant extracts and uses thereof
US12/866,389	Ayurvedic formulation advocated for the prevention and management of coronary heart disease
11020384	Antiarrhythmic formulation
US11/323,093	Aspirin formulation for cardiovascular health
DE60301888T	An ayurvedic food supplement composition

Indian patent	
205/BOM/1992	Antiarrhythmic and cardioprotective substituted -1 (2H) Isoquinolines.
202041057440	Evaluation of cardioprotective activity of flowers of Butea monosperma against isoproterenol-induced cardiac hypertrophy in Wistar albino rats
202041026431	Preliminary phytochemical investigation and evaluation of cardioprotective studies of *Withania somnifera*.
201941043716	Cardioprotective effect of Bryonopsis laciniosa against myocardial infarction and left ventricular proteomic analysis of furosemide and potassium chloride treated rats.
201941043653	Evaluation of cardioprotective activity of aqueous and ethanolic extract of *Lawsonia innermis* in CaCl2 induced arrhythmia in albino rats.
201841041593	Extraction and isolation of cardioprotective amygdalin glycoside from the *Malus domestica* seeds.
3877/DELNP/2013	Vasoprotective and cardioprotective antidiabetic therapy.
273/DEL/2013	Extraction and isolation of novel cardioprotective Ursolic acid steraoyl glucoside from the leaves of Lantana camara Linn.
1059/KOL/2010	Invention of unique potent dietary food supplement (Andam-1) from Andam fruit (*Morinda citrifolia* commonly known as Non) by freeze drying method, treatment for curing cancer, cardioprotective and hyperlipidemia with other properties/potential health diseases.
684/DELNP/2009	Cardioprotective compounds.
6123/DELNP/2005	Use of 5-methoxytryptamine as a cardioprotective agent.
3395/DELNP/2005	Cardioprotective delta opioed receptor agonists and methods of using same.
201611027647	Herbal hypertensive and blood-thinning agent and method thereof.
201611030625	Preparation and characterization of antihypertensive drug microsphere for pulsatile drug delivery system.
201727011267	HGH determination for use to guide prevention of a major adverse cardiac event or a cardiovascular disease in a subject.

TABLE 6.3 *(Continued)*

Patent application number	Patent title
201811018239	A method to identify potential angiotensin-converting enzyme inhibitory (ACE Inhibitors) compounds of *Moringa oleifera* and uses thereof.
201811028256	Polyphenolic compound pycnogenol from Pinus pinaster target Angiotensin Type 1 Receptor a promising blood pressure regulator.
201841036874	Improved pharmaceutical compositions of carvedilol ethosomes for transdermal delivery.
201941010582	Impact of pharmacokinetic parameters on in vitro and in vivo studies of controlled release solid lipid microparticles of antihypertensive drug.
201941031823	An improved process for the preparation of an antihypertensive agent.
201947038073	Herbo-mineral formulation for the treatment of cardiovascular diseases and method of preparation thereof.
201911038055	A process to prepare antihypertensive concentrate from seaweed and a product thereof.
201911050880	An in-silico method to analyze polyphenols from Crataegus pinnatifida as an ACE inhibitor.
201911053826	A method for the evaluation of antihypertensive activity of Hibiscitrin targeting angiotensin-converting enzyme.
202011011489	A composition and antihypertensive product from Marine Algae.
202011020473	Antithrombotic and antihypertensive peptides.
202021053004	Novel strategies in formulation development of an immediate release tablet containing antihypertensive agents and their evaluation thereof.
202041053913	Polymer enriched bridging liquid technique for solubility and flow property enhancement of antihypertensive drugs.
202047055031	Application method of Radix paeoniae alba and Dendrobium officinale for preparing drug for treatment of hypertension.
202121011260	Acrylic copolymer and organic acid based press coated pulsatile tablet of Nifedipine.

6.7 CONCLUSION

Most of the herbal nutraceuticals are the sources of cardio-protective agents that have been thoroughly investigated and validated. However, many herbal (Sharma et al., 2020; Rane et al., 2020a; Rane et al., 2020b; Rane et al., 2020c) remedies used traditionally still warranted the careful scientific assessment for CVD, toxicological assessment, herb–drug, and herb–food interaction. In this chapter, many such validated herbal nutraceuticals against CVD are enlisted as future remedies. These findings must be thoroughly needed to be evaluated for nutraceutical product developments having multiple health benefits including prevention and treatment of CVD conditions.

CONFLICTS OF INTEREST

The authors declare that there are no conflicts of interest.

ACKNOWLEDGMENTS

The authors are thankful to Faculty of Pharmacy, DIT University, Dehradun, India for constant encouragement.

KEYWORDS

- **nutraceuticals**
- **cardiovascular diseases (CVDs)**
- **herbs**
- **phytoconstituents**
- **diagnostic measures**

REFERENCES

Adeyi, A. O.; Awosanya, S. A.; Adeyi, O. E.; James, A. S.; Adenipekun, C. O. Ganoderma Lucidum Ethanol Extract Abrogates Metabolic Syndrome in Rats: In vivo Evaluation of Hypoglycemic, Hypolipidemic, Hypotensive and Antioxidant properties. *Obes. Med.* **2021,** *22,* 100320. doi: 10.1016/j.obmed.2021.100320

Affuso, F.; Ruvolo, A.; Micillo, F.; Sacca, L.; Fazio, S. Effects of a Nutraceutical Combination (berberine, red yeast rice and policosanols) on Lipid Levels and Endothelial Function Randomized, Double-blind, Placebo-controlled Study. *Nutr. Metab. Cardiovasc. Dis.* **2010,** *20,* 656–661. doi: 10.1016/j.numecd.2009.05.017.

Afonso, L.; Kottam, A.; Reddy, V.; Penumetcha, A. Echocardiography in Infective Endocarditis: State of the Art. *Curr. Cardiol. Rep.* **2017,** *19* (12), 127. doi: 10.1007/s11886-017-0928-9

Agarwal, B.; Campen, M. J.; Channell, M. M.; Wherry, S. J.; Varamini, B.; Davis, J. G.; Baur, J. A.; Smoliga, J. M. Resveratrol for Primary Prevention of Atherosclerosis: Clinical Trial Evidence for Improved Gene Expression in Vascular Endothelium. *Int. J. Cardiol.* **2013,** *166,* 246–248. doi: 10.1016/j.ijcard.2012.09.027

Aggarwal, B. B.; Sundaram, C.; Malani, N.; Ichikawa, H. Curcumin: The Indian Solid Gold. *Adv. Exp. Med. Biol.* **2007,** *595,* 1–75. doi:10.1007/978-0-387-46401-5_1

Ahmad, U.; Ahmad, R. S.; Arshad, M. S.; Mushtaq, Z.; Hussain, S. M.; Hameed, A. H. Antihyperlipidemic Efficacy of Aqueous Extract of Stevia Rebaudiana Bertoni in Albino *Rats. Lipids Health Dis.* **2018a,** *17* (1), 175. doi: 10.1186/s12944-018-0810-9

Ahmed, M.; Azmat, A.; Khan, S. Acute Hypotensive and Diuretic Activities of Berberis Vulgaris Root Bark Aqueous Extract in Normal Rats. *Braz. J. Pharm. Sci.* **2018b,** *53* (4), e17077. doi: 10.1590/s2175-97902017000417077

Alamgeer.; Iman, S.; Asif, H.; Saleem, M. Evaluation of Antihypertensive Potential of *Ficus carica* Fruit. *Pharm. Biol.* **2017,** *55* (1), 1047–1053. doi: 10.1080/13880209.2017.1278611

Alissa, E. M.; Ferns, G. A. Functional Foods and Nutraceuticals in the Primary Prevention of Cardiovascular Diseases. *J. Nutr. Metab.* **2012,** *569486,* 1–16. doi: 10.1155/2012/569486

Álvarez-Cilleros, D.; López-Oliva, M. E.; Ramos, S.; Martín, M. A. Preventive Effect of Cocoa Flavanols Against Glucotoxicity-induced Vascular Inflammation in the Arteria of Diabetic Rats and on the Inflammatory Process in TNF-α-stimulated Endothelial Cells. *Food Chem. Toxicol.* **2020,** *146,* 111824. doi: 10.1016/j.fct.2020.111824

Alwi, I.; Santoso, T.; Suyono, S.; Sutrisna, B.; Suyatna, F. D.; Kresno, S. B.; Ernie, S. The Effect of Curcumin on Lipid Level in Patients with Acute Coronary Syndrome. *Acta Med. Indones.* **2008,** *40,* 201–210.

American Heart Association. Heart Disease and Stroke Statistics-2021 Update. *Circulation* **2021,** *143,* e254–e743. doi: 10.1161/CIR.0000000000000950

Amssayef, A.; Ajebli, M.; Eddouks, M. Aqueous Extract of Oakmoss Produces Antihypertensive Activity in L-NAME-induced Hypertensive Rats Through sGC-cGMP Pathway. *Clin. Exp. Hypertens.* **2021,** *43* (1), 49–55. doi: 10.1080/10641963.2020.1797087

An, S.; Han, J. I.; Kim, M. J.; Park, J. S.; Han, J. M.; Baek, N. I.; Chung, H. G.; Choi, M. S.; Lee, K. T.; Jeong, T. S. Ethanolic Extracts of Brassica Campestris spp. Rapa Roots Prevent High-fat Diet-induced Obesity via Beta (3)-adrenergic Regulation of White Adipocyte Lipolytic Activity. *J. Med. Food* **2010,** *13,* 406–414. doi: 10.1089/jmf.2009.1295

Annunziata, G.; Maisto, M.; Schisano, C.; Ciampaglia, R.; Narciso, V.; Tenore, G. C.; Novellino, E. Effects of Grape Pomace Polyphenolic Extract (Taurisolo ((R))) in Reducing TMAO Serum Levels in Humans: Preliminary Results from a Randomized, Placebo-Controlled, Cross-Over Study. *Nutrients* **2019,** *11,* 139. doi: 10.3390/nu11010139

Aronson, J. K. Defining 'nutraceuticals': Neither Nutritious nor Pharmaceutical. *Br. J. Clin. Pharmacol.* **2017,** *83* (1), 8–19. doi: 10.1111/bcp.12935

Arunachalam, G.; Yao, H.; Sundar, I. K.; Caito, S.; Rahman, I. SIRT1 Regulates Oxidant- and Cigarette Smoke-induced eNOS Acetylation in Endothelial Cells: Role of Resveratrol. *Biochem. Biophys. Res. Commun.* **2010,** *393,* 66–72. doi: 10.1016/j.bbrc.2010.01.080

Asano, J.; Niisato, N.; Nakajima, K.; Miyazaki, H.; Yasuda, M.; Iwasaki, Y.; Hama, T.; Dejima, K.; Hisa, Y.; Marunaka, Y. Quercetin stimulates Na+/K+/2Cl- cotransport via PTK-dependent Mechanisms in Human Airway Epithelium. *Am. J. Respir. Cell Mol. Biol.* **2009,** *41,* 688–695. doi: 10.1165/rcmb.2008-0338OC

Askari, G.; Ghiasvand, R.; Feizi, A.; Ghanadian, S. M.; Karimian, J. The Effect of Quercetin Supplementation on Selected Markers of Inflammation and Oxidative Stress. *J. Res. Med. Sci.* **2012,** *17,* 637–641.

Aslam, N.; Janbaz, K. H.; Jabeen, Q. Hypotensive and Diuretic Activities of Aqueous-ethanol Extract of Asphodelus Tenuifolius. *Bangladesh J. Pharmacol.* **2016,** *11* (4), 830–837. doi: 10.3329/bjp.v11i4.27131

Awang, K.; Abdullah, N. H.; Hadi, A. H. A.; Fong, Y. S. Cardiovascular Activity of Labdane Diterpenes from Andrographis Paniculata in Isolated Rat Hearts. *J. Biomed. Biotechnol.* **2012,** *2012,* 876458. doi: 10.1155/2012/876458

Azhdari, M.; Karandish, M.; Mansoori, A. Metabolic Benefits of Curcumin Supplementation in Patients with Metabolic Syndrome: A Systematic Review and Meta-analysis of Randomized Controlled Trials. *Phytother. Res.* **2019,** *33,* 1289–1301. doi: 10.1002/ptr.6323

Bahadoran, Z.; Mirmiran, P.; Azizi, F. Potential Efficacy of Broccoli Sprouts as a Unique Supplement for Management of Type 2 Diabetes and its Complications. *J. Med. Food* **2013,** *16,* 375–382. doi: 10.1089/jmf.2012.2559

Bahadoran, Z.; Mirmiran, P.; Hosseinpanah, F.; Hedayati, M.; Hosseinpour-Niazi, S.; Azizi, F. Broccoli Sprouts Reduce Oxidative Stress in Type 2 Diabetes: A Randomized Double-blind Clinical Trial. *Eur. J. Clin. Nutr.* **2011,** *65,* 972–977. doi: 10.1038/ejcn.2011.59

Barthelmes, L. V. J.; Nägele, M. P.; Enseleit, F.; Ferri, C.; Flammer, A. J.; Ruschitzka, F.; Sudano, I. Cocoa, Blood Pressure, and Vascular Function. *Front. Nutr.* **2017,** *4,* 36. doi: 110.3389/fnut.2017.00036

Bernatova, I. Biological Activities of (−)-epicatechin and (−)-epicatechin-containing Foods: Focus on Cardiovascular and Neuropsychological Health. *Biotechnol. Adv.* **2018,** *36* (*3*), 666–681. doi: 10.1016/j.biotechadv.2018.01.009

Bhaskar, S.; Kumar, K. S.; Krishnan, K.; Antony, H. Quercetin Alleviates Hypercholesteraemic Fiet Induced Inflammation During Progression and Regression of Atherosclerosis in Rabbits. *Nutrition* **2013,** *29,* 219–229. doi: 10.1016/j.nut.2012.01.019

Boots, A. W.; Haenen, G. R.; Bast, A. Health Effects of Quercetin: From Antioxidant to Nutraceutical. *Eur. J. Pharmacol.* **2008,** *585,* 325–337. doi: 10.1016/j.ejphar.2008.03.008

Buono, A.; Pedrotti, P.; Soriano, F.; Veas, N.; Oliva, F.; Oreglia, J.; Ammirati, E. Myocardial Infarction with Non-obstructive Coronary Arteries (MINOCA): Diagnosis, Pathogenesis, Therapy and Prognosis. *G. Ital. Cardiol.* **2019,** *20* (9), 499–511. doi: 10.1714/3207.31839

Bureau, G.; Longpre, F.; Martinoli, M. G. Resveratrol and Quercetin, Two Natural Polyphenols, Reduce Apoptotic Neuronal Cell Death Induced by Neuroinflammation. *J. Neurosci. Res.* **2008,** *86,* 403–410. doi: 10.1002/jnr.21503

Cao, X.; Luo, T.; Luo, X.; Tang, Z. Resveratrol Prevents Ang II-induced Hypertension via AMPK Activation and Rho A/ROCK Suppression in Mice. *Hypertens. Res.* **2014,** *37,* 803–810. doi: 10.1038/hr.2014.90

Casas, R.; Castro-Barquero, S.; Estruch, R.; Sacanella, E. Nutrition and Cardiovascular Health. *Int. J. Mol. Sci.* **2018,** *19* (12), 3988. doi: 10.3390/ijms19123988

Chan, V.; Fenning, A.; Iyer, A.; Hoey, A.; Brown, L. Resveratrol Improves Cardiovascular Function in DOCA-salt Hypertensive Rats. *Curr. Pharm. Biotechnol.* **2011,** *12,* 429–436. doi: 10.2174/138920111794480552

Chang, W.; Chen, L.; Hatch, G. M. Berberine as a Therapy for Type 2 Diabetes and its Complications: From Mechanism of Action to Clinical Studies. *Biochem. Cell Biol.* **2015,** *93,* 479–486. doi: 10.1139/bcb-2014-0107

Chatuphonprasert, W.; Lao-Ong, T.; Jarukamjorn, K. Improvement of Superoxide Dismutase and Catalase in Streptozotocin-nicotinamide-induced type 2-diabetes in Mice by Berberine and Glibenclamide. *Pharm. Biol.* **2013,** *52* (4), 419–427. doi: 10.3109/13880209.2013.839714

Chauhan, B.; Kumar, G.; Kalam, N.; Ansari, S. H. Current Concepts and Prospects of Herbal Nutraceutical: A Review. *J. Adv. Pharm. Technol. Res.* **2013,** *4* (1), 4–8. doi: 10.4103/2231-4040.107494

Chen, Q. W.; Shan, H.; Sun, H.; Wang, H.; Yang, B. Effects of Cyclovirobuxine D on Intracellular Ca2+ and L-type Ca2+ Current in Rat Ventricular Cardiomyocytes. *Yao Xue Xue Bao* **2004,** *39* (7), 500–503.

Chen W, Zhang S, Hu X, Chen F, Li D. A Review of Healthy Dietary Choices for Cardiovascular Disease: From Individual Nutrients and Foods to Dietary Patterns. Nutrients. **2023;** *15*(23): 4898. https://doi.org/10.3390/nu15234898.

Cheng, F.; Wang, Y.; Li, J.; Su, C.; Wu, F.; Xia, W. H.; Yang, Z.; Yu, B. B.; Qiu, Y. X.; Tao, J. Berberine Improves Endothelial Function by Reducing Endothelial Microparticles-Mediated Oxidative Stress in Humans. *Int. J. Cardiol.* **2013,** *167,* 936–942. doi: 10.1016/j.ijcard.2012.03.090

Cheong, S. H.; Kim, M. Y.; Sok, D. E.; Hwang, S. Y.; Kim, J. H.; Kim, H. R.; Lee, J. H.; Kim, Y. B.; Kim, M. R. Spirulina Prevents Atherosclerosis by Reducing Hypercholesterolemia in Rabbits Fed a High-cholesterol Diet. *J. Nutr. Sci. Vitaminol. (Tokyo)* **2010,** *56,* 34–40. doi: 10.3177/jnsv.56.34

Chiou, W. F.; Yen, M. H.; Chen, C. F. Mechanism of Vasodilatory Effect of Berberine in Rat Mesenteric Artery. *Eur. J. Pharmacol.* **1991,** *204,* 35–40. doi: 10.1016/0014-2999(91)90832-b

Cho, I. J.; Ahn, J. Y.; Kim, S.; Choi, M. S.; Ha, T. Y. Resveratrol Attenuates the Expression of HMG-CoA Reductase mRNA in Hamsters. *Biochem. Biophys. Res. Commun.* **2008,** *367,* 190–194. doi: 10.1016/j.bbrc.2007.12.140

Chong, L. K.; Udell, L. J.; Downs, B. W. Chapter 40—Intellectual Property, Branding, Trademark, and Regulatory Approvals in Nutraceuticals and Functional Foods. In *Nutraceutical and Functional Food Regulations in the United States and around the World,* 3rd ed.; 2019, 627–636.

Cifuentes, F.; Palacios, J.; Nwokocha, C. R.; Bórquez, J.; Simirgiotis, M. J.; Norambuena, I.; Chiong, M.; Paredes, A. Polyphenolic Composition and Hypotensive Effects of Parastrephia Quadrangularis (Meyen) Cabrera in Rat. *Antioxidants* **2019,** *8* (12), 591. doi: 10.3390/antiox8120591

Cordero-Herrera, I.; Martin, M. A.; Bravo, L.; Goya, L.; Ramos, S. Cocoa Flavonoids Improve Insulin Signalling and Modulate Glucose Production via AKT and AMPK in HepG2 Cells. *Mol. Nutr. Food Res.* **2013,** *57,* 974–985. doi: 10.1002/mnfr.201200500

Cordero-Herrera, I.; Martin, M. A.; Goya, L.; Ramos, S. Cocoa Flavonoids Attenuate High Glucose-induced Insulin Signalling Blockade and Modulate Glucose Uptake and Production in Human HepG2 Cells. *Food Chem. Toxicol.* **2014,** *64,* 10–19. doi: 10.1016/j.fct.2013.11.014

Corti, R.; Flammer, A. J.; Hollenberg, N. K.; Lüscher, T. F. Cocoa and Cardiovascular Health. *Circulation* **2009,** *119* (10), 1433–1441. doi: 10.1161/CIRCULATIONAHA.108.827022

Corti, R.; Perdrix, J.; Flammer, A. J.; Noll, G. Dark or White Chocolate? Cocoa and Cardiovascular Health. *Rev. Med. Suisse* **2010,** *6* (239), 499–500.

Coskun, O.; Kanter, M.; Korkmaz, A.; Oter, S. Quercetin, a Flavonoid Antioxidant, Prevents and Protects Streptozotocin-induced Oxidative Stress and Beta-cell Damage in Rat Pancreas. *Pharmacol. Res.* **2005,** *51,* 117–123. doi: 10.1016/j.phrs.2004.06.002

Cox, K. H.; Pipingas, A.; Scholey, A. B. Investigation of the Effects of Solid Lipid Curcumin on Cognition and Mood in a Healthy Older Population. *J. Psychopharmacol.* **2015,** *29,* 642–651. doi: 10.1177/0269881114552744

Damián MR, Cortes-Perez NG, Quintana ET, Ortiz-Moreno A, Garfias Noguez C, Cruceño-Casarrubias CE, Sánchez Pardo ME, Bermúdez-Humarán LG. Functional Foods,

Nutraceuticals and Probiotics: A Focus on Human Health. *Microorganisms.* **2022;** *10*(5): 1065. https://doi.org/10.3390/microorganisms10051065

Das, L.; Bhaumik, E.; Raychaudhuri, U.; Chakraborty, R. Role of Nutraceuticals in Human Health. *J. Food Sci. Technol.* **2012,** *49* (2), 173–183. doi: 10.1007/s13197-011-0269-4

Deedwania, P.; Srikanth, S. Diabetes and Vascular Disease. *Expert Rev. Cardiovasc. Ther.* **2008,** *6* (1), 127–38. doi: 10.1586/14779072.6.1.127

DeFilippis, A. P.; Chapman, A. R.; Mills, N. L.; de Lemos, J. A.; Arbab-Zadeh, A.; Newby, L. K.; Morrow, D. A. Assessment and Treatment of Patients with Type 2 Myocardial Infarction and Acute Nonischemic Myocardial Injury. *Circulation* **2019,** *140* (20), 1661–1678. doi: 10.1161/CIRCULATIONAHA.119.040631

Delmas, D.; Jannin, B.; Latruffe, N. Resveratrol: Preventing properties Against Vascular Alterations and Ageing. *Mol. Nutr. Food Res.* **2005,** *49,* 377–395. doi: 10.1002/mnfr.200400098

Dolinsky, V. W.; Chakrabarti, S.; Pereira, T. J.; Oka, T.; Levasseur, J.; Beker, D.; Zordoky, B. N.; Morton, J. S.; Nagendran, J.; Lopaschuk, G. D.; Davidge, S. T.; Dyck, J. R. B. Resveratrol Prevents Hypertension and Cardiac Hypertrophy in Hypertensive Rats and Mice. *Biochim. Biophys. Acta.* **2013,** *1832,* 1723–1733. doi: 10.1016/j.bbadis.2013.05.018

Duran, C. G.; Burbank, A. J.; Mills, K. H.; Duckworth, H. R.; Aleman, M. M.; Kesic, M. J.; Peden, D. B.; Pan, Y.; Zhou, H.; Hernandez, M. L. A Proof-of-concept Clinical Study Examining the NRF2 Activator Sulforaphane Against Neutrophilic Airway Inflammation. *Respir. Res.* **2016,** *17,* 89. doi: 10.1186/s12931-016-0406-8

Dutta S, Ali K, Dash S, Giri B. Role of Nutraceuticals on Health Promotion and Disease Prevention: *A Review.* JDDT [Internet]. 14Jul.2018 [cited 22Apr.2024];*8*(4):42–7. Available from: https://jddtonline.info/index.php/jddt/article/view/1759

Dzau, V. J.; Antman, E. M.; Black, H. R.; Hayes, D. L.; Manson, J. E.; Plutzky, J.; Popma, J. J.; Stevenson, W. The Cardiovascular Disease Continuum Validated: Clinical Evidence of Improved Patient Outcomes, Part I: Pathophysiology and Clinical Trial Evidence (Risk Factors Through Stable Coronary Artery Disease). *Circulation* **2006,** *114,* 2850–2870. doi: 10.1161/CIRCULATIONAHA.106.655688

Eid, H. M.; Nachar, A.; Thong, F.; Sweeney, G.; Haddad, P. S. The Molecular Basis of the Antidiabetic Action of Quercetin in Cultured Skeletal Muscle Cells and Hepatocytes. *Pharmacogn. Mag.* **2015,** *11,* 74–81. doi:10.4103/0973-1296.149708

Ekor, M. The Growing Use of Herbal Medicines: Issues Relating to Adverse Reactions and Challenges in Monitoring Safety. *Front. Pharmacol.* **2013,** *4,* 177. doi: 10.3389/fphar.2013.00177

El-Hilaly, J.; Amarouch, M.; Morel, N.; Lyoussi, B.; Quetin-Leclercq, J. Ajuga iva Water Extract Antihypertensive Effect on Stroke-prone Spontaneously Hypertensive Rats, Vasorelaxant Effects ex vivo and in vitro Activity of Fractions. *J. Ethnopharmacol.* **2021,** *270,* 113791. doi: 10.1016/j.jep.2021.113791

El-Newary, S. A. The Hypolipidemic Effect of Portulaca oleracea L. Stem on Hyperlipidemic Wister Albino Rats', *Ann. Agric. Sci.* **2016,** *61* (1), 111–124. doi: 10.1016/j.aoas.2016.01.002

Embuscado, M. E. Spices and Herbs: Natural Sources of Antioxidants-a Mini Review. *J. Funct. Foods* **2015,** *18,* 811–819. doi: 10.1016/j.jff.2015.03.005

Endale, M.; Park, S. C.; Kim, S.; Kim, S. H.; Yang, Y.; Cho, J. Y.; Rhee, M. H. Quercetin Disrupts Tyrosine-phosphorylated Phosphatidylinositol 3-kinase and Myeloid Differentiation Factor-88 Association, and Inhibits MAPK/AP-1 and IKK/NF-kappaB-induced

Inflammatory Mediators Production in RAW 264.7 Cells. *Immunobiology* **2013,** *218,* 1452–1467. doi: 10.1016/j.imbio.2013.04.019

Erdman J. W. Jr.; Carson, L.; Kwik-Uribe, C.; Evans, E. M.; Allen, R. R. Effects of Cocoa Flavanols on Risk Factors for Cardiovascular Disease. *Asia Pac. J. Clin. Nutr.* **2008,** *17* (Suppl 1), 284–287.

Fan, E.; Zhang, L.; Jiang, S.; Bai, Y. Beneficial Effects of Resveratrol on Atherosclerosis. *J. Med. Food* **2008,** *11,* 610–614. doi: 10.1089/jmf.2007.0091

Feng, M.; Zou, Z.; Zhou, X.; Hu, Y.; Ma, H.; Xiao, Y.; Li, X.; Ye, X. Comparative Effect of Berberine and its Derivative 8-cetylberberine on Attenuating Atherosclerosis in ApoE-/- mice. *Int. Immunopharmacol.* **2017,** *43,* 195–202. doi: 10.1016/j.intimp.2016.12.001

Galleano, M.; Oteiza, P. I.; Fraga, C. G. Cocoa, Chocolate and Cardiovascular Disease. *J. Cardiovasc. Pharmacol.* **2009,** *54* (6), 483–490. doi: 110.1097/FJC.0b013e3181b76787

Garami, Z.; Alexandrov, A. V. Neurosonology. *Neurol. Clin.* **2009,** *27* (1), 89–10*8,* viii. doi: 10.1016/j.ncl.2008.09.010

Ghayur, M. N.; Gilani, A. H.; Afridi, M. B.; Houghton, P. J. Cardiovascular Effects of Ginger Aqueous Extract and its Phenolic Constituents are Mediated Through Multiple Pathways. *Vascul. Pharmacol.* **2005,** *43* (4), 234–241. doi: 10.1016/j.vph.2005.07.003

Ghosh, S.; Banerjee, S.; Sil, P. C. The Beneficial Role of Curcumin on Inflammation, Diabetes and Neurodegenerative Disease: A Recent Update. *Food Chem. Toxicol.* **2015,** *83,* 111–124. doi: 10.1016/j.fct.2015.05.022

Gnoni, G. V.; Paglialonga, G. Resveratrol Inhibits Fatty acid and Triacylglycerol Synthesis in Rat Hepatocytes. *Eur. J. Clin. Investig.* **2009,** *39,* 211–218. doi:10.1111/j.1365-2362.2008.02077.x

Gómez-Guzmán, M.; Jiménez, R.; Sánchez, M.; Zarzuelo, M. J.; Galindo, P.; Quintela, A. M.; López-Sepúlveda, R.; Romero, M.; Tamargo, J.; Vargas, F.; Pérez-Vizcaíno, F.; Duarte, J. Epicatechin Lowers Blood Pressure, Restores Endothelial Function, and Decreases Oxidative Stress and Endothelin-1 and NADPH Oxidase Activity in DOCA-salt Hypertension. *Free Radic. Biol. Med.* **2012,** *52* (1), 70–79. doi: 10.1016/j.freeradbiomed.2011.09.015

Gopalan, D.; Delcroix, M.; Held, M. Diagnosis of Chronic Thromboembolic Pulmonary Hypertension. *Eur. Respir. Rev.* **2017,** *26* (143), 160108. doi: 10.1183/16000617.0108-2016

Granato, D.; Barba, F. J.; Bursac Kovacevic, D.; Lorenzo, J. M.; Cruz, A. G.; Putnik, P. Functional Foods: Product Development, Technological Trends, Efficacy Testing, and Safety. *Annu. Rev. Food. Sci. Technol.* **2020,** *11,* 93–118. doi: 10.1146/annurev-food-032519-051708

Grassi, D.; Desideri, G.; Necozione, S.; Lippi, C.; Casale, R.; Properzi, G.; Blumberg, J. B.; Ferri, C. Blood Pressure is Reduced and Insulin Sensitivity Increased in Glucose-intolerant, Hypertensive Subjects After 15 Days of Consuming High-polyphenol Dark Chocolate. *J. Nutr.* **2008,** *138,* 1671–1676. doi: 10.1093/jn/138.9.1671

Grassi, D.; Lippi, C.; Necozione, S.; Desideri, G.; Ferri, C. Short-term Administration of Dark Chocolate is Followed by a Significant Increase in Insulin Sensitivity and a Decrease in Blood Pressure in Healthy Persons. *Am. J. Clin. Nutr.* **2005,** *81,* 611–614. doi: 10.1093/ajcn/81.3.611

Gu, X. S.; Wang, Z. B.; Ye, Z.; Lei, J. P.; Li, L.; Su, D. F.; Zheng, X. Resveratrol, an Activator of SIRT*1,* Upregulates AMPK and Improves Cardiac Function in Heart Failure. *Genet. Mol. Res.* **2014,** *13,* 323–335. doi: 10.4238/2014.January.17.17

Guan, H.; Lin, Y.; Bai, L.; An, Y.; Shang, J.; Wang, Z.; Zhao, S.; Fan, J.; Liu, E. Dietary Cocoa Powder Improves Hyperlipidemia and Reduces Atherosclerosis in apoE Deficient Mice through the Inhibition of Hepatic Endoplasmic Reticulum Stress. *Mediators Inflamm.* **2016,** *2016,* 1937572. doi: 10.1155/2016/1937572

Gupta, P. K.; Di-Pette, D. J.; Supowit, S. C. Protective Effect of Resveratrol Against Pressure Overload-induced Heart Failure. *Food Sci. Nutr.* **2014,** *2,* 218–229. doi: 10.1002/fsn3.92

Hajavi, J.; Abbas Momtazi, A.; Johnston, T. P.; Banach, M.; Majeed, M.; Sahebkar, A. Curcumin: A Naturally Occurring Modulator of Adipokines in Diabetes. *J. Cell. Biochem.* **2017,** *118* (12), 4170–4182. doi: 10.1002/jcb.26121

Hamedifard, Z.; Milajerdi, A.; Reiner, Z.; Taghizadeh, M.; Kolahdooz, F.; Asemi, Z. The Effects of Spirulina on Glycemic Control and Serum Lipoproteins in Patients with Metabolic Syndrome and Related Disorders: A Systematic Review and Meta-analysis of Randomized Controlled Trials. *Phytother. Res.* **2019,** *33,* 2609–2621. doi: 10.1002/ptr.6441

Han, J. H.; Lee, H. J.; Kim, T. S.; Kang, M. H. The Effect of Glutathione S-transferase M1 and T1 Polymorphisms on Blood Pressure, Blood Glucose, and Lipid Profiles Following the Supplementation of Kale (Brassica oleracea acephala) Juice in South Korean Subclinical Hypertensive Patients. *Nutr. Res. Pract.* **2015,** *9,* 49–56. doi: 10.4162/nrp.2015.9.1.49

Hasler, C. M. Functional Foods: Benefits, Concerns and Challenges-A Position Paper from the American Council on Science and Health. *J. Nutr.* **2002,** *132* (12), 3772–3781. doi: 10.1093/jn/132.12.3772

Hatcher, H.; Planalp, R.; Cho, J.; Torti, F. M.; Torti, S. V. Curcumin: From Ancient Medicine to Current Clinical Trials. *Cell. Mol. Life Sci.* **2008,** *65,* 1631–1652. doi: 10.1007/s00018-008-7452-4

He, R.; Alashi, A.; Malomo, S. A.; Girgih, A. T.; Chao, D.; Ju, X.; Aluko, R. E. Antihypertensive and Free Radical Scavenging Properties of Enzymatic Rapeseed Protein Hydrolysates. *Food Chem.* **2013,** *141,* 153–159. doi: 10.1016/j.foodchem.2013.02.087

Hertog, M. G.; Feskens, E. J.; Hollman, P. C.; Katan, M. B.; Kromhout, D. Dietary Antioxidant Flavonoids and Risk of Coronary Heart Disease: The Zutphen Elderly Study. *Lancet* **1993,** *342,* 1007–1011. doi: 10.1016/0140-6736(93)92876-u

Hoeper, M. M.; Bogaard, H. J.; Condliffe, R.; Frantz, R.; Khanna, D.; Kurzyna, M.; Langleben, D.; Manes, A.; Satoh, T.; Torres, F.; Wilkins, M. R.; Badesch, D. B. Definitions and Diagnosis of Pulmonary Hypertension. *J. Am. Coll. Cardiol.* **2013,** *62* (25 Suppl), D42–D50. doi: 10.1016/j.jacc.2013.10.032

Hollenberg, N. K. Red Wine Polyphenols Enhance Endothelial Nitric Oxide Synthase Expression and Subsequent Nitric Oxide Release from Endothelial Cells. *Curr. Hypertens. Rep.* **2003,** *5,* 287–288. doi: 10.1161/01.cir.0000034445.31543.43

Hu, D.; Liu, X.; Wang, Y.; Chen, S. Cyclovirobuxine D Ameliorates Acute Myocardial Ischemia by KATP Channel Opening, Nitric Oxide Release and Anti-thrombosis. *Eur. J. Pharmacol.* **2007,** *569* (1–2), 103–109. doi: 10.1016/j.ejphar.2007.04.038

Huffman, M. D.; Prabhakaran, D.; Osmond, C.; Fall, C. H. D.; Tandon, N.; Lakshmy, R.; Ramji, S.; Khalil, A.; Gera, T.; Prabhakaran, P.; Biswas, S. K. D.; Reddy, K. S.; Bhargava, S. K.; Sachdev, H. S. Incidence of Cardiovascular Risk Factors in an Indian Urban Cohort. *J. Am. Coll. Cardiol.* **2011,** *57* (17), 1765–1774. doi: 10.1016/j.jacc.2010.09.083

Hung, C. H.; Chan, S. H.; Chu, P. M.; Tsai, K. L. Quercetin is a Potent Anti-atherosclerotic Compound by Activation of SIRT1 Signaling under ox LDL Stimulation. *Mol. Nutr. Food Res.* **2015,** *59,* 1905–1917. doi: 10.1002/mnfr.201500144

Iizuka, B.; Yamagishi, N.; Honma, N.; Hayashi, N. Cardio Vascular Disease Associated with Ulcerative Colitis. *Nihon Rinsho* **1999,** *57* (11), 2540–2545.

Izzo, R.; de Simone, G.; Giudice, R.; Chinali, M.; Trimarco, V.; De Luca, N.; Trimarco, B. Effects of Nutraceuticals on Prevalence of Metabolic Syndrome and on Calculated Framingham Risk Score in Individuals with Dyslipidemia. *J. Hypertens.* **2010,** *28,* 1482–1487. doi: 10.1097/HJH.0b013e3283395208

Javed, I.; Rahman, Z. U.; Khan, M. Z.; Muhammad, F.; Aslam, B.; Iqbal, Z.; Sultan, J. I.; Ahmad, I. Antihyperlipidaemic Efficacy of *Trachyspermum ammi* in Albino Rabbits. *Acta Vet. Brno.* **2009,** *78* (2), 229–236. doi: 10.2754/avb200978020229

Jeyasekaran, P. Awareness on Functional Foods and Nutraceuticals Among Educated Home Makers Through Booklet Distribution. *J. Med. Nutr. Nutraceut.* **2015,** *4,* 39–46. doi: 10.4103/2278-019X.146160

Jiang, Z.; Fu, L.; Xu, Y.; Hu, X.; Yang, H.; Zhang, Y.; Luo, H.; Gan, S.; Tao, L.; Liang, G.; Shen, X. Cyclovirobuxine D Protects Against Diabetic Cardiomyopathy by Activating Nrf2-Mediated Antioxidant Responses. *Sci. Rep.* **2020**, *10,* 6427. doi: 10.1038/s41598-020-63498-3

José, S. C.; Albuquerque, B. R.; Aguiar, J.; Corrêa, R. C. G.; Gonçalves, J. L.; Granato, D.; Pereira, J. A. M.; Barros, L.; Ferreira, I. C. F. R. Food Bioactive Compounds and Emerging Techniques for Their Extraction: Polyphenols as a Case Study. *Foods* **2021,** *10* (1), 37. doi: 10.3390/foods10010037

Kaur, G.; Meena, C. Amelioration of Obesity, Glucose Intolerance, and Oxidative Stress in High-fat Diet and Low-dose Streptozotocin-induced Diabetic Rats by Combination Consisting of "curcumin with piperine and quercetin". *ISRN Pharmacol.* **2012,** *2012,* 957283. doi: 10.5402/2012/957283

Kempuraj, D.; Madhappan, B.; Christodoulou, S.; Boucher, W.; Cao, J.; Papadopoulou, N.; Cetrulo, C. L.; Theoharides, T. C. Flavonols Inhibit Proinflammatory Mediator Release, Intracellular Calcium Ion Levels and Protein Kinase C Theta Phosphorylation in Human Mast Cells. *Br. J. Pharmacol.* **2005,** *145,* 934–944. doi: 10.1038/sj.bjp.0706246

Keservani, R. K.; Sharma, A. K. Flavonoids: Emerging Trends and Potential Health Benefits. *J. Chinese Pharma. Sci.* **2014**, *23* (12), 815.

Keservani, R. K.; Kesharwani, R. K.; Vyas, N.; Jain, S.; Raghuvanshi, R.; Sharma, A. K. Nutraceutical and Functional Food as Future Food: A Review. *Der Pharmacia Lettre.* **2010a**, *2* (1), 106–116.

Keservani, R. K.; Sharma, A. K.; Kesharwani, R. K. Nutraceutical and Functional Foods for Cardiovascular Health. In *Food Process Engineering*; Apple Academic Press, CRC Press, 2016; pp. 291–312; pp. 257–278. ISBN: 9781771884020.

Keservani, R. K.; Sharma, A. K.; Kesharwani, R. K. An Overview and Therapeutic Applications of Nutraceutical and Functional Foods. *Rec. Adv. Drug Deliv. Technol.* **2017,** 160–201. ISBN:9781522507543.

Keservani, R. K.; Sharma, A. K.; Kesharwani, R. K.; Eds. *Nutraceuticals and Dietary Supplements: Applications in Health Improvement and Disease Management*; CRC Press, 2020. ISBN: 9781771888738.

Keservani, R. K.; Kesharwani, R. K.; Sharma, A. K.; Vyas, N.; Chadoker, A. Nutritional Supplements: An Overview. *Int. J. Curr. Pharm. Rev Res.* **2010b**, *1* (1), 59–75.

Khan, M.; Shobha, J. C.; Mohan, I. K.; Naidu, M. U.; Sundaram, C.; Singh, S.; Kuppusamy, P.; Kutala, V. K. Protective Effect of Spirulina Against Doxorubicin-induced Cardiotoxicity. *Phytother. Res.* **2005,** *19,* 1030–1037. doi: 10.1002/ptr.1783

Kim, H. P.; Mani, I.; Iversen, L.; Ziboh, V. A. Effects of Naturally-occurring Flavonoids and Biflavonoids on Epidermal Cyclooxygenase and Lipoxygenase from

Guinea-pigs. *Prostaglandins Leukot. Essent. Fatty Acids* **1998,** *58,* 17–24. doi: 10.1016/s0952-3278(98)90125-9

Kim, J. H.; Kang, M. J.; Choi, H. N.; Jeong, S. M.; Lee, Y. M.; Kim, J. I. Quercetin Attenuates Fasting and Postprandial Hyperglycemia in Animal Models of Diabetes Mellitus. *Nutr. Res. Pract.* **2011,** *5,* 107–111. doi: 10.4162/nrp.2011.5.2.107

Kleemann, R.; Verschuren, L.; Morrison, M.; Zadelaar, S.; van Erk, M. J.; Wielinga, P. Y.; Kooistra, T. Anti-inflammatory, Anti-proliferative and Anti-atherosclerotic Effects of Quercetin in Human in vitro and in vivo Models. *Atherosclerosis* **2011,** *218,* 44–52. doi: 10.1016/j.atherosclerosis.2011.04.023

Kluknavsky, M.; Balis, P.; Puzserova, A.; Radosinska, J.; Berenyiova, A.; Drobna, M.; Lukac, S.; Muchova, J.; Bernatova, I. (-)-Epicatechin Prevents Blood Pressure Increase and Reduces Locomotor Hyperactivity in Young Spontaneously Hypertensive Rats. *Oxid. Med. Cell Longev.* **2016,** *2016,* 6949020. doi: 10.1155/2016/6949020

Ko, W. H.; Yao, X. Q.; Lau, C. W.; Law, W. I.; Chen, Z. Y.; Kwok, W.; Ho, K.; Huang, Y. Vasorelaxant and Antiproliferative Effects of Berberine. *Eur. J. Pharmacol.* **2000,** *399,* 187–196. doi: 10.1016/s0014-2999(00)00339-3

Kovaříková, A.; Chen, K. K. Activities of Newer Glycosides of *Adonis Vernalis* L. *Life Sci.* **1965,** *4* (1), 41–43. doi: 10.1016/0024-3205(65)90030-5

Kumar, A. S.; Sinha, N. Cardiovascular Disease in India: A 360-degree Overview. *Med. J. Armed Forces India* **2020,** *76* (1), 1–3. doi: 10.1016/j.mjafi.2019.12.005

Kurosawa, T.; Itoh, F.; Nozaki, A.; Nakano, Y.; Katsuda, S.; Osakabe, N.; Tsubone, H.; Kondo, K.; Itakura, H. Suppressive Effect of Cocoa Powder on Atherosclerosis in Kurosawa and Kusanagi-hypercholesterolemic Rabbits. *J. Atheroscler. Thromb.* **2005,** *12,* 20–28. 10.5551/jat.12.20

Lahlou, S.; Carneiro-Leão, R. F.; Leal-Cardoso, J. H.; Toscano, C. F. Cardiovascular Effects of the Essential oil of *Mentha x villosa* and its Main Constituent, Piperitenone Oxide, in Normotensive Anaesthetised Rats: Role of the Autonomic Nervous System. *Planta Med.* **2001,** *67* (7), 638–643. doi: 10.1055/s-2001-17352

Lahlou, S.; Carneiro-Leão, R. F. L.; Leal-Cardoso, J. H. Cardiovascular Effects of the Essential Oil of Mentha x Villosa in DOCA-salt-hypertensive Rats. *Phytomedicine* **2002,** *9* (8), 715–720. doi: 10.1078/094471102321621313

Lahlou, S.; Figueiredo, A. F.; Magalhães, P. J. C.; Leal-Cardoso, J. H.; Gloria, P. D. Cardiovascular Effects of Methyleugenol, a Natural Constituent of Many Plant Essential Oils, in Normotensive rats. *Life Sci.* **2004,** *74* (19), 2401–2412. doi: 10.1016/j.lfs.2003.09.063

Lahlou, S.; Galindo, C. A. B.; Leal-Cardoso, J. H.; Fonteles, M. C.; Duarte, G. P. Cardiovascular Effects of the Essential Oil of Alpinia Zerumbet Leaves and its Main Constituent, Terpinen-4-ol, in Rats: Role of the Autonomic Nervous System. *Planta Med.* **2002,** *68* (12), 1097–1102. doi: 10.1055/s-2002-36336

Lahlou, S.; Interaminense, L. F. L.; Leal-Cardoso, J. H.; Morais, S. M.; Duarte, G. P. Cardiovascular Effects of the Essential oil of Ocimum Gratissimum Leaves in Rats: Role of the Autonomic Nervous System. *Clin. Exp. Pharmacol. Physiol.* **2004,** *31* (4), 219–225. doi: 10.1111/j.1440-1681.2004.03976.x

Lahlou, S.; Leal-Cardoso, J. H.; Magalhães, P. J.; Coelho-de-Souza, A. N.; Duarte, G. P. Cardiovascular Effects of the Essential Oil of Croton Nepetaefolius in Rats: Role of the Autonomic Nervous System. *Planta Med.* **1999,** *65* (6), 553–557. doi: 10.1055/s-1999-14025

Lan, J.; Zhao, Y.; Dong, F.; Yan, Z.; Zheng, W.; Fan, J.; Sun, G. Meta-analysis of the Effect and Safety of Berberine in the Treatment of Type 2 Diabetes Mellitus, Hyperlipemia and Hypertension. *J. Ethnopharmacol.* **2015,** *161,* 69–81. doi: 10.1016/j.jep.2014.09.049

Larson, A. J.; Symons, J. D.; Jalili, T. Therapeutic Potential of Quercetin to Decrease Blood Pressure: Review of Efficacy and Mechanisms. *Adv. Nutr.* **2012,** *3,* 39–46. doi: 10.3945/an.111.001271

Latic, N.; Erben, R. G. Vitamin D and Cardiovascular Disease, with Emphasis on Hypertension, Atherosclerosis, and Heart Failure. *Int. J. Mol. Sci.* **2020,** *21* (18), 6483. doi: 10.3390/ijms21186483

Lee, H.; Lee, Y.; Kim, S. U.; Kim, H. C. Metabolic Dysfunction-Associated Fatty Liver Disease and Incident Cardiovascular Disease Risk: A Nationwide Cohort Study. *Clin. Gastroenterol. Hepatol.* **2021,** *19* (10), 2138–2147. doi: 10.1016/j.cgh.2020.12.022

Lee, K. M.; Hwang, M. K.; Lee, D. E.; Lee, K. W.; Lee, H. J. Protective Effect of Quercetin Against Arsenite-induced COX-2 Expression by Targeting PI3K in Rat Liver Epithelial cells. *J. Agric. Food Chem.* **2010,** *58,* 5815–5820. doi: 10.1021/jf903698s

Li, H. L.; Liu, C.; de Couto, G.; Ouzounian, M.; Sun, M.; Wang, A. B.; Huang, Y.; He, C. W.; Shi, Y.; Chen, X.; Nghiem, M. P.; Liu, Y.; Chen, M.; Dawood, F.; Fukuoka, M.; Maekawa, Y.; Zhang, L.; Leask, A.; Ghosh, A. K.; Kirshenbaum, L. A.; Liu, P. P. Curcumin Prevents and Reverses Murine Cardiac Hypertrophy. *J. Clin. Investig.* **2008,** *118,* 879–893. doi: 10.1172/JCI32865

Li, K.; Yao, W.; Zheng, X.; Liao, K. Berberine Promotes the Development of Atherosclerosis and Foam Cell Formation by Inducing Scavenger Receptor A Expression in Macrophage. *Cell Res.* **2009,** *19,* 1006–1017. doi: 10.1038/cr.2009.76

Li, Z.; Geng, Y. N.; Jiang, J. D.; Kong, W. J. Antioxidant and Anti-inflammatory Activities of Berberine in the Treatment of Diabetes Mellitus. *Evid. Based Complement. Altern. Med.* **2014,** *2014,* 289264. doi: 10.1155/2014/289264

Liang, K. W.; Yin, S. C.; Ting, C. T.; Lin, S. J.; Hsueh, C. M.; Chen, C. Y.; Hsu, S. L. Berberine Inhibits Platelet-derived Growth Factor-induced Growth and Migration Partly through an AMPK-dependent Pathway in Vascular Smooth Muscle Cells. *Eur. J. Pharmacol.* **2008,** *590,* 343–354. doi: 10.1016/j.ejphar.2008.06.034

Lin, C.; Zhang, L.; Li, B.; Zhang, F.; Shen, Q.; Kong, G.; Wang, X.; Cui, S.; Dai, R.; Cao, W.; Zhang, P. Selenium-Containing Protein from Selenium-Enriched *Spirulina platensis*Attenuates High Glucose-Induced Calcification of MOVAS Cells by Inhibiting ROS-Mediated DNA Damage and Regulating MAPK and PI3K/AKT Pathways. *Front. Physiol.* **2020,** *11,* 791. doi: 10.3389/fphys.2020.00791

Lokhande, P. D.; Dhaware, B. S.; Jagdale, S. C.; Chabukswar, A. R.; Mulkalwar, S. A. Cardiac Activity of Isolated Constituents of Inula Racemosa. *J. Herb. Pharmacother.* **2006,** *3–4,* 81–88. doi: 10.1300/J157v06n03_03

Lokhande, S. S. Role of Nutraceuticals in Various Diseases: A Comprehensive Review. *Asian J. Pharm. Res.* **2018,** *8* (4), 236–240. doi: 10.5958/2231-5691.2018.00040.0

Lu, J.; Ren, D. F.; Xue, Y. L.; Sawano, Y.; Miyakawa, T.; Tanokura, M. Isolation of an Antihypertensive Peptide from Alcalase Digest of *Spirulina platensis*. *J. Agric. Food Chem.* **2010,** *58,* 7166–7171. 10.1021/jf100193f

Lu, X. L.; Zhao, C. H.; Yao, X. L.; Zhang, H. Quercetin Attenuates High Fructose Feeding-induced Atherosclerosis by Suppressing Inflammation and Apoptosis via ROS-regulated PI3K/AKT Signaling Pathway. *Biomed. Pharmacother.* **2017,** *85,* 658–671. doi: 10.1016/j.biopha.2016.11.077

Maham, M.; Sarrafzadeh-Rezaei, F. Cardiovascular Effects of Adonis Aestivalis in Anesthetized Sheep. *Vet. Res. Forum Summer* **2014,** *5* (3), 193–199.

Manivannan, J.; Balamurugan, E.; Silambarasan, T.; Raja, B. Diosgenin Improves Vascular Function by Increasing Aortic eNOS Expression, Normalize Dyslipidemia and ACE Activity in Chronic Renal Failure Rats. *Mol. Cell Biochem.* **2013,** *384* (1–2), 113–120. doi: https://doi.org/10.1007/s11010-013-1788-2

Marin-Neto, J. A.; Maciel, B. C.; Secches, A. L.; Gallo Junior, L. Cardiovascular Effects of Berberine in Patients with Severe Congestive Heart Failure. *Clin. Cardiol.* **1988,** *11,* 253–260. doi: 10.1002/clc.4960110411

Marunaka, Y.; Marunaka, R.; Sun, H.; Yamamoto, T.; Kanamura, N.; Inui, T.; Taruno, A. Actions of Quercetin, a Polyphenol, on Blood Pressure. *Molecules* **2017,** *22,* 209. doi:10.3390/molecules22020209

Marunaka, Y.; Niisato, N.; Miyazaki, H.; Nakajima, K.; Taruno, A.; Sun, H.; Marunaka, R.; Okui, M.; Yamamoto, T.; Kanamura, N.; Kogiso, H.; Ikeuchi, Y.; Kashio, M.; Hosogi, S.; Nakahari, T. Quercetin is a Useful Medicinal Compound Showing Various Actions Including Control of Blood Pressure, Neurite Elongation and Epithelial Ion Transport. *Curr. Med. Chem.* **2018,** *25* (37), 4876–4887. doi: 10.2174/0929867323666160919095043

Marzo, A.; Ghirardi, P.; Croce, G.; Marchetti, G. Absorption, Distribution and Excretion of K-strophanthoside (3H) Administered Rectally to Guinea-pigs. *Naunyn Schmiedebergs Arch. Pharmacol.* **1973,** *279* (1), 19–29. doi: 10.1007/BF00502064.

Mashour, N. H.; Lin, G. I.; Frishman, W. H. Herbal Medicine for the Treatment of Cardiovascular Disease: Clinical Considerations. *Arch. Intern. Med.* **1998,** *158* (20), 2225–2234. doi:10.1001/archinte.158.20.2225

Matta, V. K.; Pasala, P. K.; Netala, S.; Pandrinki, S.; Konduri, P. Anti-hypertensive Activity of the Ethanolic Extract of Lantana Camara Leaves on High Salt Loaded Wistar Albino Rats. *Phcog. J.* **2015,** *7* (5), 289–295. doi: 10.5530/pj.2015.5.7

Matziouridou, C.; Marungruang, N.; Nguyen, T. D.; Nyman, M.; Fåk, F. Lingonberries Reduce Atherosclerosis in Apoe-/- Mice in Association with Altered Gut Microbiota Composition and Improved Lipid Profile. *Mol Nutr Food Res.* **2016,** *60* (5), 1150–1160. doi: 10.1002/mnfr.201500738

Mbikay, M.; Sirois, F.; Simoes, S.; Mayne, J.; Chretien, M. Quercetin-3-glucoside Increases Low-Density Lipoprotein Receptor (LDLR) Expression, Attenuates Proprotein Convertase Subtilisin/kexin 9 (PCSK9) Secretion, and Stimulates LDL Uptake by Huh7 Human Hepatocytes in Culture. *FEBS Open Bio.* **2014,** *4,* 755–762. doi: 10.1016/j.fob.2014.08.003

Mehrinfar, R.; Frishman, W. H. Flavanol-rich Cocoa: A Cardioprotective Nutraceutical. *Cardiol. Rev.* **2008,** *16* (3), 109–115. doi: 110.1097/CRD.0b013e31815d95e2

Meng, B.; Li, J.; Cao, H. Antioxidant and Antiinflammatory Activities of Curcumin on Diabetes Mellitus and its Complications. *Curr. Pharm. Des.* **2013,** *19,* 2101–2113.

Mizoguchi, H.; Matsubara, H. Cardio-pulmonary Disorders Associated with Collagen Vascular Disease. *Nihon Rinsho* **2009,** *67* (3), 543–549.

Morimoto, T.; Sunagawa, Y.; Kawamura, T.; Takaya, T.; Wada, H.; Nagasawa, A.; Komeda, M.; Fujita, M.; Shimatsu, A.; Kita, T.; Hasegawa, K. The Dietary Compound Curcumin Inhibits p300 Histone Acetyltransferase Activity and Prevents Heart Failure in Rats. *J. Clin. Investig.* **2008,** *118,* 868–878. doi: 10.1172/JCI33160

Nabavi, S. F.; Thiagarajan, R.; Rastrelli, L.; Daglia, M.; Sobarzo-Sanchez, E.; Alinezhad, H.; Nabavi, S. M. Curcumin: A Natural Product for Diabetes and its Complications. *Curr. Top. Med. Chem.* **2015,** *15,* 2445–2455. doi: 10.2174/1568026615666150619142519

Nasirian, F.; Mesbahzadeh, B.; Maleki, S. A.; Mogharnasi, M.; Kor, N. M. The Effects of Oral Supplementation of *Spirulina platensis* Microalgae on Hematological Parameters in Streptozotocin-induced Diabetic rats. *Am. J. Transl. Res.* **2017,** *9* (12), 5238–5244.

Nasri, H.; Baradaran, A.; Shirzad, H.; Rafieian-Kopaei, M. New Concepts in Nutraceuticals as Alternative for Pharmaceuticals. *Int. J. Prev. Med.* **2014,** *5* (12), 1487–1499. PMID: 25709784

Novaes, R. D.; PeluzioMdo, C.; Maldonado, I. R. Resveratrol Causes Antiatherogenic Effects in an Animal Model of Atherosclerosis. *Arq. Bras. Cardiol.* **2012,** *98,* 571–572. doi: 10.1590/s0066-782x2012005000006

Oharomari, L. K.; de Moraes, C.; Navarro, A. M. Exercise Training but not Curcumin Supplementation Decreases Immune Cell Infiltration in the Pancreatic Islets of a Genetically Susceptible Model of Type 1 Diabetes. *Sports Med. Open* **2017,** *3,* 15. doi: 10.1186/s40798-017-0082-3

Olmez, E.; Ilhan, M. Evaluation of the Alpha-adrenoceptor Antagonistic Action of Berberine in Isolated Organs. *Arzneimittelforschung* **1992,** *42,* 1095–1097.

Palomaki, A.; Pohjantahti-Maaroos, H.; Wallenius, M.; Kankkunen, P.; Aro, H.; Husgafvel, S.; Pihlava, J. M.; Oksanen, K. Effects of Dietary Cold-pressed Turnip Rapeseed Oil and Butter on Serum Lipids, Oxidized LDL and Arterial Elasticity in Men with Metabolic Syndrome. *Lipids Health Dis.* **2010,** *9,* 137. doi: 10.1186/1476-511X-9-137

Pan, W.; Yu, H.; Huang, S.; Zhu, P. Resveratrol Protects against TNF-alpha-Induced Injury in Human Umbilical Endothelial Cells through Promoting Sirtuin-1-Induced Repression of NF-KB and p38 MAPK. *PLoS One* **2016,** *11,* e0147034. doi: 10.1371/journal.pone.0147034

Panchal, S. K.; Poudyal, H.; Brown, L. Quercetin Ameliorates Cardiovascular, Hepatic, and Metabolic Changes in Diet-induced Metabolic Syndrome in Rats. *J. Nutr.* **2012,** *142,* 1026–1032. doi:10.3945/jn.111.157263

Pang, B.; Zhao, L. H.; Zhou, Q.; Zhao, T. Y.; Wang, H.; Gu, C. J.; Tong, X. L. Application of Berberine on Treating type 2 Diabetes Mellitus. *Int. J. Endocrinol.* **2015,** *2015,* 905749. doi: 10.1155/2015/905749

Paramo, J. A. A New Favourable Effect of Cocoa on Atherosclerosis? *Cardiovasc. Res.* **2008,** *79,* 3–4. doi: 10.1093/cvr/cvn103

Pawar, Y. B.; Munjal, B.; Arora, S.; Karwa, M.; Kohli, G.; Paliwal, J. K.; Bansal, A. K. Bioavailability of a Lipidic Formulation of Curcumin in Healthy Human Volunteers. *Pharmaceutics* **2012,** *4,* 517–530. doi:10.3390/pharmaceutics4040517

Pelaez-Soto, A.; Fernandez-Espinar, M. T.; Roig, P.; Gil, J. V. Evaluation of the Ability of Polyphenol Extracts of Cocoa and Red Grape to Promote the Antioxidant Response in Yeast Using a Rapid Multiwell Assay. *J. Food Sci.* **2017,** *82,* 324–332. doi: https://doi.org/10.1111/1750-3841.13602

Peredo-Escarcega, A. E.; Guarner-Lans, V.; Perez-Torres, I.; Ortega-Ocampo, S.; Carreon-Torres, E.; Castrejon-Tellez, V.; Diaz-Diaz, E.; Rubio-Ruiz, M. E. The Combination of Resveratrol and Quercetin Attenuates Metabolic Syndrome in Rats by Modifying the Serum Fatty Acid Composition and by Upregulating SIRT 1 and SIRT 2 Expression in White Adipose Tissue. *Evid. Based Complement. Alternat. Med.* **2015,** *2015,* 474032. doi: 10.1155/2015/474032

Peter, J.; Max, W. 3-epi-periplogenin: Ein Neues Cardenolid Aus *Adonis vernalis*. *Phytochemistry* **1980,** *19* (10), 2193–2197. doi: https://doi.org/10.1016/S0031-9422(00)82222-0

Petrie, J. R.; Guzik, T. J.; Touyz, R. M. Diabetes, Hypertension, and Cardiovascular Disease: Clinical Insights and Vascular Mechanisms. *Can. J. Cardiol.* **2018,** *34* (5), 575–584. doi: 10.1016/j.cjca.2017.12.005

Powell-Wiley, T. M.; Poirier, P.; Burke, L. E.; Després, J.; Gordon-Larsen, P.; Lavie, C. J.; Lear, S. A.; Ndumele, C. E.; Neeland, I. J.; Sanders, P.; St-Onge, M. Obesity and Cardiovascular Disease: A Scientific Statement from the American Heart Association. *Circulation* **2021,** *143,* e984–e1010. doi: 10.1161/CIR.0000000000000973

Rabadan-Chavez, G. M.; Reyes-Maldonado, E.; Quevedo-Corona, L.; Paniagua-Castro, N.; Escalona-Cardoso, G.; Jaramillo-Flores, M. E. The Prothrombotic State Associated with Obesity-induced Hypertension is Reduced by Cocoa and its Main Flavanols. *Food Funct.* **2016,** *7,* 4880–4888. doi: 10.1039/c6fo01165a

Raiola, A.; Errico, A.; Petruk, G.; Monti, D. M.; Barone, A.; Rigano, M. M. Bioactive Compounds in Brassicaceae Vegetables with a Role in the Prevention of Chronic Diseases. *Molecules.* **2017,** *23* (1), 15. doi: 10.3390/molecules23010015

Ramirez-Sanchez, I.; Taub, P. R.; Ciaraldi, T. P.; Nogueira, L.; Coe, T.; Perkins, G.; Hogan, M.; Maisel, A. S.; Henry, R. R.; Ceballos, G.; Villarreal, F. 2 (-)-Epicatechin Tich Cocoa Mediated Modulation of Oxidative Stress Regulators in Skeletal Muscle of Heart Failure and type 2 Diabetes Patients. *Int. J. Cardiol.* **2013,** *168,* 3982–3990. doi: 10.1016/j.ijcard.2013.06.089

Rane, B. R.; Bharath, M. S.; Patil, R. R.; Keservani, R. K., Jain, A. S. Novel Approaches in Nutraceuticals. In *Enhancing the Therapeutic Efficacy of Herbal Formulations through Novel Drug Delivery Systems*; Kesharwani, R. K., Keservani, R. K., Sharma, A. K., Eds.; IGI Global International Publisher, Pennsylvania: USA, 2020b; Chapter *11,* pp 241–266. ISBN13: 9781799844532.

Rane, B. R., Patil, A. S., Keservani, R. K., Jain, A. S. Novel Approaches in Herbal Formulation. In *Enhancing the Therapeutic Efficacy of Herbal Formulations through Novel Drug Delivery Systems*; Kesharwani, R. K., Keservani, R. K., Sharma, A. K. IGI Global International Publisher: Pennsylvania, USA, 2020a; Chapter *2,* pp 43–68. ISBN13: 9781799844532.

Rane, B. R., Tadavi, S. A., Keservani, R. K. Naturopathy. In *Herbal Product Development*; Sharma, A. K., Keservani, R. K., Gautam, S. P., Eds.; Apple Academic Press, CRC Press, Taylor & Francis Group, 2020c; Chapter 1*2,* pp 321–347. ISBN: 9781771888776.

Ren, Z.; Wang, L.; Cui, J.; Huoc, Z.; Xue, J.; Cui, H.; Mao, Q.; Yang, R. Resveratrol Inhibits NF-kB Signaling Through Suppression of p65 and I kappa B Kinase Activities. *Pharmazie* **2013,** *68,* 689–694.

Rimbaud, S.; Ruiz, M.; Piquereau, J.; Mateo, P.; Fortin, D.; Veksler, V.; Garnier, A.; Ventura-Clapier, R. Resveratrol Improves Survival, Hemodynamic and Energetics in a Rat Model of Hypertension Leading to Heart Failure. *PLoS One* **2011,** *6,* e26391. doi: 10.1371/journal.pone.0026391

Rivera, L.; Moron, R.; Zarzuelo, A.; Galisteo, M. Long-term Resveratrol Administration Reduces Metabolic Disturbances and Lowers Blood Pressure in Obese Zucker Rats. *Biochem. Pharmacol.* **2009,** *77,* 1053–1063. doi: 10.1016/j.bcp.2008.11.027

Roberfroid, M. B. A European Consensus of Scientific Concepts of Functional Foods. *Nutrition* **2000,** *16,* 689–691. doi: 10.1016/s0899-9007(00)00329-4

Roghani-Dehkordi, F.; Roghani, M.; Baluchnejadmojarad, T. Diosgenin Mitigates Streptozotocin Diabetes-induced Vascular Dysfunction of the Rat Aorta: The Involved

Mechanisms. *J. Cardiovasc. Pharmacol.* **2015,** *66* (6), 584–592. doi: 10.1097/FJC.0000000000000308

Roos, B.; Wood, S.; Bremner, D.; Bashir, S.; Betancor, M. B.; Fraser, W. D.; Duthie, S. J.; Horgan, G. W.; Sneddon, A. A. The Nutritional and Cardiovascular Health Benefits of Rapeseed Oil-fed Farmed Salmon in Humans are not Decreased Compared with Those of Traditionally Farmed Salmon: A Randomized Controlled Trial. *Eur. J. Nutr.* **2021,** *60* (4), 2063–2075. doi: 10.1007/s00394-020-02396-w

Rostami, A.; Khalili, M.; Haghighat, N.; Eghtesadi, S.; Shidfar, F.; Heidari, I.; Ebrahimpour-Koujan, S.; Eghtesadi, M. High-cocoa Polyphenol-rich Chocolate Improves Blood Pressure in Patients with Diabetes and Hypertension. *ARYA Atheroscler.* **2015,** *11,* 21–29. PMID: 26089927

Rubattu, S.; Di Castro, S.; Cotugno, M.; Bianchi, F.; Mattioli, R.; Baima, S.; Stanzione, R.; Madonna, M.; Bozzao, C.; Marchitti, S.; Gelosa, P.; Sironi, L.; Pignieri, A.; Maldini, M.; Giusti, A. M.; Nardini, M., Morelli, G. Protective Effects of Brassica Oleracea Sprouts Extract toward Renal Damage in High-salt-fed SHRSP: Role of AMPK/PPARalpha/UCP2 Axis. *J. Hypertens.* **2015,** *33,* 1465–1479. doi: 10.1097/HJH.0000000000000562

Rungseesantivanon, S.; Thenchaisri, N.; Ruangvejvorachai, P.; Patumraj, S. Curcumin Supplementation Could improve Diabetes-induced Endothelial Dysfunction Associated with Decreased Vascular Superoxide Production and PKC Inhibition. *BMC Complement. Altern. Med.* **2010,** *10,* 57. doi: 10.1186/1472-6882-10-57

Sadeghi, A.; Seyyed Ebrahimi, S. S.; Golestani, A.; Meshkani, R. Resveratrol Ameliorates Palmitate-Induced Inflammation in Skeletal Muscle Cells by Attenuating Oxidative Stress and JNK/NF-KB Pathway in a SIRT1-Independent Mechanism. *J. Cell. Biochem.* **2017,** *118* (9), 2654–2663. doi: 10.1002/jcb.25868

Sadek, K. M.; Lebda, M. A.; Nasr, S. M.; Shoukry, M. *Spirulina platensis* Prevents Hyperglycemia in Rats by Modulating Gluconeogenesis and Apoptosis via Modification of Oxidative Stress and MAPK-pathways. *Biomed. Pharmacother.* **2017,** *92,* 1085–1094. 10.1016/j.biopha.2017.06.023

Šamec, D.; Urlić, B.; Salopek-Sondi, B. Kale (Brassica oleracea var. acephala) as a Superfood: Review of the Scientific Evidence Behind the Statement. *Crit. Rev. Food Sci. Nutr.* **2019,** *59* (15), 2411–2422. doi: 10.1080/10408398.2018.1454400

Santos-Parker, J. R.; Strahler, T. R.; Bassett, C. J.; Bispham, N. Z.; Chonchol, M. B.; Seals, D. R. Curcumin Supplementation Improves Vascular Endothelial Function in Healthy Middle-aged and Older Adults by Increasing Nitric Oxide Bioavailability and Reducing Oxidative Stress. *Aging (Albany NY)* **2017,** *9,* 187–208. doi:10.18632/aging.101149

Saqib, F., Ali, A.; Ahmedah, H. T.; Irimie, C. A.; Toma, S. I.; Popovici, B. E.; Moga, M.; Irimie, M. Cardioprotective, Hypotensive and Toxicological Studies of Populus Ciliata (Wall. ex Royle). *Biomed. Pharmacother.* **2021,** *142,* 112065. doi: 10.1016/j.biopha.2021.112065

Saqib, F.; Janbaz, K. H. Ethnopharmacological Basis for Folkloric Claims of Anagallis Arvensis Linn. (Scarlet Pimpernel) as Prokinetic, Spasmolytic and Hypotensive in Province of Punjab, Pakistan. *J. Ethnopharmacol.* **2021,** *267,* 113634. doi: 10.1016/j.jep.2020.113634

Sato, T.; Yamamoto, C.; Fujiwara, Y.; Kaji, T. Biological Activities of Exogenous Polysaccharides via Controlling Endogenous Proteoglycan Metabolism in Vascular Endothelial Cells. *Yakugaku Zasshi* **2008,** *128* (5), 717–23. doi: 10.1248/yakushi.128.717

Satoskar, R. R.; Shah, S. J.; Shenoy, S. G. Evaluation of Anti-inflammatory Property of Curcumin (diferuloyl methane) in Patients with Postoperative Inflammation. *Int. J. Clin. Pharmacol. Ther. Toxicol.* **1986,** *24,* 651–654.

Saxena, A.; Ng, E. Y. K.; Lim, S. T. Imaging Modalities to Diagnose Carotid Artery Stenosis: Progress and Prospect. *Biomed. Eng. Online* **2019,** *18* (1), 66. doi: 10.1186/s12938-019-0685-7

Schulz, H. U.; Schurer, M.; Bassler, D.; Weiser, D. Investigation of Pharmacokinetic Data of Hypericin, Pseudohypericin, Hyperforin and the Flavonoids Quercetin and Isorhamnetin Revealed from Single and Multiple Oral Dose Studies with a Hypericum Extract Containing Tablet in Healthy Male Volunteers. *Arzneimittelforschung* **2005,** *55,* 561–568. doi:10.1055/s-0031–1296905

Sengupta, S.; Koley, H.; Dutta, S.; Bhowal, J. Hypocholesterolemic Effect of Spirulina Platensis (SP) Fortified Functional Soy Yogurts on Diet-induced Hypercholesterolemia. *J. Funct. Foods* **2018,** *48,* 54–64. 10.1016/j.jff.2018.07.007

Sharma, A. K.; Keservani, R. K.; Gautam, S. P. *Herbal Product Development*; Apple Academic Press, CRC Press, Taylor & Francis Group, 2020. ISBN: 9781771888776.

Sharma, D. K.; Manral, A.; Saini, V.; Singh, A.; Srinivasan, B. P.; Tiwari, M. Novel Diallyldisulfide Analogs Ameliorate Cardiovascular Remodeling in Rats with L-NAME-induced Hypertension. *Eur. J. Pharmacol.* **2012,** *691* (1–3), 198–208. doi: 10.1016/j.ejphar.2012.07.022

Shen, L. L.; Wang, X. M.; He, B. L. Advance of Resveratrol in Treating Diabetes Mellitus. *Zhongguo Zhong Xi Yi Jie He Za Zhi* **2013,** *33,* 279–281.

Singh, P., Kesharwani, R. K., & Keservani, R. K. Antioxidants and Vitamins: Roles in Cellular Function and Metabolism. In *Sustained energy for Enhanced Human Functions and Activity*; Academic Press, 2017a; pp 385–407. ISBN:9780128054130.

Singh, P.; Kesharwani, R. K.; Keservani, R. K. Protein, Carbohydrates, and Fats: Energy Metabolism. In *Sustained Energy for Enhanced Human Functions and Activity*; Academic Press, 2017b; pp 103–115. ISBN:9780128054130.

Smoliga, J. M.; Baur, J. A.; Hausenblas, H. A. Resveratrol and Health–a Comprehensive Review of Human Clinical Trials. *Mol. Nutr. Food Res.* **2011,** *55,* 1129–1141. doi: 10.1002/mnfr.201100143

Soares, M. C. M. S.; Damiani, C. E. N.; Moreira, C. M.; Stefanon, I.; Vassallo, D. V. Eucalyptol, an Essential Oil, Reduces Contractile Activity in Rat Cardiac Muscle. *Braz. J. Med. Biol. Res.* **2005,** *38* (3), 453–461. doi: 10.1590/S0100-879X2005000300017

Sommella, E.; Conte, G. M.; Salviati, E.; Pepe, G.; Bertamino, A.; Ostacolo, C.; Sansone, F.; Prete, F. D.; Aquino, R. P.; Campiglia, P. Fast Profiling of Natural Pigments in Different Spirulina (Arthrospira platensis) Dietary Supplements by DI-FT-ICR and Evaluation of their Antioxidant Potential by Pre-Column DPPH-UHPLC Assay. *Molecules* **2018,** *23,* 1152. doi: 10.3390/molecules23051152

Song, R.; Li, W. Q.; Dou, J. L.; Li, L.; Hu, Y. J.; Guo, J. Z.; Lu, D.; Zhang, G.; Sun, L. Resveratrol Reduces Inflammatory Cytokines via Inhibiting Nuclear Factor-kappaB and Mitogen-activated Protein Kinase Signal Pathway in a Rabbit Atherosclerosis Model. *Zhonghua Xin Xue Guan Bing Za Zhi* **2013,** *41,* 866–869.

Sosnowska, B.; Penson, P.; Banach, M. The Role of Nutraceuticals in the Prevention of Cardiovascular Disease. *Cardiovasc. Diagn. Ther.* **2017,** *7* (Suppl 1), S21–S31. doi: 10.21037/cdt.2017.03.20

Stewart, L. K.; Wang, Z.; Ribnicky, D.; Soileau, J. L.; Cefalu, W. T.; Gettys, T. W. Failure of Dietary Quercetin to Alter the Temporal Progression of Insulin Resistance among Tissues of C57BL/6J Mice uring the Development of Diet-induced Obesity. *Diabetologia* **2009,** *52,* 514–523. doi: 10.1007/s00125-008-1252-0

Strimpakos, A. S.; Sharma, R. A. Curcumin: Preventive and Therapeutic Properties in Laboratory Studies and Clinical Trials. *Antioxid. Redox Signal.* **2008,** *10,* 511–545. doi: 10.1089/ars.2007.1769

Szabó, K.; Gesztelyi, R.; Lampé, N.; Kiss, R.; Remenyik, J.; Pesti-Asbóth, G.; Priksz, D.; Szilvássy, Z.; Juhász, B. Fenugreek (*Trigonella foenum-Graecum*) Seed Flour and Diosgenin Preserve Endothelium-dependent Arterial Relaxation in a Rat Model of Early-stage Metabolic Syndrome. *Int. J. Mol. Sci.* **2018,** *19* (3), 798. doi: 10.3390/ijms19030798

Szkudelska, K.; Szkudelski, T. Resveratrol, Obesity and Diabetes. *Eur. J. Pharmacol.* **2010,** *635,* 1–8. doi: 10.1016/j.ejphar.2010.02.054

Szulinska, M.; Gibas-Dorna, M.; Miller-Kasprzak, E.; Suliburska, J.; Miczke, A.; Walczak-Galezewska, M.; Stelmach-Mardas, M.; Walkowiak, J.; Bogdanski, P. Spirulina Maxima Improves Insulin Sensitivity, Lipid Profile, and Total Antioxidant Status in Obese Patients with Well-treated Hypertension: A Randomized Double-blind Placebo-controlled Study. *Eur. Rev. Med. Pharmacol. Sci.* **2017,** *21,* 2473–2481.

Tang, Y. H.; Bao, M. W.; Yang, B.; Zhang, Y.; Zhang, B. S.; Zhou, Q.; Chen, J. L.; Huang, C. X. Curcumin Attenuates Left Ventricular Dysfunction and Remodelling in Rabbits with Chronic Heart Failure. *Zhonghua Xin Xue Guan Bing Za Zhi* **2009,** *37,* 262–267.

Tanghe, A.; Celie, B.; Shadid, S.; Rietzschel, E.; Roodt, J. A.; Reesink, K. D.; Heyman, E.; Calders, P. Acute Effects of Cocoa Flavanols on Blood Pressure and Peripheral Vascular Reactivity in Type 2 Diabetes Mellitus and Essential Hypertension: A Protocol for an Acute, Randomized, Double-Blinded, Placebo-Controlled Cross-Over Trial. *Front. Cardiovasc. Med.* **2021,** *8,* 602086. doi: 10.3389/fcvm.2021.602086

Tapia, E.; Soto, V.; Ortiz-Vega, K. M.; Zarco-Marquez, G.; Molina-Jijon, E.; Cristobal-Garcia, M.; Santamaria, J.; Garcia-Nino, W. R.; Correa, F.; Zazueta, C.; Pedraza-Chaverri, J. Curcumin Induces Nrf2 Nuclear Translocation and Prevents Glomerular Hypertension, Hyperfiltration, Oxidant Stress, and the Decrease in Antioxidant Enzymes in 5/6 Nephrectomized Rats. *Oxid. Med. Cell. Longev.* **2012,** *2012,* 269039. doi:10.1155/2012/269039.

Tata, C. M.; Sewani-Rusike, C. R.; Oyedeji, O. O.; Gwebu, E. T.; Mahlakata, F.; Nkeh-Chungag, B. N. Antihypertensive Effects of the Hydro-ethanol Extract of Senecio Serratuloides DC in Rats. *BMC Complement Altern. Med.* **2019,** *19,* 52. doi: 10.1186/s12906-019-2463-2

Vallianou, N. G.; Evangelopoulos, A.; Kazazis, C. Resveratrol and Diabetes. *Rev. Diabet. Stud.* **2013,** *10,* 236–242. doi: 10.1900/RDS.2013.10.236

Vareed, S. K.; Kakarala, M.; Ruffin, M. T.; Crowell, J. A.; Normolle, D. P.; Djuric, Z.; Brenner, D. E. Pharmacokinetics of Curcumin Conjugate Metabolites in Healthy Human Subjects. *Cancer Epidemiol. Biomark. Prev.* **2008,** *17,* 1411–1417. doi: 10.1158/1055-9965.EPI-07-2693

Vo, T. S.; Kim, S. K. Down-regulation of Histamine-induced Endothelial Cell Activation as Potential Anti Atherosclerotic Activity of Peptides from Spirulina Maxima. *Eur. J. Pharm. Sci.* **2013,** *50,* 198–207. doi: 10.1016/j.ejps.2013.07.001

Voloshyna, I.; Hussaini, S. M.; Reiss, A. B. Resveratrol in Cholesterol Metabolism and Atherosclerosis. *J. Med. Food* **2012,** *15,* 763–773. doi: 10.1089/jmf.2012.0025

Wang, G. Y.; Wang, Y. M.; Zhang, L. N.; Li, Q.; Yue, H.; Song, C. M.; Feng, J. K.; Wang, N. Effect of Resveratrol on Heart Function of Rats with Adriamycin-induced Heart Failure. *Zhongguo Zhong Yao Za Zhi* **2007,** *32,* 1563–1565.

Wang, Q.; Zhang, M.; Liang, B.; Shirwany, N.; Zhu, Y.; Zou, M. H. Activation of AMP-activated Protein Kinase is Required for Berberine-induced Reduction of Atherosclerosis in Mice:

The Role of Uncoupling Protein 2. *PLoS One* **2011,** *6,* e25436. doi: 10.1371/journal.pone.0025436

Weisberg, S. P.; Leibel, R.; Tortoriello, D. V. Dietary Curcumin Significantly Improves Obesity-associated Inflammation and Diabetes in Mouse Models of Diabesity. *Endocrinology* **2008,** *149,* 3549–3558. doi: 10.1210/en.2008-0262

Wickenberg, J.; Ingemansson, S. L.; Hlebowicz, J. Effects of Curcuma Longa (turmeric) on Postprandial Plasma Glucose and Insulin in Healthy Subjects. *Nutr. J.* **2010,** *9,* 43. doi: 10.1186/1475-2891-9-43

Wigati, D.; Anwar, K.; Sudarsono.; Nugroho, A. E. Hypotensive Activity of Ethanolic Extracts of Morinda citrifolia L. Leaves and Fruit in Dexamethasone-Induced Hypertensive Rat. *J. Evid. Based Complementary Altern. Med.* **2017,** *22* (1), 107–113. doi: 10.1177/2156587216653660

Windecker, S.; Stortecky, S.; Meier, B. Paradoxical Embolism. *J. Am. Coll. Cardiol.* **2014,** *64* (4), 403–415. doi: 10.1016/j.jacc.2014.04.063

Wong, B.; Kruse, G.; Kutikova, L., Ray, K. K., Mata, P.; Bruckert, E. Cardiovascular Disease Risk Associated with Familial Hypercholesterolemia: A Systematic Review of the Literature. *Clin. Ther.* **2016,** *38* (7), 1696–709. doi: 10.1016/j.clinthera.2016.05.006

Wongcharoen, W.; Jai-Aue, S.; Phrommintikul, A.; Nawarawong, W.; Woragidpoonpol, S.; Tepsuwan, T.; Sukonthasarn, A.; Apaijai, N.; Chattipakorn, N. Effects of Curcuminoids on Frequency of Acute Myocardial Infarction After Coronary Artery Bypass Grafting. *Am. J. Cardiol.* **2012,** *110,* 40–44. doi: 10.1016/j.amjcard.2012.02.043

Wongeakin, N.; Bhattarakosol, P.; Patumraj, S. Molecular Mechanisms of Curcumin on Diabetes-Induced Endothelial Dysfunctions: Txnip, ICAM-*1,* and NOX2 Expressions. *BioMed Res. Int.* **2014,** *2014,* 161346. doi: 10.1155/2014/161346

Wu, L.; Noyan Ashraf, M. H.; Facci, M.; Wang, R.; Paterson, P. G.; Ferrie, A.; Juurlink, B. H. Dietary Approach to Attenuate Oxidative Stress, Hypertension, and Inflammation in the Cardiovascular System. *Proc. Natl. Acad. Sci. USA* **2004,** *101,* 7094–7099. doi: 10.1073/pnas.0402004101

Wu, M.; Wang, J. Advance on Study in Anti-atherosclerosis Mechanism of Berberine. *Zhongguo Zhong Yao Za Zhi* **2008,** *33,* 2013–2016. doi: 10.1007/s11655-010-0188-7

Wu, M.; Wang, J.; Liu, L. T. Advance of Studies on Anti-atherosclerosis Mechanism of Berberine. *Chin. J. Integr. Med.* **2010,** *16,* 188–192. doi: 10.1007/s11655-010-0188-7

Wu, X.; Li, Y.; Wang, Q.; Li, W.; Feng, Y. Effects of Berberine and Pomegranate Seed Oil on Plasma Phospholipid Metabolites Associated with Risks of type 2 Diabetes Mellitus by U-HPLC/Q-TOF-MS. *J. Chromatogr. B Analyt. Technol. Biomed. Life Sci.* **2015,** *1007,* 110–120. doi: 10.1016/j.jchromb.2015.11.008

Xiao, L.; Liu, L.; Guo, X.; Zhang, S.; Wang, J.; Zhou, F.; Liu, L.; Tang, Y.; Yao, P. Quercetin Attenuates High Fat Diet-induced Atherosclerosis in Apolipoprotein E Knockout Mice: A Critical Role of NADPH Oxidase. *Food Chem. Toxicol.* **2017,** *105,* 22–33. doi: 10.1016/j.fct.2017.03.048

Xu, M. G.; Wang, J. M.; Chen, L.; Wang, Y.; Yang, Z.; Tao, J. Berberine-induced Upregulation of Circulating Endothelial Progenitor Cells is Related to Nitric Oxide Production in Healthy Subjects. *Cardiology* **2009,** *112,* 279–286. doi: 10.1159/000157336

Xuan, W.; Wu, B.; Chen, C.; Chen, B.; Zhang, W.; Xu, D.; Bin, J.; Liao, Y. Resveratrol Improves Myocardial Ischemia and Ischemic Heart Failure in Mice by Antagonizing the Detrimental Effects of Fractalkine. *Crit. Care Med.* **2012,** *40,* 3026–3033. doi: 10.1097/CCM.0b013e31825fd7da

Xue, F.; Nie, X.; Shi, J.; Liu, Q.; Wang, Z.; Li, X.; Zhou, J.; Su, J.; Xue, M.; Chen, W. D.; Wang, Y. D. Quercetin Inhibits LPS-Induced Inflammation and ox-LDL-Induced Lipid Deposition. *Front. Pharmacol.* **2017,** *8,* 40. doi:10.3389/fphar.2017.00040

Yao, Y.; Wang, W.; Li, M.; Ren, H.; Chen, C.; Wang, J.; Wang, W. E.; Yang, J.; Zeng, C. Curcumin Exerts its Anti-hypertensive Effect by Down-regulating the AT1 Receptor in Vascular Smooth Muscle Cells. *Sci. Rep.* **2016,** *6,* 25579. doi:10.1038/srep25579

Yasuda, A.; Natsume, M.; Sasaki, K.; Baba, S.; Nakamura, Y.; Kanegae, M.; Nagaoka, S. Cacao Procyanidins Reduce Plasma Cholesterol and Increase Fecal Steroid Excretion in Rats Fed a High-cholesterol Diet. *Biofactors* **2008,** *33,* 211–223. doi: 10.1002/biof.5520330307

Zeggwagh, N.; Moufid, A.; Michel, J. B.; Eddouks, M. Hypotensive Effect of *Chamaemelum nobile* Aqueous Extract in Spontaneously Hypertensive Rats. *Clin. Exp. Hypertens.* **2009,** *31* (5), 440–450. doi: 10.1080/10641960902825453

Zeggwagh, N. A.; Michel, J. B.; Eddouks, M. Vascular Effects of Aqueous Extract of Chamaemelum Nobile: In vitro Pharmacological Studies in Rats. *Clin. Exp. Hypertens.* **2013,** *35* (3), 200–206. doi: 10.3109/10641963.2012.712179

Zeng, X. H.; Zeng, X. J.; Li, Y. Y. Efficacy and Safety of Berberine for Congestive Heart Failure Secondary to Ischemic or Idiopathic Dilated Cardiomyopathy. *Am. J. Cardiol.* **2003,** *92,* 173–176. doi: 10.1016/s0002-9149(03)00533-2

Zhang, D. W.; Fu, M.; Gao, S. H.; Liu, J. L. Curcumin and Diabetes: A Systematic Review. *Evid. Based Complement. Alternat. Med.* **2013,** *2013,* 636053. doi: 10.1155/2013/636053

Zhang, P.; Li, Y.; Du, Y.; Li, G.; Wang, L.; Zhou, F. Resveratrol Ameliorated Vascular Calcification by Regulating Sirt-1 and Nrf2. *Transplant. Proc.* **2016,** *48,* 3378–3386. doi: 10.1016/j.transproceed.2016.10.023

Zhang, Q.; Li, Y.; Chen, L. Effect of Berberine in Treating type 2 Diabetes Mellitus and Complications and its Relevant Mechanisms. *Zhongguo Zhong Yao Za Zhi* **2015,** *40,* 1660–1665.

Zhang, X. D.; Ren, H. M.; Liu, L. Effects of Different Dose Berberine on Hemodynamic Parameters and [Ca2+] I of Cardiac Myocytes of Diastolic Heart Failure Rat Model. *Zhongguo Zhong Yao Za Zhi* **2008,** *33,* 818–821.

Zhang, Y.; Li, X.; Zou, D.; Liu, W.; Yang, J.; Zhu, N.; Huo, L.; Wang, M.; Hong, J.; Wu, P.; et al. Treatment of type 2 Diabetes and Dyslipidemia with the Natural Plant Alkaloid Berberine. *J. Clin. Endocrinol. Metab.* **2008,** *93,* 2559–2565. doi: 10.1210/jc.2007-2404

Zhou, Z.; Xu, J.; Lan, T. Protective Effect of Berberine on Isolated Perfused Heart in Heart Failure. *Hua Xi Yi KeDa Xue Xue Bao* **2001,** *32,* 417–418.

Zordoky, B. N.; Robertson, I. M.; Dyck, J. R. Preclinical and Clinical Evidence for the Role of Resveratrol in the Treatment of Cardiovascular Diseases. *Biochim. Biophys. Acta* **2015,** *1852,* 1155–1177. doi: 10.1016/j.bbadis.2014.10.016

CHAPTER 7

HERBAL NUTRACEUTICALS: SAFE AND POTENT THERAPEUTICS IN COMBATING CANCER

ALKA LOHANI[1], SHAHBAZ KHAN[2], SUSHIL KUMAR[1], and NAVNEET VERMA[2]

[1]*School of Pharmaceutical Sciences, IFTM University Moradabad, Uttar Pradesh, India*

[2]*Pharmacy Academy, IFTM University Moradabad, Uttar Pradesh, India*

ABSTRACT

Nutraceuticals is a wide-ranging term that refers to any product generated from food sources that provides additional health advantages beyond the fundamental nutritional content present in foods. As non-specific biological therapies, nutraceuticals can be used to improve overall health, manage symptoms, and prevent a variety of illnesses. The phrase "nutraceutical" is a combination of the words "nutrient," which refers to a nutritive dietary component, and "pharmaceutical," which refers to a medical drug. Nutraceuticals, on the other hand, are edibles that are used as medicines in the form of tablets, capsules, or liquids to deliver physiological benefits. Nutraceuticals, which give health advantages and are an alternative to contemporary medicine, have gained popularity in recent years. The use of nutraceuticals can reduce or eliminate the need for traditional pharmaceuticals while lowering the risk of negative or side effects. Obesity, osteoporosis, cancer, diabetes, allergies, and dental problem are some of the modern-day ailments, with which the entire world is fighting. Nutraceuticals are made up mostly of

Herbals As Nutraceuticals: Their Role in Healthcare.
Raj K. Keservani, Rajeshwar Kamal Kant Arya, and Rajesh K. Kesharwani (Eds.)

nutrients, herbals, and dietary supplements, which make them useful in maintaining health, fighting disease, and improving the overall quality of life.

7.1 INTRODUCTION

According to the World Health Organization (WHO), "Health refers to a state of complete mental, physical, and social well-being rather than the absence of disease or infirmity" (WHO, 2022). Although lifestyle choices can contribute to good health, they can also harm it. Optimal nutrition, in particular, is an essential component of good health. Inadequate and excessive nutrition can lower immunity, increase susceptibility to diseases that are preventable, influence physical and mental development, and have long-term consequences for future generations. Over the last few decades, rapid changes in the lifestyle and diet have occurred as a result of urbanization, industrialization, economic progress, and market globalization, having a substantial impact on health and nutrition. Chronic diseases/conditions (like diabetes, heart disease, stroke, obesity, arthritis, asthma, hypertension, cancer, etc.) are becoming more serious sources of disability and early death in the world (Joint WHO and FAO Expert Consultation, 2003). The scientific community is constantly challenged to find new ways to prevent disease through the application of food biotechnology findings.

In recent decades, nutrition science has been focused on the detection and understanding of deficiencies. Nutraceuticals and functional foods, in addition to being nutritionally rich, have specific health-promoting properties. New nutritional trends and the desire to meet social and health requirements have prompted this concept. Dr. Stephen DE Felice, President of the Foundation for Innovation in Medicine (FIM) in Cranford, New Jersey, USA, coined the term "nutraceutical" in 1989 from the word "nutrition" and "pharmaceutical." Nutraceuticals, according to Dr. DeFelice, are "foods or components of foods that offer medicinal or health advantages, including illness prevention or therapy" (DeFelice, 2022). Nutraceuticals may be defined as "A product obtained from food that is normally offered in medicinal forms not commonly associated with food, proved to have a physiological benefit or provide protection against chronic disease." Nutraceuticals are the class of biologically active substances that are derived from natural origin using non-denaturing techniques in order to maintain their original qualities without the use of chemicals. A nutraceutical is an extract of a food that contains a higher concentration of nutrients than the food itself, resulting in a larger health benefit than it would from the natural food. These products are appealing because of their natural origins, high bioavailability, and the fact that they may normally be taken for a long time without causing negative effects (Keservani et al., 2010a, 2010b, 2017, 2020).

7.2 CANCER

The term "cancer" covers a wide range of illnesses that can affect any part of the body. Cancer is characterized by abnormal cell growth and its ability to invade and spread throughout the body. Other terminologies used for cancer are malignant tumors and neoplasms. The rapid development of aberrant cells that grow beyond their normal borders and can later infect neighboring sections of the body and migrate to other organs are a distinguishing feature of cancer. This process is known as metastasis. The most common cause of death is cancer metastases. As a genetic disease, cancer is caused by alterations in the genes that influence how our cells function, especially in terms of growth and division. As research progresses, it is becoming clear that nutrition plays a significant role in cancer. The American Institute for Cancer Research and the World Cancer Research Fund estimate that good eating habits, regular exercise, and maintaining a healthy weight prevent 30–40% of all cancers (Glade, 1999). An important part of cancer treatment is nutrition. Consuming healthy foods prior to, during, and after treatment can help to feel better and stay stronger. Here, nutraceuticals come into the picture that can actually help the person in the fight against cancer.

7.3 WHY NUTRACEUTICALS?

For many of us, getting appropriate nutrients from regular diet is impossible. We live in a highly toxic environment that is polluted and contaminated with pesticides, disrupting our bodies' ability to manage. Initially, cancer was spreading to different parts of the body and has now infiltrated our entire population. Use of natural sources is more rational, rather than relying on antibiotics that have lost their potency, to improve our body's system. Because medicines are not natural to the body, they often have negative effects; however, good quality supplementation that can be absorbed and utilized by the body may genuinely strengthen and provide vigor to our bodies.

7.4 CONCEPT OF NUTRACEUTICALS

Recently, a proposal has been made to redefine the term "nutraceuticals" in order to clearly distinguish them from dietary supplements and medications. Nutraceuticals are defined as: (1) phytocomplexes for plant-based foods, and (2) pools of secondary metabolites for animal-based foods. Both are highly concentrated and supplied in a pharmaceutically acceptable form. They have the potential to have positive health impacts, such as disease prevention and/or therapy (Santini et al., 2018). Nutraceuticals are thus a new category that blurs the line between medications and food, and might be seen of as the progression of the food supplement notion.

From the consumer's point of view, nutraceuticals may offer the following benefits:

- It's possible that our diet's health value will improve.
- The health-promoting effect could be enhanced.
- May assist us in avoiding medical disorders.
- Doing something for oneself may provide a psychological advantage.
- May be seen as being more "natural" than conventional medication, with less negative side effects.
- May provide food for special-needs populations (e.g., nutrient-dense foods for the elderly).

7.5 HERBAL NUTRACEUTICALS

By sustaining health and combating nutritionally induced acute and chronic ailments, herbal nutraceuticals are used to promote good health, longevity, and wellbeing. Herbal (Sharma et al., 2020; Rane et al., 2020a) nutraceuticals are bioactive phytochemicals obtained as secondary or tertiary metabolites from plants (Dillard and German, 2000). Food extract supplements provide a wide range of therapeutic benefits, making them a valuable addition to the healthcare system and an excellent preventative measure against a variety of illnesses, including cancer (Das et al., 2012). Herbal (Rane et al., 2020b, 2020c) nutraceuticals have been found to detoxify environmental and ingested carcinogen agents by stimulating antioxidant enzymes, DNA repair pathways, and tumor suppressor proteins, as well as fight cancer progression by inhibiting epithelial mesenchymal transition, angiogenesis, and metastasis and promoting apoptosis (Wargovich et al., 2010; Ranzato et al., 2014). Herbal nutraceuticals, in addition to their anticancer actions, have also been documented to boost the therapeutic efficacy of chemotherapies against hypoxic malignancies, and ensure that these cytotoxic treatments don't have harmful side effects.

Herbal nutraceuticals are categorized into distinct classes based on their chemical structure and function, such as cannabinoids, saponins, flavonoids, carotenoids, sitosterols, anthocyanins, terpenes, and isothiocyanates (Nalini et al., 2020).

7.6 HERBAL NUTRACEUTICALS IN COMBATING CANCER

7.6.1 POLYPHENOLS

There has been increased research interest in polyphenols since it was discovered that they provide a number of health benefits. Polyphenols have the potential to be used in the development of anticancer drugs. Polyphenols are water soluble (Quideau et al., 2011) and according to their chemical structure, polyphenols are grouped into five major categories: flavonoids (Keservani and Sharma, 2014; Keservani et al., 2016a, 2016b), phenolic acids, lignans, stilbenes, and other polyphenols. Among of all these polyphenols, phenolic acids and flavonoids are at the top.

Herbal polyphenols may exert anticancer effects via different mechanisms, including the following (Hazafa et al., 2020):

a. Antioxidant enzyme activity
b. Cell cycle signaling modulation
c. Removal of anticancer agents
d. Apoptosis and cell cycle arrest

7.6.1.1 FLAVONOIDS

Flavonoids are naturally occurring substances with different phenolic structures found in fruits, vegetables, seeds, flowers, and beverages. The quantity of anticancerous flavonoids varies in different foods. Following are some flavonoids that have anticancer activity.

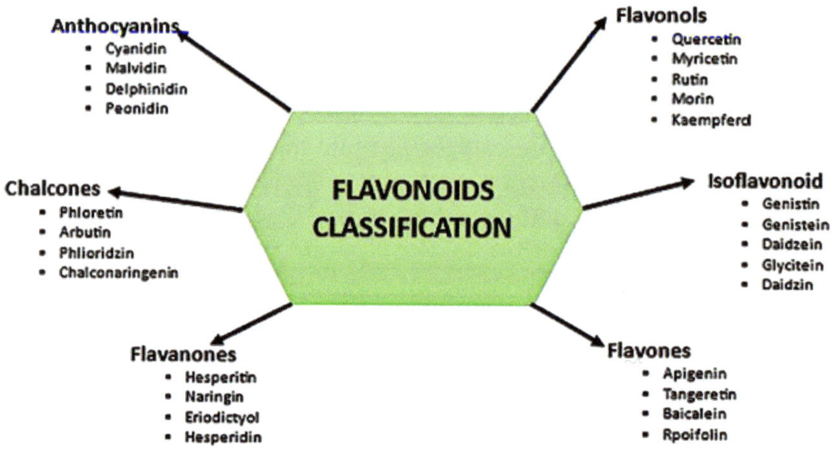

FIGURE 7.1 Classification of flavonoids.

7.6.1.1.1 Quercetin

Since many years ago, humans have relied on nutraceuticals as complementary therapies to treat various types of cancer and influence cellular signaling (Martin, 2006). A natural compound such as quercetin has been used in the treatment of cancer as an alternative treatment. Quercetin ($C_{15}H_{10}O_7$), also known as 3, 3′, 4′, 5, 7-pentahydroxyflavone, is polyphenol flavonoid that occurs naturally in a variety of vegetables, fruits, seeds, as well as various medicinal plants, including green tea, etc. (Iwashina, 2000). Onions, grapes, apples, berries, citrus fruits, broccoli tea, and capers are all high in quercetin (Bischoff, 2008). The most important property of quercetin exhibit is its antioxidant property (Iacopetta et al., 2017). Additionally, quercetin inhibits the spread of cancers like those of the lung, breast, colon, prostate, liver, and cervical (Yi et al., 2017). Quercetin has anticancer properties due to its ability to bind to the cellular receptors and proteins and these anticancer properties are exerted through a variety of mechanisms, including cellular signaling and the ability to suppress enzymes responsible for the activation of carcinogen (Jung et al., 2018; Bevara et al., 2018). Shortly after consumption of quercetin-rich vegetables, metabolites of quercetin reportedly accumulate in tissues. Several researchers have reported that cancer cell proliferation has been suppressed by quercetin, such as breast cancer (Choi et al., 2001), oral cancer (ElAttar and Virji, 1999), pancreatic cancer

(Angst et al., 2013), lung cancer (Zheng et al., 2012), and thyroid cancer (Altundag et al., 2016).

7.6.1.1.2 Apigenin

Apigenin is a flavonoid found throughout the plant kingdom that has been used for millennia as a traditional medicine. It is chemically called as 4′,5,7-trihydroxyflavone, and has a molecular formula $C_{15}H_{10}O_5$ and molecular weight 270.24. Apigenin is found in many fruits and herbs like chamomile, basil, peppermint, endives, clove, etc. Apples, celery, and some spices such as oregano, grapefruit, tarragon, cilantro, wheat sprouts, parsley, onions, oranges, and plant-derived beverages are also high in apigenin.

This natural bioflavonoid has been discovered to have little cytotoxicity, making its anticancer activities exceedingly targeted and precise. Apigenin has gotten a lot of attention recently because of its anticancer capabilities and low toxicity (Yan et al., 2017). Apigenin has been demonstrated to reduce a range of human malignancies in vitro and in vivo by a variety of biological mechanisms, including cell cycle arrest, cell migration and invasion suppression, cell apoptosis and autophagy, and immune stimulation (Cardenas et al., 2016).

7.6.1.1.3 Genistein

Genistein ($C_{15}H_{10}O_5$) is a natural flavonoid from isoflavone family that is derived from soy-rich products and accounts for approximately 60% of total isoflavones in soybeans. The principal food source of genistein is a variety of plants such as kudzu, fava beans, psoralea, and lupin.

Genistein has been shown to possess anti-tumor activity (via induction of apoptosis, cell proliferation suppression, and cell cycle regulation), as well as inhibiting angiogenesis in hormone-related and hormone-unrelated cancer cells and modulating antioxidant properties (Thangavel et al., 2019). Moderate doses of genistein have been shown to inhibit malignancies/cancer of the cervix (Kim et al., 2009), prostate (Morito et al., 2001), breast, brain, and colon (Sakamoto et al., 2010). It has also been discovered that genistein increases the sensitivity of some cells to radiotherapy (De and Hilakivi-Clarke, 2006). It has been reported that genistein is a potential anti-leukemia agent (Zhou et al., 2021).

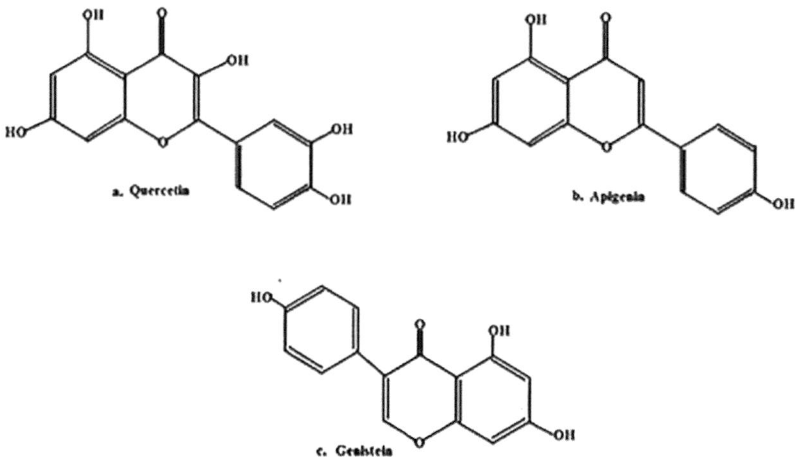

FIGURE 7.2 Structure of: (a) quercetin; (b) apigenin; and (c) genistein.

7.6.1.2 PHENOLIC ACIDS

Phenolic acids are a varied and well-studied group of plant polyphenols that are produced as a by-product of the phenylpropanoid pathway by shikimic acid, as aby-product during monolignol pathway, and by the breakdown of cell wall polymers like lignin. Plant phenolic acids are an important part of the human diet because they have powerful antioxidant properties as well as other health advantages.

7.6.1.2.1 Ferulic Acid

Ferulic acid (4-hydroxy-3-methoxycinnamic acid) is among the most prevalent phenolic acids in plants, and it's plentiful in foods like maize, bran, navy beans, wheat bran, artichokes, eggplant, and beets (Rosazza et al., 1995). Ferulic acid has been shown in research investigations to inhibit cell proliferation and invasion in Hela and Caski cells. Its ability to inhibit autophagy and trigger cell cycle arrest in human cervical carcinoma cells make it an effective anticancer agent (Gao et al., 1998).

7.6.1.2.2 Gallic Acid

Gallic acid, also known as 3,4,5-trihydroxybenzoic acid, is a phenolic acid that can be found in many fruits and medicinal plants. Tea, grapes, berries, blueberries, walnuts, apples, and other fruits, as well as wine, contain a large amount of gallic acid (Ma et al., 2003). Gallic acid is found in several hard wood plant species such as oak (*Quercus robur*), chestnut (*Castanea sativa L.*), and others (Eyles et al., 2004). Gallic acid has significant antioxidant and anticancer capabilities, as it affects a number of pharmacological and metabolic processes. Because of its anticancer properties, this chemical is an essential biomolecule for medicinal purposes. Gallic acid has been shown to promote apoptosis in cancer cell types due to its pro-oxidant properties (Yen and Duh, 2002). Gallic acid's preventive function in chemically induced carcinogenesis has also been confirmed by research (Senapathy et al., 2011). Gallic acid is shown to have selective cytotoxicity for malignant cells while causing very little harm in normal cells (Chia et al., 2010). Gallic acid is a desirable addition to vitamins and dietary supplements because of this feature, which helps to reduce the risk of cancer (Golumbic and Mattill, 1942).

7.6.1.2.3 p-Coumaric Acid

p-Coumaric acid is 4-hydroxy-cinnamic acid, which is a hydroxy derivative of cinnamic acid and can be found in diverse of fruits (like pineapple, berries, apples, grapes, oranges, tomatoes, and pears), vegetables (like onions, beans, and potatoes), and cereals (like maize, oats, and wheat). According to a research, p-Coumaric acid exerts its anticancer effects in SNU-16 gastric cancer cells via altering the expression of particular microRNAs (Jang et al., 2020).

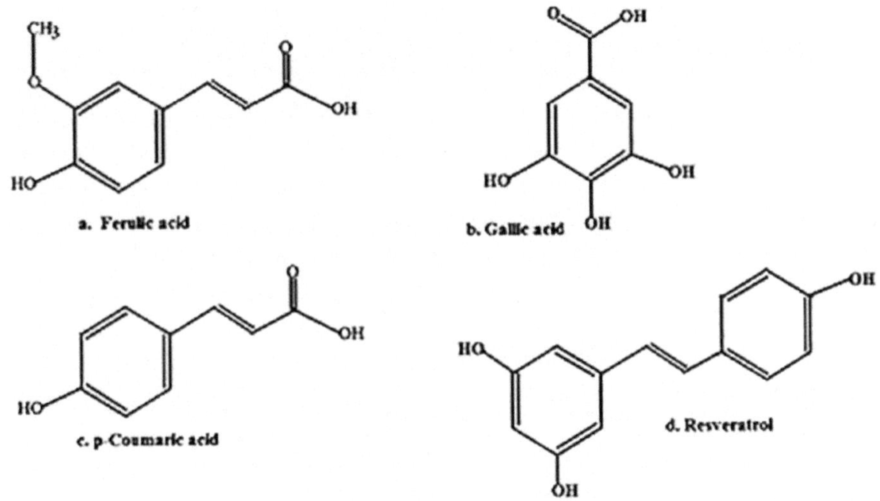

FIGURE 7.3 Structure of: (a) ferulic acid; (b) gallic acid; (c) p-coumaric acid; and (d) resveratrol.

7.6.1.3 STILBENES

7.6.1.3.1 Resveratrol

Resveratrol (3,5,4′-trihydroxy-trans-stilbene) is a stilbenoid that is produced by numerous plants. It has been detected in over 70 plant species, with the skin and seeds of grapes, raspberries, blueberries, mulberries, and peanuts being the most common. It can also be present in trace amounts in red wines and other foods consumed by humans. The nutrient resveratrol is considered to be a potential preventative and therapeutic agent for several forms of cancer. Research has validated resveratrol's cancer-fighting abilities in in vitro and in vivo, finding that it can prevent the initiation, promotion, and progression of cancer (Salehi et al., 2018). In 1997, Jang et al. discovered that resveratrol suppressed carcinogenesis in a mouse skin cancer model, resulting in a wealth of research publications followed.

In-vitro studies have demonstrated that resveratrol is cytotoxic to a number of human tumor cells, including myeloid and lymphoid cancer cells, as well as cells from the breast, colon, skin, ovary, cervix, stomach, prostate, liver, pancreas, and thyroid (Tome-Carneiro et al., 2013; Kundu and Letters, 2008).

7.6.1.4 OTHER POLYPHENOLS

7.6.1.4.1 Curcumin

Scientists worldwide have been intrigued by the many biological properties of curcumin; a polyphenol obtained from the rhizomes of *Curcuma longa.* Curcumin is known for its inflammatory properties, antioxidant properties, antiviral properties, and antibacterial properties, the most well-known of which is its anticancer efficacy. Chemically curcumin is (1E,6E)-1,7-bis(4-hydroxy-3-methoxyphenyl)-1,6-heptadiene-3,5-dione, with a chemical formula $C_{21}H_{20}O_6$ and molecular weight 368.38. It also shows anticancer properties by targeting many cells signaling pathways, including growth factors, transcription factors, cytokines, and genes that modulate cellular proliferation and apoptosis (Giordano and Tommonaro, 2019). Curcumin lowers the negative effects of chemotherapy and radiotherapy, enhancing the quality of life of patients (Mansouri et al., 2020).

7.6.1.5 SAPONINS

A saponin is a group of naturally occurring plant glycosides whose composition is complex, chemically diverse, and has the ability to create stable, soap-like foams in aqueous solutions. Dammaranes, lupanes, tirucallanes, lanostanes, taraxasteranes, hopanes, cucurbitane, oleananes, ursanes, cycloartanes, and steroids are the distinguished classes of saponins. A wide variety of higher plants contain saponins in their roots, tubers, leaves, flowers, and seeds. Saponins have an anticancer effect on numerous cancer cells, according to recent studies. With IC_{50} values as low as 0.2 mM, some saponins suppress tumor cell growth by cell cycle arrest and death. Meanwhile, saponins, when used in conjunction with traditional tumor therapy procedures, improve therapeutic outcomes (Man et al., 2010).

7.7 CONCLUSION

Chemo-preventive medicines are in high demand as an early interventional strategy for lowering cancer incidence risk. The only treatments currently accessible for cancer are chemotherapy, radiation, and surgery. Considering their side effects, alternative or adjuvant therapies are urgently needed. Phytochemicals are relatively harmless and readily available in the diet. Nutraceuticals are non-specific biological therapies that help promote well-being, manage symptoms, and prevent a variety of diseases. It is vital to conduct more research to identify the most relevant nutraceutical targets so that tailored clinical studies can produce consistent outcomes for cancer prevention and therapy management.

KEYWORDS

- **nutrients**
- **nutraceuticals**
- **disease**
- **food**
- **medicine**

REFERENCES

Altundag, E. M.; Kasacı, T.; Yılmaz, A. M. Quercetin-Induced Cell Death in Human Papillary Thyroid Cancer (B-CPAP) Cells. *J. Thy. Res.* **2016,** 1–14.

Angst, E.; Park, J.; Moro, A.; Lu, Q.; Lu, X.; Li, G.; Pancreas, J. K. The Flavonoid Quercetin Inhibits Pancreatic Cancer Growth In Vitro and In Vivo. *Pancreas* **2013,** *42* (2), 223–225.

Bevara, G. B.; Kumar, A. N.; Koteshwaramma, K. L.; Badana, A.; Kumari, S.; Malla, R. R. C-Glycosyl Flavone from Urginea Indica Inhibits Proliferation & Angiogenesis & Induces Apoptosis via Cyclin-Dependent Kinase 6 in Human Breast, Hepatic & Colon Cancer Cell Lines. *Ind. J. Med. Res.* **2018,** *147* (2), 155–158.

Bischoff, S. C. Quercetin: Potentials in the Prevention and Therapy of Disease. *Curr. Opin. Clin. Nutr. Met. Care* **2008,** *11* (6), 733–740.

Cardenas, H.; Arango, D.; Nicholas, C.; Duarte, S.; Nuovo, G. J.; He, W.; Voss, O. H.; Gonzalez-Mejia, M.; Guttridge, D. C.; Grotewold, E.; Doseff, A. I. Dietary Apigenin Exerts Immune-Regulatory Activity In Vivo by Reducing NF-κB Activity, Halting Leukocyte Infiltration and Restoring Normal Metabolic Function. *Int. J. Mol. Sci.* **2016,** *17* (3), 323–328.

Chia, Y. C.; Rajbanshi, R.; Calhoun, C.; Chiu, R. H. Anti-Neoplastic Effects of Gallic Acid, a Major Component of Toona Sinensis Leaf Extract, on Oral Squamous Carcinoma Cells. *Molecules* **2010,** *15* (11), 8377–8389.

Choi, J. A.; Kim, J. Y.; Lee, J. Y.; Kang, C. M.; Kwon, H. J.; Yoo, Y. D.; Kim, T. W.; Lee, Y. S.; Lee, S. J. Induction of Cell Cycle Arrest and Apoptosis in Human Breast Cancer Cells by Quercetin. *Inter. J. Oncol.* **2001,** *19* (4), 837–844.

Das, L.; Bhaumik, E.; Raychaudhuri, U.; Chakraborty, R. Role of Nutraceuticals in Human Health. *J. Food Sci. Technol.* **2012,** *49* (2), 173–183.

De, A. S.; Hilakivi-Clarke, L. Timing of Dietary Estrogenic Exposures and Breast Cancer Risk. *Ann. NY Acad. Sci.* **2006,** *1089*, 14–35.

DeFelice, S. L. Nutrition Stymied: The Nutraceutical Solution. In: *Proceedings of the Plenary Lecture XXV National Congress of the Italian Chemical Society-SCI-The University of Calabria*, Calabria, Italy, 2014 (9).

Dillard, C.; German, J. B. Phytochemicals: Nutraceuticals and Human Health. *J. Sci. Food Agric.* **2000,** *80* (12), 1744–1756.

ElAttar, T. M.; Virji, A. S. Modulating Effect of Resveratrol and Quercetin on Oral Cancer Cell Growth and Proliferation. *Anti-Cancer Drug* **1999,** *10* (2), 187–193.

Eyles, A.; Davies, N. W.; Mitsunaga, T.; Mihara, R.; Mohammed, C. Role of Eucalyptus Globulus Wound Wood Extractives: Evidence of Superoxide Dismutase-like Activity. *Forest Pathol.* **2004,** *34* (4), 225–232.

Gao, R.; Yuan, Z.; Zhao, Z.; Gao, X. Mechanism of Pyrogallol Autoxidation and Determination of Superoxide Dismutase Enzyme Activity. *Bioelectrochem. Bioenerg.* **1998,** *45* (1), 41–45.

Giordano, A.; Tommonaro, G. Curcumin and Cancer. *Nutrients* **2019,** *11* (10), 2376–2379.

Glade, M. J. Food, Nutrition, and the Prevention of Cancer: A Global Perspective: American Institute for Cancer Research/World Cancer Research Fund, American Institute for Cancer Research. *Nutrition* **1999,** *15* (6), 523–526.

Golumbic, C.; Mattill, H. A. The Antioxidant Properties of Gallic Acid and Allied Compounds. *Oil Soap.* **1942,** *19* (8), 144–145.

Hazafa, A.; Rehman, K. U.; Jahan, N.; Jabeen, Z. The Role of Polyphenol (Flavonoids) Compounds in the Treatment of Cancer Cells. *Nutr. Cancer* **2020,** *72* (3), 386–397.

Iacopetta, D.; Grande, F.; Caruso, A.; Mordocco, R. A.; Plutino, M. R.; Scrivano, L.; Ceramella, J.; Muia, N.; Saturnino, C.; Puoci, F.; Rosano, C.; Sinicropi, M. S. New Insights for the Use of Quercetin Analogs in Cancer Treatment. *Fut. Med. Chem.* **2017,** *9* (17), 2011–2028.

Iwashina, T. The Structure and Distribution of the Flavonoids in Plants. *J. Plant Res.* **2000,** *113* (3), 287–299.

Jang, M. G.; Ko, H. C.; Kim, S. J. Effects of P-Coumaric Acid on MicroRNA Expression Profiles in SNU-16 Human Gastric Cancer Cells. *Genes Genom.* **2020,** *42* (7), 817–825.

Joint WHO; FAO Expert Consultation. Diet, Nutrition and the Prevention of Chronic Diseases? World Health Organization Technical Report Series, 2003; 27; 916 (i–viii); pp 1–49.

Jung, K. H.; Rumman, M.; Yan, H.; Cheon, M. J.; Choi, J. G.; Jin, X.; Park, S.; Oh, M. S.; Hong, S. S. An Ethyl Acetate Fraction of Artemisia Capillaris (ACE-63) Induced Apoptosis and Anti-Angiogenesis via Inhibition of PI3K/AKT Signaling in Hepatocellular Carcinoma. *Phytother. Res.* **2018,** *32* (10), 2034–2046.

Keservani, R. K.; Sharma, A. K. Flavonoids: Emerging Trends and Potential Health Benefits. *J. Chin. Pharm. Sci.* **2014,** *23* (12), 815.

Keservani, R. K.; Kesharwani, R. K.; Vyas, N.; Jain, S.; Raghuvanshi, R.; Sharma, A. K. Nutraceutical and Functional Food as Future Food: A Review. *Der Pharmacia Lettre* **2010a,** *2* (1), 106–116.

Keservani, R. K.; Kesharwani, R. K.; Sharma, A. K.; Vyas, N.; Chadoker, A. Nutritional Supplements: An Overview. *Int. J. Curr. Pharm. Rev. Res.* **2010b,** *1* (1), 59–75.

Keservani, R. K.; Sharma, A. K.; Kesharwani, R. K. Nutraceutical and Functional Foods for Cardiovascular Health. In: *Food Process Engineering*; Apple Academic Press, CRC Press, 2016a; pp. 257–278, 291–312.

Keservani, R. K.; Sharma, A. K.; Kesharwani, R. K. Medicinal effect of Nutraceutical Fruits for the Cognition and Brain Health. *Scientifica* **2016b,** Article ID 3109254, 10 pages. DOI: 10.1155/2016/3109254

Keservani, R. K.; Sharma, A. K.; Kesharwani, R. K. An Overview and Therapeutic Applications of Nutraceutical and Functional Foods. *Recent Adv. Drug Deliv. Technol.* **2017,** 160–201.

Keservani, R. K.; Sharma, A. K.; Kesharwani, R. K. (Eds.). *Nutraceuticals and Dietary Supplements: Applications in Health Improvement and Disease Management*; CRC Press, 2020.

Kim, S. H.; Kim, S. H.; Kim, Y. B.; Jeon, Y. T.; Lee, S. C.; Song, Y. S. Genistein Inhibits Cell Growth by Modulating Various Mitogen-Activated Protein Kinases and AKT in Cervical Cancer Cells. *Ann. NY Acad. Sci.* **2009,** *1171*, 495–500.

Kundu, J.; Letters, Y. S. C. Cancer Chemo Preventive and Therapeutic Potential of Resveratrol: Mechanistic Perspectives. *Cancer Lett.* **2008,** *269* (2), 243–261.

Ma, J.; Luo, X.; Protiva, P.; Yang, H.; C. M. J. Bioactive Novel Polyphenols from the Fruit of Manilkara Zapota (Sapodilla). *ACS Pub.* **2003,** *66* (7), 983–986.

Man, S.; Gao, W.; Zhang, Y.; Huang, L.; Liu, C. Chemical Study and Medical Application of Saponins as Anti-Cancer Agents. *Fitoterapia* **2010,** *81* (7), 703–714.

Mansouri, K.; Rasoulpoor, S.; Daneshkhah, A.; Abolfathi, S.; Salari, N.; Mohammadi, M.; Rasoulpoor, S.; Shabani, S. Clinical Effects of Curcumin in Enhancing Cancer Therapy: A Systematic Review. *BMC Cancer* **2020,** *20* (1), 1–12.

Martin, K. R. Targeting Apoptosis with Dietary Bioactive Agents. *Exp. Biol. Med.* **2006,** *231* (2), 117–129.

Morito, K.; Hirose, T.; Kinjo, J.; Hirakawa, T.; Okawa, M.; Nohara, T.; Ogawa, S.; Inoue, S.; Muramatsu, M.; Masamune, Y. Interaction of Phytoestrogens with Estrogen Receptors α and β. *Bio. Pharm. Bull.* **2001,** *24* (4), 351–356.

Nalini, D.; Selvaraj, J.; Kumar, G. S. Herbal Nutraceuticals: Safe and Potent Therapeutics to Battle Tumor Hypoxia. *J. Can. Res. Clin. Oncol.* **2020,** *146* (1).

Quideau, S.; Deffieux, D.; Douat-Casassus, C.; Pouysogu, L. Plant Polyphenols: Chemical Properties, Biological Activities, and Synthesis. *Angew. Chem. Inter. Ed.* **2011,** *50* (3), 586–621.

Rane, B. R.; Patil, A. S.; Keservani, R. K.; Jain, A. S. Novel Approaches in Herbal Formulation, In: *Enhancing the Therapeutic Efficacy of Herbal Formulations through Novel Drug Delivery Systems*; Kesharwani, R. K., Keservani, R. K., Sharma, A. K., Eds.; IGI Global International Publisher: Pennsylvania, 2020a; Chapter 2, pp 43–68.

Rane, B. R.; Bharath, M. S.; Patil, Rutuja, R.; Keservani, R. K.; Jain, A. S. Novel Approaches in Nutraceuticals, In: *Enhancing the Therapeutic Efficacy of Herbal Formulations Through Novel Drug Delivery Systems*; Kesharwani, R. K., Keservani, R. K., Sharma, A. K., Eds.; IGI Global International Publisher: Pennsylvania, 2020b; Chapter 11, pp 241–266.

Rane, B. R.; Tadavi, S. A.; Keservani, R. K. Naturopathy. In: *Herbal Product Development*; Sharma, A. K.; Keservani, R. K.; Gautam, S. P., Eds.; Apple Academic Press, CRC Press, Taylor & Francis Group, 2020c; Chapter 12, pp 321–347.

Ranzato, E.; Martinotti, S.; Myriam Calabrese, C.; Calabrese, G. Role of Nutraceuticals in Cancer Therapy. *J. Food Res.* **2014,** *3* (4), 1–18.

Rosazza, J.; Huang, Z.; Dostal, L.; Volm, T.; Rousseau, B. Biocatalytic Transformations of Ferulic Acid: An Abundant Aromatic Natural Product. *J. Ind. Microbiol. Biot.* **1995,** *15* (6), 457–471.

Sakamoto, T.; Horiguchi, H.; Oguma, E.; Kayama, F. Effects of Diverse Dietary Phytoestrogens on Cell Growth, Cell Cycle and Apoptosis in Estrogen-Receptor-Positive Breast Cancer Cells. *J. Nutr. Biochem.* **2010,** *21* (9), 856–864.

Salehi, B.; Mishra, A. P.; Nigam, M.; Sener, B.; Kilic, M.; Sharifi-Rad, M.; Valere, P.; Fokou, T.; Martins, N.; Sharifi-Rad. J. Resveratrol: A Double-Edged Sword in Health Benefits. *Biomedicines* **2018,** *6* (3), 91–97.

Santini, A.; Cammarata, S. M.; Capone, G.; Ianaro, A.; Tenore, G. C.; Pani, L.; Novellino, E. Nutraceuticals: Opening the Debate for a Regulatory Framework. *Br. J. Clin. Pharmacol.* **2018,** *84* (4), 659–672.

Senapathy, G. J.; GiftsonSenapathy, J.; Jayanthi, S.; Viswanathan, P.; Umadevi, P.; Nalini, N. Effect of Gallic Acid on Xenobiotic Metabolizing Enzymes in 1,2-Dimethyl Hydrazine Induced Colon Carcinogenesis in Wistar Rats–A Chemo Preventive Approach. *Food Chem. Toxicol.* **2011,** *49* (4), 887–892.

Sharma, A. K.; Keservani, R. K.; Gautam Surya, P. *Herbal Product Development*; Apple Academic Press, CRC Press, Taylor & Francis Group, 2020.

Thangavel, P.; Puga-Olguin, A.; Rodriguez-Landa, J. F.; Zepeda, R. C.; Castelazo Ayala, L. Genistein as Potential Therapeutic Candidate for Menopausal Symptoms and Other Related Diseases. *Mol.* **2019,** *24* (21), 3892–3899.

The WHO Definition of "Health"- PubMed. https://pubmed.ncbi.nlm.nih.gov/4607284/ (accessed 19 Jan 2022).

Tome-Carneiro, J.; Larrosa, M.; Gonzalez-Sarrias, A.; A. Tomas-Barberan, F.; Teresa Garcia-Conesa, M.; Carlos Espin, J. Resveratrol and Clinical Trials: The Crossroad from In Vitro Studies to Human Evidence. *Curr. Pharm. Des.* **2013,** *19* (34), 6064–6093.

Wargovich, M. J.; Morris, J.; Brown, V.; Ellis, J.; Logothetis, B.; Weber, R. Nutraceutical Use in Late-Stage Cancer. *Cancer Metastasis Rev*. **2010,** *29* (3), 503–510.

Yan, X.; Qi, M.; Li, P.; Zhan, Y.; Shao, H. Apigenin in Cancer Therapy: Anti-Cancer Effects and Mechanisms of Action. *Cell Biosci*. **2017,** *7* (1), 1–12.

Yen, G. C.; Duh, P. D. Antioxidant and Pro-Oxidant Properties of Ascorbic Acid and Gallic Acid. *Food Chem*. **2002,** *79* (3), 307–313.

Yi, L.; Zongyuan, Y.; Cheng, G.; Lingyun, Z.; Guilian, Y.; Wei, G. Quercetin Enhances Apoptotic Effect of Tumor Necrosis Factor-Related Apoptosis-Inducing Ligand (TRAIL) in Ovarian Cancer Cells Through Reactive Oxygen Species (ROS). *Cancer Sci*. **2014,** *105* (5), 520–527.

Zheng, S.; Li, Y.; Jiang, D.; Zhao, J.; Ge, J.-F. Anticancer Effect and Apoptosis Induction by Quercetin in the Human Lung Cancer Cell Line A-549. *Mol. Med. Rep*. **2012,** *5* (3), 822–826.

Zhou, H. r.; Shen, J. z.; Fu, H. y.; Zhang, F. Genistein-Induced Anticancer Effects on Acute Leukemia Cells Involve the Regulation of Wnt Signaling Pathway Through H4K20me1 Rather Than DNA Demethylation. *Curr. Med. Sci*. **2021,** *41* (5), 869–879.

CHAPTER 8

NUTRACEUTICAL HERBS AS IMMUNO-INFLAMMATORY AGENTS

RAHUL TIWARI[1], BHAVNA KUMAR[2], RUPA GUPTA[3], DHEERAJ BISHT[4], RAJESHWAR KAMAL KANT ARYA[4], RITU TOMAR[2], and NEERAJ KUMAR SETHIYA[2]

[1]*Department of Pharmacology, Rajiv Academy for Pharmacy, Mathura, Uttar Pradesh, India*

[2]*Faculty of Pharmacy, DIT University Dehradun, Uttarakhand, India*

[3]*Amity Institute of Pharmacy, Amity University, Gurugram, Haryana, India*

[4]*Department of Pharmaceutical Sciences, Sir J.C. Bose Technical Campus Bhimtal, Kumaun University, Nainital, India*

ABSTRACT

Immuno-inflammatory agents have the ability to transform the immune system by improving immune defences to boost reaction of body against several infectious or exogenous injuries, or suppressing the abnormal immune response caused any immune disorders. Currently, nutraceuticals products, which have positive control and prevent the influencing risk factors associated with immuno-inflammation, are being recognized through various studies. The use of medicinal plants, as referred in traditional medicine, is a possible alternative method to prevent onsets of immuno-inflammation. In the recent past, many researches on bioactive plants demonstrated their significance as nutraceutical food for the prevention of immuno-inflammation and has been customized. Several studies pertaining in-vitro and/or

Herbals As Nutraceuticals: Their Role in Healthcare.
Raj K. Keservani, Rajeshwar Kamal Kant Arya, and Rajesh K. Kesharwani (Eds.)

in-vivo activity of medicinal plants has explored their possibilities to prevent immuno-inflammation through various pharmacological screening models. Therefore, in this chapter, the authors tried to customize some of the selected recent plant-based nutraceutical research findings which would be useful against immuno-inflammation. In this context, databases were obtained from online search engine from various sources and were consulted; and relevant studies were selected to summarize this chapter. Further, patents, involving technologies related to treatment and prevention of immuno-inflammation, were reviewed and presented in this chapter.

8.1 INTRODUCTION

Biological responses of immune system may get triggered by various agents, that is, pathogenic materials, dead cells, and lethal substances or irradiation (Medzhitov et al., 2010). This is followed by initiating the process of healing by elimination of injurious stimuli through self-defense mechanism (Nathan et al., 2010; Ferrero-Milliani et al., 2007). The factors that are responsible for inflammation may cause acute or chronic inflammatory conditions to various parts of the body like heart, kidney, liver, pancreas, brain, intestinal tract, lungs, and reproductive system that may lead to tissue damage or can cause disease. In acute inflammatory condition, the cellular and molecular processes and their interaction gets affected that leads to infection or injury. When the acute inflammatory condition is uncontrolled, it may lead to cause various chronic inflammatory diseases (Zhou et al., 2016). Further, at tissue site, the local immune, vascular, and inflammatory cell responses get activated against infection or injury and can be characterized by various conditions like rubor, heat, pain, swelling, and loss in tissue functioning (Takeuchi et al., 2010). There is existence of many microcirculatory events during the inflammatory process that comprises change in vascular permeability, recruitment of leukocyte, and its accumulation and release of various inflammatory mediators (Ferrero-Milliani et al., 2007; Chertov et al., 2000). A chemical signaling cascade is initiated by the body in response to tissue injury that leads to stimulation of responses which are intended for the healing of affected tissues through recruitment of leukocyte chemotaxis from the general circulation to the damaged site for inflammatory responses stimulation (Jabbour et al., 2009). Taking together, inflammatory response is a coordinated activation of various signaling pathways responsible for regulation of inflammatory mediators in resident tissue cells and inflammatory cells (Shurin et al., 2007).

8.2 CAUSES OF INFLAMMATION

The causes of inflammation are depicted in the table below.

Parameters	Non-infectious	Infectious
Physical	Burn, injury, trauma, and ionizing radiation	Bacteria
Chemical	Glucose, fatty acid, toxins, and irritants	Viruses
Biological	Damaged or dead cells	Other micro-organisms
Psychological	Excitement	

8.3 IMMUNO-INFLAMMATION

Both immune and inflammation are dependent conditions that cannot be differentiated easily from one another and are generally referred to as immune-inflammation. Additionally, immune-mediated inflammatory disease (IMID) is defined as a group of disorder that lacks a definitive etiology, but are characterized by common inflammatory pathways that lead to inflammation which get triggered by dysregulation of the normal immune response. IMID is characterized by immune dysregulation, and one underlying manifestation of this immune dysregulation is the inappropriate activation of inflammatory cytokines, which may lead to pathological consequences (Koopman et al., 2017). A pleiotropic anti-inflammatory cytokine, that is, interleukin (IL)-10 gets secreted by several activated immune cells (e.g., T cells and macrophages), which leads to the inhibition of pro-inflammatory cytokine synthesis or release. Presentation of antigen enhances uptake of pathogens by macrophages or monocytes (Sabat et al., 2010). TGF-β and IL-6, in combination or alone, lead to the increase in production and release of IL-10 into circulation, which, in turn, elicits suppressive effects on the pro-inflammatory response (McGeachy et al., 2007). According to studies, there is reduction in activity of serum IL-10 among patients signifying a state of proinflammation and pro-atherogenicity (Smith et al., 2001). For the establishment and maintenance of health and tissue homeostasis, both innate and adaptive immunities are equally required and both include cellular and humoral immunity factors. The antigen-specific production of antibody for the neutralization of toxins and pathogens, facilitation of allergic reactions and autoimmunity, production of immune specific memory cells, and stimulation of cytokine secretion are referred to as humoral immunity. However, elimination of pathogens by macrophages or natural killer cells, elimination of intracellular bacteria through the initiation of cytotoxic T-cell mediated apoptosis, and stimulation of tissue cells, for example, fibroblasts, for the secretion of cytokines, which further leads to modulate the responses of innate and adaptive immune system, are referred to as cellular immunity (Nicholson, 2016). The initiation of inflammation by pathogens, tissue damage, or trauma is characteristically the self-limiting factor and it leads to the restoration of tissue and re-establishment of homeostatic condition of tissue followed by the elimination of the cause. The alteration in the inflammatory responses is supposed to be the main source for the induction and development of IMID (Feghali and Wright, 1997). IMID are further referred to as the collection of unrelated diseases that mutually share the common inflammatory pathways and are activated by or lead to the functional dysregulation of both innate and adaptive immune system (Kuek et al., 2007). The main cause is still not clear; but, the genetic susceptibility and environmental factors like infection and trauma may lead to the induction of these conditions, that comprises of, but are not restricted to, psoriasis, rheumatoid arthritis, inflammatory bowel diseases (IBD), systemic lupus erythematosus, sjogren syndrome, and type-1 diabetes (Kruger et al., 2005; Martin et al., 2014; McGinley et al., 2018). The IMID co-existing with another systemic inflammatory/autoimmune disease is not rare, for example: 20–30% of patients suffering from psoriasis will ultimately develop psoriatic arthritis (Wu et al., 2012). Although the pathogenesis of IMID is not yet completely understood and clarified, the comorbid condition shows that an inflammatory cytokine network which is misbalanced is involved in the disease processes (Tan et al., 2017; Thilagar et al., 2018). Inflammation (Fig. 8.1) is a reaction to stimulation through the attack of pathogens or endogenous signs and signals which can repair

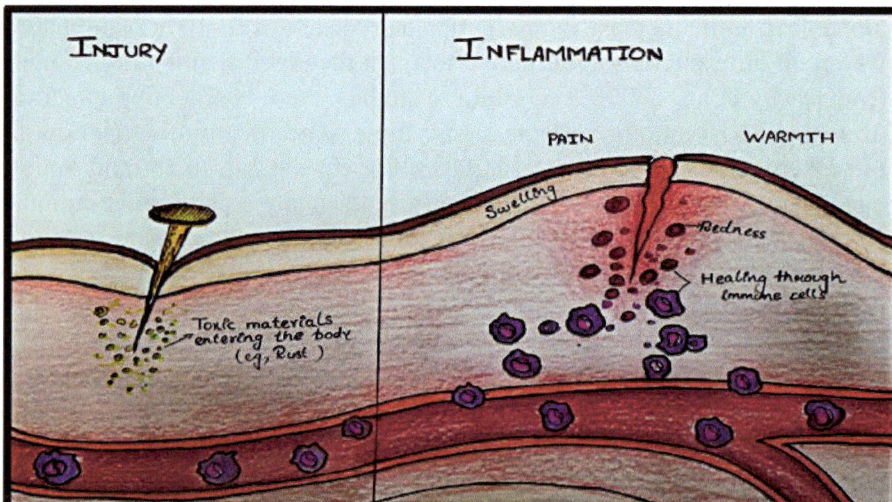

FIGURE 8.1 Mechanism of inflammation.

the tissue or may restore the pathology, when the response goes abandoned. The mechanism behind can well be understood in the context of inflammation during physiological immune responses and when the pathology is constantly evolving (Editorial, 2017).

8.4 TYPES OF IMMUNE RESPONSE

Figure 8.2 represents the different types of immune responses.

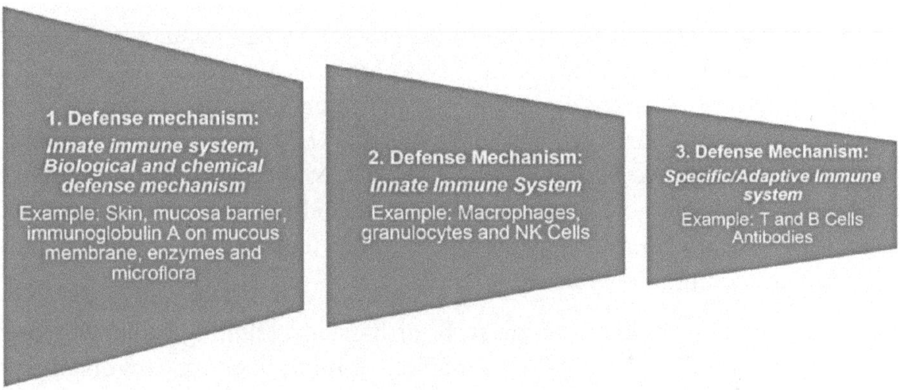

FIGURE 8.2 Functional representation of immune response.

8.4.1 INNATE IMMUNE SYSTEM

The mechanism for innate immune system is in a nonspecific manner which is through direct cellular phagocytosis and by damaging the invading microbes via mast cells, macrophages, and neutrophils. Basically, the presence of a foreign antigen, which is a prerequisite for engaging the innate immune response, is not amended by any kind of "memory" with an antigen. The advancement has provided that the innate immune system includes the repertoire of bacterial, viral, and foreign antigens that can be immediately recognized upon revelation; thus, in this respect, it is dissimilar from the adaptive immune response.

8.4.2 MADAPTIVE/SELECTIVE IMMUNE SYSTEM

The adaptive immune response ensures the improvement established on immunologic "memory" of an already-encountered antigen. This will be a highly dependent antigen performance and also based on cell activation, cell differentiation, and cell memory. Here, the cell memory deliberates a long-lasting immunologic protection unlike that for the innate mechanism. When there is a repetition of the exposure of same antigen, it leads to give a quick and effective immune response. The cells accountable for the adaptive system are antigen-presenting cells (APC), B cells, and T cells. There is another category as professional APC (includes macrophages, dendritic cells, and B lymphocytes) which are equipped with special surface receptors.

8.5 ROLE OF IMMUNE SYSTEM

The immune system generally provides protection and defence to the body through several mechanisms. The gut is regarded as a place for 70–80% of the immune system. It provides almost 400 m^2 contact area including intestine, a simultaneously extensive lymphoid tissue (GALT, gut-associated lymphoid tissue) that forms the largest immune organ in our body. The three barriers theory of immune systems for protection to body from various diseases is the most accepted and as per this theory there includes the following three barriers of the immune system:

1. The first barrier is a coarse filter barrier, including protection from potential germ through mechanical or biological force involvements.
2. The second barrier includes involvement of directional response of the body through innate immune system. In this cascade, immune response attacks all exogenous substances and germs recognized as pathogenic to eliminate invading pathogens in fastest way or initiate further combat creating inflammation.
3. The third barrier includes the adaptive immune system, mainly comprised of specialized cells including antibodies, which are able to detect pathogens and infected cells and response to specific molecular patterns on cell surfaces.

8.6 DIAGNOSTIC TOOL FOR IMMUNO-INFLAMMATION

Disease	Diagnostic tool/test	References
Rheumatoid arthritis	Digital X-ray radiogrammetry and anti-CCP antibody	Niewold et al. (2007); An and Zhang (2019)
HIV	Rapid test and CD_4 count	Ford et al. (2018)
Tuberculosis	Mantoux test	Nayak and Acharjya (2012)
Leukemia	Flow cytometry	Weir and Borowitz (2001)
Allergy	Lymphocyte proliferation test	Räsänen et al. (1991)
IBD	Colonoscopy	Nett et al. (2014)

8.7 CURRENT THERAPEUTIC APPROACHES AND ASSOCIATED CHALLENGES

The following are the various therapeutic approaches associated with immuno-inflammatory conditions:

1. Development of radiopharmaceuticals which targets the specific biomarkers of inflammation (Fig. 8.3), infection, and tumor immunology is one such approach (Signore et al., 2019). Various radiolabeled molecules such as monoclonal antibodies (MoAbs), cytokines, peptides, and their fragments, as well as radiolabeled immune cells, have been suggested to detect pathological inflammatory sites. The amalgamation of radiological imaging modalities like computed tomography (CT), magnetic resonance imaging (MRI), and ultrasound (US), as well as NM techniques, containing single photon emission CT (SPECT) and positron emission tomography (PET), will be used for analyzing target molecules and specify the location, to make a timely diagnosis, and this have been known to be the best therapy in a diversity of pathologies. Thus, numerous radiopharmaceuticals could be used relatively for the complex pathways involved in diseases (Signore et al., 2018) It can be understood by the following example: Radiolabeling white blood cell (WBC), using both 99mTechnetium (99mTc) or 111Indium (111In), is the nuclear medicine gold standard investigation in the diagnosis of an infection. Its application will cover almost most of the field associated with infective diseases comprising of indications such as osteomyelitis, and for the estimation of prosthetic joint infections or implantable cardiac device (Chianelli et al., 2008; Ruf et al., 2010).
2. Another approach for the therapeutic treatment of immuno-inflammation is by immune potentiates, which holds a promising approach where the enhancement of the immune response can be achieved either for foreign or for endogenous antigen. Various infectious diseases can be treated by the immune activation—most commonly used are the antiviral and antibacterial. However, it has been reported that in many situations, the pathogens get mutated and shows resistance to these drugs. On the other hand, these pathogens can elude or even blunt the host immune response. The host immunity can be raised or restored for providing a solution to treat chronic infections that are difficult to cure (Isorce

et al., 2015). Although the immune potentiates are beneficial in many ways to treat infectious diseases, cancer, and allergy, immune activation is a dual-edged sword, and the development of novel immune adjuvants will get hampered by poor safety and tolerability. Hence, to uncouple the immune-mediated efficacy from the toxicity, new concepts of novel immune agonists have emerged—the most studied being toll-like receptors (TLRs) (Kawai and Akira, 2010). The treatment of infectious diseases, such as hepatitis C, by the use of oral TLR7 agonists {R848, ANA773 (now RG-7795)} and PF-4878691 have been reported in past literatures (Bergmann et al., 2011; Fidock et al., 2011). Till so far, the clinical applications of immune agonists have been primarily limited to local delivery for minimizing the immune-related toxicities.

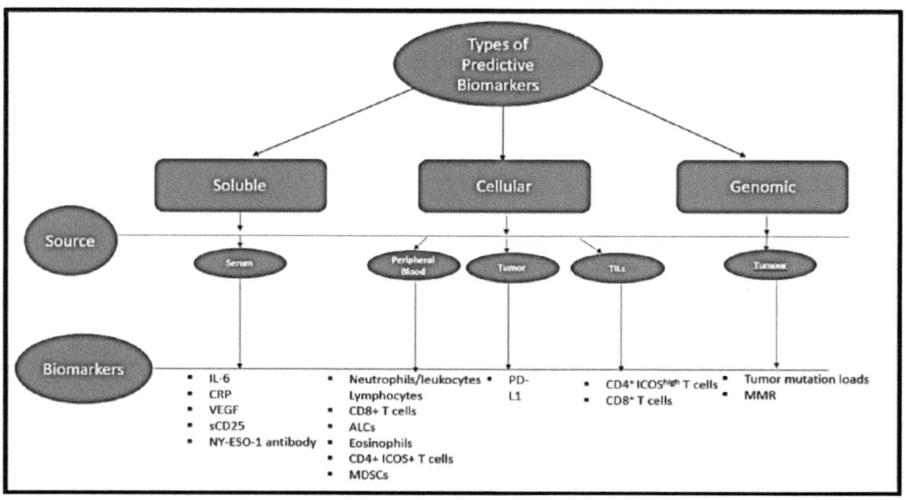

FIGURE 8.3 Biomarkers for immunotherapy.

8.8 NUTRACEUTICAL HERBS IN IMMUNO-INFLAMMATION

Several medicinal plants have been studied in the past to investigate the prevention and treatment from immuno-inflammatory conditions through series of in-vitro and in-vivo studies. The outcome of the such studies has been mentioned in Table 8.1.

TABLE 8.1 Medicinal Plants Against Immuno-Inflammation.

Family	Plant name	Extract/derived compound	In-vitro/in-vivo models	Mechanism of actions	References
Zingiberaceae	*Curcuma longa*	Curcumin	U937 human monocytes	Strong affinity of curcumin to various inflammatory mediators (ERK, PKC, P38 MAP Kinase, NfKB, and Lipoxygenase) and its affinity toward chemokines/cytokines and TNF-α was found to be ineffective.	Venkata et al. (2021)
Theaceae	*Camellia sinensis*	Catechin	RAW267.4 cells	Decreased the protein levels of Akt, mitogen-activated protein kinase (MAPK), and cyclooxygenase (COX)-1 or -2.	Yamaguchi and Levy (2016)
		Epigallocatechin-3-gallate	RAW 264.7 cells	NO, COX-2, IL-6, IL-1β, and TNF-α. OTEE had a higher inhibition activity toward NO, COX-2, IL-6, and IL-1β than EGCG; the reverse was seen for TNF-α. However, both OTEE and EGCG suppressed production of NO, COX-2, IL-6, IL-1β, and TNF-α.	Novilla et al. (2017)
Ranunculaceae	*Coptis chinensis*	Berberine	RAW267.4 cells	Down-regulate T-helper cells cytokines [Th1 (TNF-α, IL-2), and Th2 (IL-4)] production.	Bae and Cheon (2016)
	Nigella sativa	Thymoquinone	Monocytic leukemia (THP1)	Pro-inflammatory cytokines (IL-1-α, IL1-β, IL2, IL4, IL6, IL8, IL10, IL12, IL17A, IFN-γ, TNF-α, and GM-CSF).	Bordoni et al. (2019)
Papaveraceae	*Chelidonium majus*	Chelerythrine	HeLa cell	Inhibit PGE-2 release by regulating COX-2.	Dvořák et al. (2006)
Piperaceae	*Piper longum*	Piperine	HaCaT keratinocyte cells	Reduce level of pro inflammatory cytokines IL-1β, IL-6, and TNF-α. Down regulate expression of COX-2, NOS-2, and NF-κB. Inhibit eicosanoide generation by inhibiting phospholipase A2 and TXA-2 synthase activity.	Jaisin et al. (2020)
Amaryllidaceae	*Lycoris radiata*	Lycorine	RAW264.7	Inhibit iNOS and COX-2.	Kang et al. (2012)

TABLE 8.1 *(Continued)*

Family	Plant name	Extract/derived compound	In-vitro/in-vivo models	Mechanism of actions	References
Apiaceae	*Angelica sinensis*	Z-ligustilide	RAW 264.7 cells	Inhibit iNOS and COX-2 induction by regulating the NF-kB and MAPK signal pathways.	Chao and Lin (2011)
	Centella asiatica	Madecassoside	RAW 264.7 cells	Decrease pro-inflammatory cytokines such as IL-1, IL-2, IL-4, IL-6, and IFN-γ through inhibition of NF-kB activation.	Park et al. (2017)
Araliaceae	*Panax ginseng*	Ginsan	RAW264.7 cells	Inhibit COX-2, iNOS and TNF-α in a dose-dependent manner.	Han et al. (2018)
Fabaceae	*Glycine max*	Genistein	RAW264.7 cells	Significant decrease in COX-2 gene expression.	Lin and Wu (2021)
Fabaceae	*Glycyrrhiza inflata*	Licochalcone E	Kunming mice	Inhibit COX-2 synthesis.	Cui et al. (2008); Bisht et al. (2022)
Burseraceae	*Boswellia carter*	11-keto-β-boswellic acid	APC $^{Min/+}$ male mice	Decrease pro-inflammatory cytokines such as IL-1, IL-2, IL-4, IL-6, and IFN-γ through inhibition of NF-kB activation.	Wang et al. (2014)
Cucurbitaceae	*Luffa cylindrica*	Echinocystic acid	RAW 264.7	Reduce level of IL-1α, IL-6, and TNF-α, as well as their effect on complement pathway though the inhibition of C3 convertase. Inhibits adenosine deaminase activity.	Kao et al. (2012)
Acanthaceae	*Andrographis paniculata*	14-deoxy-andrographolide	Balb/c mice	Inhibit NO production by activated peritoneal macrophages.	Sheeja et al. (2006)
Rubiaceae	*Uncaria rhynchophylla*	Rhynchophylline	Sprague-Dawley rats	Blocking the activation of NF-κB, ERK, and p38 MAPKs.	Wu and Xiao (2019)
Boraginaceae	*Lithospermum erythrorhyzon*	Furylhydroquinones	HaCaT cells	IL-6 inhibition by TNF-α stimulation.	Ahn et al. (2019)

8.9 PATENTS

Several molecules and processes have been filed for Intellectual Property Right protection for the prevention and treatment of immuno-inflammation and have been mentioned in Table 8.2.

TABLE 8.2 Useful Patents Filled with Reference to Natural Product or Herbal Formulation for Management of Immuno-Inflammation.

Patent number	Patent title
	International patent
US11/201,416	Dietary supplement formulation for controlling inflammation and cancer
EP05734279.2A	Compositions and method for alleviating inflammation and oxidative stress in a mammal
EP08860496A	Compositions comprising hydroxytyrosol and chondroitin and use thereof for the treatment, co-treatment, or prevention of inflammatory disorders
US11/737,740	Methods and related systems and formulations to normalize and improve human body chemistry and healing ability
US09/512,673	Herbal composition for reducing inflammation and methods of using same
US14/251,074	Anti-inflammatory compounds
US09/825,838	Anti-inflammatory herbal composition and method of use
CA2864816A	A novel herbal formulation for the modulation of immune system of HIV-infected patients and a process of preparation thereof
US13/465,656	Novel herbal formulation for the modulation of immune system of HIV-infected patients and a process of preparation thereof
US11/737,740	Methods and related systems and formulations to normalize and improve human body chemistry and healing ability
US11/318,869	Methods for recommending a program to improve or maintain pro-inflammatory immune health
US10/378,572	Method for treatment of inflammation and pain in mammals
US11/287,901	Herbal composition for treatment and maintenance of hormone-dependent conditions, osteoporosis, circulatory conditions, and for use as an immunostimulant
US11/148,407	Herbal compositions for prevention and treatment of rheumatic and inflammatory diseases and method of preparing the same
EP07734177A	A pharmaceutical composition useful as an immunomodulating agent and a process for the preparation thereof
US12/593,581	Synergistic anti-inflammatory and antioxidant dietary supplement compositions
US14/735,360	Synergistic anti-inflammatory compositions comprising *Boswellia serrata* extract

TABLE 8.2 *(Continued)*

Patent number	Patent title
	International patent
EP10157239.4A	Pharmaceutical composition presenting anti-inflammatory and anti-histaminic properties
US10/561,713	Inflammatory disease treatment
US12/919,941	Activation of innate and adaptive immune responses by a ginseng extract
PCT/IB2006/053540	Herbal composition for inflammatory disorders
US09/773,280	Dietary supplements for treating inflammation-related diseases
CA002643270A	Pharmaceutical composition and functional food for suppressing and relieving itching and inflammation
US10/938,093	Cytokine modulators and related method of use
US09/842,428	Immunity-enhancing supplements for lung support
US14/534,074	Anti-inflammatory composition
US09/102,355	Herbal extract composition and method with immune-boosting capability
US10/852,718	Immune system reconstructor composition
US12/079,853	Herbal composition for inflammatory disorders
US13/933,764	Botanical composition and method for treating pain and discomfort of various conditions
PCT/US2004/011886	Immunostimulatory agents in botanicals
EP08014201A	Combination therapy of beta-blockers and non-steroidal anti-inflammatory drugs (NSAID)
US13/157,040	Phytochemical combinations that regulate pathological immunity
EP06827402A	Flavanols and b-type procyanidins and inflammation
EP13155586.4A	D-fagomine for the control of inflammatory processes related to an overactivation of the humoral immune response
US11/683,268	Anti-inflammatory activity of eggshell membrane and processed eggshell membrane preparations
US10/421,202	Extracts of celery seed for the prevention and treatment of pain, inflammation, and gastrointestinal irritation
US11/572,769	Use of at least one conjugated triene-containing fatty acid for preparing a medicine for treating inflammation
Indian patent	
202127019169	Nanocarriers for lung inflammation therapy
202121018989	Dual application topical formulations for treating pain and inflammation
202117008984	Formulations containing lipophilic extracts of spicy edible plants useful in controlling pain and inflammation.
202038049073	Use of anatabine to treat inflammation and methods of synthesizing anatabine

TABLE 8.2 *(Continued)*

Patent number	Patent title
	International patent
202017048049	Indene derivatives useful in treating pain and inflammation
202017039445	Probiotic formulations for the treatment and alleviation of metabolic and oxidative stress, inflammation, and neurodegeneration
202037037704	Fasting-mimicking diet (FMD) but not water-only fasting promotes reversal of inflammation and IBD pathology
202017032804	Compositions and methods for treating pulmonary edema or lung inflammation
202138029141	Quinoline derivatives for the treatment of inflammatory diseases
202141023941	Process of preparation of anti-inflammatory and analgesic herbal formulation and product thereof
202147019477	Compositions and methods for treating inflammatory neurological disorders
2191/DEL/1996	An anti-inflammatory, antiarthritic, and a vasculodilator herbal oil from Momordica charantia l, (bitter gourd)"
791/DEL/1996	Oral anti-spasmodic anti-inflammatory composition comprising pitofenone, fenperinium, and NSAID
202127019169	Nanocarriers for lung inflammation therapy
202047029396	Method and system for improved motion robustness during measurement of localized oral inflammation
202028012607	A purified extract (atc2) isolated from *Pseudolysimachion rotundum* var. Subintegrum containing abundant amount of active ingredient, the preparation thereof, and the composition comprising the same as an active ingredient for preventing or treating inflammation, allergy, and asthma
563/MUM/2002	Anti-inflammatory phytoconstituent derivatives
595/MUM/2001	Process for preparation of anti-inflammatory phytoconstituent derivatives

8.10 CONCLUSION

Immuno-inflammatory condition is of much interest and can receive much more attention for its prevention and treatment. There are several natural products including medicinal and nutraceutical plants which finds way to success. Here, some evidence-based studies are presented for easy understanding of such medicinal and nutraceutical plants for the management of immuno-inflammation. Further, more studies are required for testing the clinical effectivity at molecular level through series of experimentation in near future.

Additionally, interaction with drug and food is needed to be understood during on-course of time of such medicinal and nutraceutical plants.

KEYWORDS

- **nutraceuticals**
- **immuno-inflammatory**
- **herbs**
- **phytoconstituents**
- **diagnostic measures**

REFERENCES

Ahn, J.; Chae, H. S.; Chin, Y. W.; Kim, J. Furylhydroquinones and Miscellaneous Compounds from the Roots of *Lithospermum erythrorhizon* and Their Anti-Inflammatory Effect in HaCaT Cells. *Nat. Prod. Res.* **2019,** *33* (12), 1691–1698. DOI: 10.1080/14786419.2018.1431632

An, H.; Zhang, J. Diagnostic Accuracy of Digital X-ray Radiogrammetry on Hand Bone Loss for Patients with Rheumatoid Arthritis. *Medicine* **2019,** *98* (39), 17280. DOI: 10.1097/MD.0000000000017280

Bae, Y. A.; Cheon, H. G. Activating Transcription Factor-3 Induction Is Involved in the Anti-Inflammatory Action of Berberine in RAW264. 7 Murine Macrophages. *Korean J. Physiol. Pharmacol.* **2016,** *20* (4), 415–424. DOI: 10.4196/kjpp.2016.20.4.415

Bergmann, J. F.; Bruijne, J.; Hotho, D. M.; Knegt, R. J.; Boonstra, A.; Weegink, C. J.; Vliet, A. A.; Wetering, J.; Fletcher, S. P.; Bauman, L. A.; Rahimy, M.; Appleman, J. R.; Freddo, J. L.; Janssen, H. L. A.; Reesink, H. W. Randomised Clinical Trial: Anti-Viral Activity of ANA773, an Oral Inducer of Endogenous Interferons Acting via TLR7, in Chronic HCV. *Aliment. Pharmacol. Ther.* **2011,** *34* (4), 443–453. DOI: 10.1111/j.1365–2036.2011.04745.x

Bisht, D.; Rashid, M.; Arya, R. K. K.; Kumar, D.; Chaudhary, S. K.; Rana, V. S.; Sethiya, N. K. Revisiting Liquorice (*Glycyrrhiza glabra* L.) as Anti-Inflammatory, Antivirals and Immunomodulators: Potential Pharmacological Applications with Mechanistic Insight. *Phytomed. Plus* **2022,** *2* (1), 100206. DOI: 10.1016/j.phyplu.2021.100206

Bordoni, L.; Fedeli, D.; Nasuti, C.; Maggi, F.; Papa, F.; Wabitsch, M.; De-Caterina, R.; Gabbianelli, R. Antioxidant and Anti-Inflammatory Properties of *Nigella sativa* Oil in Human Pre-Adipocytes. *Antioxidants* **2019,** *8* (2), 51. DOI: 10.3390/antiox8020051

Chao, W. W.; Lin, B. F. Bioactivities of Major Constituents Isolated from *Angelica sinensis* (Danggui). *Chin. Med.* **2011,** *6* (1), 1–7. DOI: 10.1186/1749–8546-6-29

Chertov, O.; Yang, D.; Howard, O.; Oppenheim, J. J. Leukocyte Granule Proteins Mobilize Innate Host Defense and Adaptive Immune Responses. *Immunol. Rev.* **2000,** *177*, 68–78. DOI: 10.1034/j.1600–065x.2000.17702.x

Chianelli, M.; Boerman, O. C.; Malviya, G.; Galli, F.; Oyen, W. J.; Signore, A. Receptor Binding Ligands to Image Infection. *Curr. Pharm. Des.* **2008,** *14*, 3316–3325. DOI: 10.2174/138161208786549416.

Cui, Y.; Ao, M.; Hu, J.; Yu, L. Anti-Inflammatory Activity of Licochalcone A Isolated from *Glycyrrhiza inflata. Z. Naturforsch. C. J. Biosci.* **2008,** *63* (5–6), 361–365. DOI: 10.1515/znc-2008–5-609

Dvořák, Z.; Vrzal, R.; Maurel, P.; Ulrichová, J. Differential Effects of Selected Natural Compounds with Anti-Inflammatory Activity on the Glucocorticoid Receptor and NF-κB in HeLa Cells. *Chem. Biol. Interact.* **2006,** *159* (2), 117–128. DOI: 10.1016/j.cbi.2005.10.105

Editorial. Current View on Inflammation. *Nat. Immunol.* **2017,** *18*, 825. DOI: 10.1038/ni.3798

Feghali, C. A.; Wright, T. M. Cytokines in Acute and Chronic Inflammation. *Front. Biosci.* **1997,** *2*, d12–d26. DOI: 10.2741/a171

Ferrero-Miliani, L.; Nielsen, O.; Andersen, P.; Girardin, S. Chronic Inflammation: Importance of NOD2 and NALP3 in Interleukin-1β Generation. *Clin. Exp. Immunol.* **2007,** *147*, 227–235. DOI: 10.1111/j.1365–2249.2006.03261.x.

Fidock, M. D.; Souberbielle, B. E.; Laxton, C.; Rawal, J.; Delpuech-Adams, O.; Corey, T. P.; Colman, P.; Kumar, V.; Cheng, J. B.; Wright, K.; Srinivasan, S.; Rana, K.; Craig, C.; Horscroft, N.; Perros, M.; Westby, M.; Webster, R.; Ryst, E. The Innate Immune Response, Clinical Outcomes, and Ex Vivo HCV Antiviral Efficacy of a TLR7 Agonist (PF-4878691). *Clin. Pharmacol. Ther.* **2011,** *89*, 821–829. DOI: 10.1038/clpt.2011.60

Ford, N.; Shubber, Z.; Jarvis, J. N.; Chiller, T.; Greene, G.; Migone, C.; Vitoria, M.; Doherty, M.; Meintjes, G. CD4 Cell Count Threshold for Cryptococcal Antigen Screening of HIV-Infected Individuals: A Systematic Review and Meta-Analysis. *Clin. Infect. Dis.* **2018,** *66* (suppl-2), S152–S159. DOI: 10.1093/cid/cix1143

Han, S. Y.; Kim, J.; Kim, E.; Kim, S. H.; Seo, D. B.; Kim, J. H.; Shin, S. S.; Cho, J. Y. AKT-Targeted Anti-Inflammatory Activity of *Panax ginseng* Calyx Ethanolic Extract. *J. Ginseng Res.* **2018,** *42* (4), 496–503. DOI: 10.1016/j.jgr.2017.06.00310

Isorce, N.; Lucifora, J.; Zoulim, F.; Durantel, D. Immune-Modulators to Combat Hepatitis B Virus Infection: From IFN-a to Novel Investigational Immunotherapeutic Strategies. *Antiviral. Res.* **2015,** *122*, 69–81. DOI: 10.1016/j.antiviral.2015.08.008

Jabbour, H. N.; Sales, K. J.; Catalano, R. D.; Norman, J. E. Inflammatory Pathways in Female Reproductive Health and Disease. *Reproduction* **2009,** *138*, 903–919. DOI: 10.1530/REP-09–0247

Jaisin, Y.; Ratanachamnong, P.; Wongsawatkul, O.; Watthammawut, A.; Malaniyom, K.; Natewong, S. Antioxidant and Anti-Inflammatory Effects of Piperine on UV-B-Irradiated Human HaCaT Keratinocyte Cells. *Life Sci.* **2020,** *263*, 118607. DOI: 10.1016/j.lfs.2020.118607

Kang, J.; Zhang, Y.; Cao, X.; Fan, J.; Li, G.; Wang, Q. I.; Diao, Y.; Zhao, Z.; Luo, L.; Yin, Z. Lycorine Inhibits Lipopolysaccharide-Induced iNOS and COX-2 Up-Regulation in RAW264. 7 Cells Through Suppressing P38 and STATs Activation and Increases the Survival Rate of Mice After LPS Challenge. *Int. Immunopharmacol.* **2012,** *12* (1), 249–256. DOI: 10.1016/j.intimp.2011.11.018

Kao, T. H.; Huang, C. W.; Chen, B. H. Functional Components in *Luffa cylindrica* and Their Effects on Anti-Inflammation of Macrophage Cells. *Food Chem.* **2012,** *135* (2), 386–395. DOI: 10.1016/j.foodchem.2012.04.128

Kawai, T.; Akira, S. The Role of Pattern-Recognition Receptors in Innate Immunity: Update on Toll-Like Receptors. *Nat. Immunol.* **2010,** *11* (5), 373–384. DOI: 10.1038/ni.1863

Koopman, F. A.; Maanen, M. A.; Vervoordeldonk, M. J.; Tak, P. P. Balancing the Autonomic Nervous System to Reduce Inflammation in Rheumatoid Arthritis. *J. Intern. Med.* **2017,** *282* (1), 64–75.

Kruger, J.; Wirtz, D.; Miller, D. T. Counterfactual Thinking and the First Instinct Fallacy. *J. Pers. Soc. Psychol.* **2005,** *88* (5), 725–735. DOI: 10.1037/0022–3514.88.5.725

Kuek, A.; Hazleman, B. L.; Ostor, A. J. Immune-Mediated Inflammatory Diseases (IMIDs) and Biologic Therapy: A Medical Revolution. *Postgrad. Med. J.* **2007,** *83*, 251–260. DOI: 10.1136/pgmj.2006.052688

Lin, Y.; Wu, S. Vegetable Soybean (Glycine max (L.) Merr.) Leaf Extracts: Functional Components and Antioxidant and Anti-Inflammatory Activities. *J. Food Sci.* **2021,** *86* (6), 2468–2480. DOI: 10.1111/1750–3841.15765

Martin, J. C.; Baeten, D. L.; Josien, R. Emerging Role of IL-17 and Th17 Cells in Systemic Lupus Erythematosus. *Clin. Immunol.* **2014,** *154*, 1–12. DOI: 10.1016/j.clim.2014.05.004

McGeachy, M. J.; Bak-Jensen, K. S.; Chen, Y.; Tato, C. M.; Blumenschein, W.; McClanahan, T.; Cua, D. J. TGF-[Beta] and IL-6 Drive the Production of IL-17 and IL-10 by T Cells and Restrain TH-17 Cell-Mediated Pathology. *Nat. Immunol.* **2007,** *8*, 1390–1397. DOI: 10.1038/ni1539

McGinley, A. M.; Edwards, S. C.; Raverdeau, M.; Mills, K. H. G. Th17cells, Gammadelta T Cells and Their Interplay in EAE and Multiple Sclerosis. *J. Autoimmun.* **2018,** *87*, 97–108. DOI: 10.1016/j.jaut.2018.01.001

Medzhitov, R. Inflammation 2010: New Adventures of an Old Flame. *Cell* **2010,** *140*, 771–776. DOI: 10.1016/j.cell.2010.03.006

Nathan, C.; Ding, A. Nonresolving Inflammation. *Cell* **2010,** *140*, 871–882. DOI: 10.1016/j.cell.2010.02.029

Nayak, S.; Acharjya, B. Mantoux Test and Its Interpretation. *Indian Dermatol. Online J.* **2012,** *3* (1), 2–6. DOI: 10.4103/2229–5178.93479

Nett, A.; Velayos, F.; McQuaid, K. Quality Bowel Preparation for Surveillance Colonoscopy in Patients with Inflammatory Bowel Disease Is a Must. *Gastrointest. Endosc. Clin. N. Am.* **2014,** *24* (3), 379–392. DOI: 10.1016/j.giec.2014.03.004

Nicholson, L. B. The Immune System. *Essays Biochem.* **2016,** *60*, 275–301. DOI: 10.1042/EBC20160017

Niewold, T. B.; Harrison, M. J.; Paget, S. A. Anti-CCP Antibody Testing as a Diagnostic and Prognostic Tool in Rheumatoid Arthritis. *Q. J. Med.* **2007,** *100*, 193–201. DOI: 10.1093/qjmed/hcm015

Novilla, A.; Djamhuri, D. S.; Nurhayati, B.; Rihibiha, D. D.; Afifah, E.; Widowati, W. Anti-Inflammatory Properties of Oolong Tea (*Camellia sinensis*) Ethanol Extract and Epigallocatechin Gallate in LPS-Induced RAW 264.7 Cells. *Asian Pac. J. Trop. Biomed.* **2017,** *7* (11), 1005–1009. DOI: 10.1016/j.apjtb.2017.10.002

Park, J. H.; Choi, J. Y.; Son, D. J.; Park, E. K.; Song, M. J.; Hellström, M.; Hong, J. T. Anti-Inflammatory Effect of Titrated Extract of *Centella asiatica* in Phthalic Anhydride-Induced Allergic Dermatitis Animal Model. *Int. J. Mol. Sci.* **2017,** *18* (4), 738. DOI: 10.3390/ijms18040738

Räsänen, L.; Sainio, H.; Lehto, M.; Reunala, T. Lymphocyte Proliferation Test as a Diagnostic Aid in Chromium Contact Sensitivity. *Contact Dermatitis* **1991,** *25* (1), 25–29. DOI: 10.1111/j.1600–0536.1991.tb01768.x

Ruf, J.; Oeser, C.; Amthauer, H. Clinical Role of Anti-Granulocyte MoAb versus Radiolabelled White Blood Cells. *Q. J. Nucl. Med. Mol. Imaging* **2010,** *54*, 599–616.

Sabat, R.; Grutz, G.; Warszawska, K.; Kirsch, S.; Witte, E.; Wolk, K.; Geginat, J. Biology of Interleukin-10. *Cytokine Growth Factor Rev.* **2010,** *21*, 331–344. DOI: 10.1016/j.cytogfr.2010.09.002

Sheeja, K.; Shihab, P. K.; Kuttan, G. Antioxidant and Anti-Inflammatory Activities of the Plant *Andrographis paniculata* Nees. *Immunopharmacol. Immunotoxicol.* **2006,** *28* (1), 129–140. DOI: 10.1080/08923970600626007

Shurin, M. R.; Smolkin, Y. S. Immune Mediated Diseases: From Theory to Therapy (Advances in Experimental Medicine and Biology). In: *Proceedings from the 2nd International Congress of Immune Mediated Diseases Held in Moscow*, Russia on September 10–15, Springer, 2007.

Signore, A.; Anzola, K. L.; Auletta, S.; Varani, M.; Petitti, A.; Pacilio, M.; Galli, F.; Lauri, C. Current Status of Molecular Imaging in Inflammatory Disorders. *Curr. Pharm. Des.* **2018,** *24*, 743–753. DOI: 10.2174/1381612824666180130115153

Signore, A.; Lauri, C.; Auletta, S.; Anzola, K.; Galli, F.; Casali, M.; Versari, A.; Glaudemans, A. W. J. M. Immuno-Imaging to Predict Treatment Response in Infection, Inflammation and Oncology. *J. Clin. Med.* **2019,** *8*, 681. DOI: 10.3390/jcm8050681

Smith, D. A.; Irving, S. D.; Sheldon, J.; Cole, D.; Kaski, J. C. Serum Levels of the Antiinflammatory Cytokine Interleukin-10 are Decreased in Patients with Unstable Angina. *Circulation* **2001,** *104*, 746–749. DOI: 10.1161/hc3201.094973

Takeuchi, O.; Akira, S. Pattern Recognition Receptors and Inflammation. *Cell* **2010,** *140*, 805–820. DOI: 10.1016/j.cell.2010.01.022

Tan, Y.; Qi, Q.; Lu, C.; Niu, X.; Bai, Y.; Jiang, C.; Wang, Y.; Zhou, Y.; Lu, A.; Xiao, C. Cytokine Imbalance as a Common Mechanism in Both Psoriasis and Rheumatoid Arthritis. *Mediat. Inflamm.* **2017,** *2017*, 2405291. DOI: 10.1155/2017/2405291

Thilagar, S.; Theyagarajan, R.; Sudhakar, U.; Suresh, S.; Saketharaman, P.; Ahamed, N. Comparison of Serum Tumor Necrosis Factor-Alpha Levels in Rheumatoid Arthritis Individuals with and Without Chronic Periodontitis: A Biochemical Study. *J. Indian Soc. Periodontol.* **2018,** *22*, 116–121. DOI: 10.4103/jisp.jisp_362_17

Venkata, M.; Sripathy, R.; Anjana, D.; Somashekara, N.; Krishnaraju, A.; Krishanu, S.; Murali, M.; Verma. S. R.; Ramchand, C. N. In Silico, In Vitro and In Vivo Assessment of Safety and Anti-Inflammatory Activity of Curcumin. *Am. J. Infect. Dis.* **2012,** *8* (1), 26. DOI: 10.3844/ajidsp.2012.26.33

Wang, R.; Wang, Y.; Gao, Z.; Qu, X. The Comparative Study of Acetyl-11-Keto-Beta-Boswellic Acid (AKBA) and Aspirin in the Prevention of Intestinal Adenomatous Polyposis in APCMin/+ Mice. *Drug Discov. Ther.* **2014,** *8* (1), 25–32. DOI: 10.5582/ddt.8.25

Weir, E. G.; Borowitz, M. J. Flow Cytometry in the Diagnosis of Acute Leukemia. *Semin Hematol.* **2001,** *38* (2), 124–138. DOI: 10.1016/s0037-1963 (01)90046-0

Wu, J. J.; Nguyen, T. U.; Poon, K. Y.; Herrinton, L. J. The Association of Psoriasis with Autoimmune Diseases. *J. Am. Acad. Derm.* **2012,** *67*, 924–930. DOI: 10.1016/j.jaad.2012.04.039

Wu, L. Z.; Xiao, X. M. Evaluation of the Effects of *Uncaria rhynchophylla* Alkaloid Extract on LPS-Induced Preeclampsia Symptoms and Inflammation in a Pregnant Rat Model. *Braz. J. Med. Biol. Res.* **2019,** *52* (6), e8273. DOI: 10.1590/1414-431X20198273

Yamaguchi, M.; Levy, R. M. The Combination of β-Caryophyllene, Baicalin and Catechin Synergistically Suppresses the Proliferation and Promotes the Death of RAW267. 4 Macrophages In Vitro. *Int. J. Mol. Med.* **2016,** *38* (6), 1940–1946. DOI: 10.3892/ijmm.2016.2801

Zhou, Y.; Hong, Y.; Huang, H. Triptolide Attenuates Inflammatory Response in Membranous Glomerulo Nephritis Rat via Downregulation of NF-κB Signaling Pathway. *Kidney Blood Pressure Res.* **2016,** *41*, 901–910. DOI: 10.1159/000452591

CHAPTER 9

NUTRACEUTICALS: AN EMERGING CONCEPT FOR SKIN CARE

ALKA LOHANI[1], RAMANDEEP KAUR[2], and NAVNEET VERMA[2]

[1]*School of Pharmaceutical Sciences, IFTM University, Moradabad, Uttar Pradesh, India*

[2]*Pharmacy Academy, IFTM University, Moradabad, Uttar Pradesh, India*

ABSTRACT

People from all across the world are now obsessed not only with their health but also with their appearance and beauty. The yearning for younger-looking skin is stronger than ever. Skin conditions are linked to not only environmental factors (such as sun exposure, humidity, temperature, and chemical wear and tear) but also to an individual's age, gender, and nutritional history. In recent times, there has been a substantial surge in the development of the 'ingestible beauty' trend, and the systemic method, which involves feeding the skin from the inside out with essential nutrients, is gaining popularity among customers. It is widely acknowledged that healthy nutrition affects every organ, including the skin. As a result, the range of products relating to "beauty from within" on the market has expanded in recent years. New tactics are being developed by the food and cosmetic sectors to establish a link between nutrient consumption and healthy skin. Nutraceuticals are a promising technique for preventing, delaying, or minimizing skin aging, as well as for treating some skin problems. This chapter gives a thorough overview of nutraceuticals for skin care.

Herbals As Nutraceuticals: Their Role in Healthcare.
Raj K. Keservani, Rajeshwar Kamal Kant Arya, and Rajesh K. Kesharwani (Eds.)

9.1 INTRODUCTION

The definitions and criteria of beauty have evolved over time to reflect shifting cultural ideals. Beauty has long been valued, and the appearance of the skin is important to the majority of people (Jones, 2018). People all across the world are now obsessed with not just their health, but also their appearance and beauty. The importance of healthy skin in physical appearance cannot be overstated. The human body's outermost tissue is called skin, covering roughly 1.7 m^2 of body surface and accounting for around 10% of the average body mass. The skin's principal job is to act as a hindrance between the body and the outside world. This barrier prevents ultraviolet (UV) radiation, chemicals, allergens, and germs from entering the skin, as well as nutrient and moisture loss. The skin also helps to maintain homeostasis by controlling body temperature and blood pressure (Meyer et al., 2007). The epidermis, dermis, and hypodermis are the three layers of the skin.

People are hyperaware of and sensitive to their looks. As a result, skin appearance has piqued the interest of pharmaceutical and other cosmetic scientists in recent years. Topical preparations have been used in cosmetic dermatology and the cosmetic market is flooded with topically applied cosmetics that are utilized for enhancing skin health and looks. Because of inadequate skin penetration and retention of cosmetically active ingredients in the skin, conventional or topical cosmetic formulations are judged unsuccessful in treating various skin-related problems such as skin aging, acne, alopecia, etc.

In recent times, there has been a substantial surge in the development of the "ingestible beauty" trend, and the systemic method, which involves feeding the skin from the inside out with essential nutrients, is gaining popularity among customers (Spiro and Lockyer, 2018). It's becoming widely acknowledged that healthy nutrition affects every organ, including the skin. As a result, the range of products relating to "beauty from within" on the market has expanded in recent years. "Beauty from within" appeals to today's customers and is gaining popularity. Clinical studies have revealed that beauty is more than just skin deep; it is an integrated strategy that includes both skin care and a good diet. Nutritional status in terms of macro and micronutrients is widely known to be important for skin health and attractiveness (Park, 2015). New tactics are being developed by the food and cosmetic sectors to establish a link between nutrient consumption and healthy skin. As a result, the use of dietary supplements and components that promise to lower the risk of skin issues is on the rise. This is why the nutraceutical industry, particularly oral nutrition supplements promising to benefit skin health, has seen such rapid expansion. This chapter focuses on the use of herbal nutraceuticals for skin care.

9.2 BENEFICIAL BIOLOGICAL FUNCTIONS OF NUTRACEUTICALS FOR SKIN

Following are the various biological functions of nutraceuticals that are directly related to skin health.

9.2.1 ANTIOXIDANTS

Peroxidation attacks the skin cell membrane on a regular basis, causing oxidation. Peroxidation is a chemical reaction that involves when unsaturated fatty acids come into contact with reactive oxygen molecules (like superoxide anion, hydrogen peroxide, and hydroxyl radical). Skin lipids (a blend of sebaceous and epidermal lipids) have a distinct species and fatty acid composition. Essential lipids such as sebaleic, linoleic acid, squalene, and cholesterol are found in human skin and are oxidized by reactive oxygen species (ROS) or free radicals (Niki, 2015). Skin is continually subjected to external environmental stressors (like air pollution, UV radiation, and chemical oxidants). Environmental stress significantly raises the amount of ROS, causing significant damage to the cell structure. This assault on the structure of the skin causes the damage of essential collagen and elastic fibers and decreasing skin hydration, resulting in dyschromia, change in skin physiology, skin dullness, and wrinkles (Taniguchi et al., 2012). Antioxidants have the capacity to neutralize free radicals. Nutraceuticals derived from plants can bring antioxidant action and offer protection to the skin against the harmful effects of oxidation. Nutraceuticals that have antioxidant activity and their source are mentioned in Table 9.1.

TABLE 9.1 Some Nutraceuticals with Antioxidant Properties and Their Food Source.

Phytochemical groups/subgroups	Representative chemical constituent	Common food source
Flavones	Apigenin, luteolin	Parsley, oranges, oregano, celery, chamomile, Bbsil, bell peppers, olive oil, onions, tomatoes, grapes, green beans, cabbage, broccoli, cauliflower
Flavonols	Kaempferol, tangeretin, quercetin, myricetin, isorhamnetin	Citrus plants, broccoli, almonds, dill weed, fennel leaves, apple, berries, grapes, strawberries, broccoli, asparagus, peppers, tomatoes, cereal grains
Flavanones	Hesperidin, naringenin	Lemon, oranges, lime, grapefruit, and other citrus fruits
Flavanonols	Taxifolin	Red onion, milk thistle, and citrus fruits
Flavanols/ procyanidins	Catechin, epicatechin	Grapes, mangoes, apples, pears, plums, okra, peaches, berries, Swiss chard, peanut, green tea
Carotenoids	Lycopene, β-carotene, lutein	Butternut squash, pumpkins, carrot, sweet potato, tomatoes, dark green leafy vegetables, watermelon, pulse crops, oil seeds, citrous fruits
Amides	Avenanthramides, capsaicin	Oats, chili pepper, jalapeno, serrano pepper
Isoflavonoids	Diadzein, genistein, biochanin A	Soy, red clover, lima beans, pigeon peas, chickpeas, split peas, green tea, peanuts
Cinnamic acids	Ferulic acid	Apple seeds, eggplant, cereal grains, oatmeal, small berry fruits, rice

9.2.2 ANTI-INFLAMMATION

Inflammation, which can be acute or chronic, is a protective mechanism against the root cause of cell damage. Skin rash, red spots, or even a breakout like blisters or abscess are signs of skin inflammation. When a rash induced by inflammation is touched, it frequently stings, burns, or itches. Nutraceuticals exert anti-inflammatory properties by hindering NF-B activation, stopping the overexpression of interleukin-1, restricting phospholipase A2, cyclooxygenase-2 (COX-2), lipoxygenase, myeloperoxidase, and preventing the enzymes responsible for ROS generation and enhance the ability to scavenge ROS (Inan, 2019). Examples of nutraceutical herbs that possess anti-inflammatory activity include rosemary (*Rosmarinus officinalis*), ginger (*Zingiber officinale*), stinging nettle (*Urtica dioica*), garlic (*Allium sativum*), turmeric (*Curcuma longa*), cardamom (*Elettaria cardamomum*), black pepper (*Piper nigrum* L.), avocado (*Persea americana*), olive (*Olea europaea*), green tea (*Camellia sinensis* L.), rosemary (*R. officinalis*), etc.

9.2.3 SKIN HYDRATION

Maintaining healthy skin requires enough skin moisture and water is vitally necessary for the correct functioning of the skin. The skin's ability to hold water is primarily determined by the stratum corneum (SC), which functions as a strong barrier to the loss of water from the skin. Water retention in the SC is determined by two key components: (1) natural hygroscopic substances present in the corneocytes and (2) barrier formation against transepidermal water loss (TEWL) by the ordered arrangement of SC intercellular lipids. As a result of increased TEWL, enzymatic processes necessary for regular skin exfoliation are inhibited, and the skin appears dry and flaky. ELASTEN® (QUIRIS Healthcare, Gütersloh, Germany), a liquid nutraceutical, was studied by researchers for skin aging and health. They reported that the skin hydration, suppleness, roughness, and density were all improved by the tested nutraceutical formulation (Bolke et al., 2019).

9.3 NUTRACEUTICALS FOR SKIN CARE

Various nutraceuticals used or beneficial for skin care are listed below in the following subsections.

9.3.1 HYALURONIC ACID

Hyaluronic acid is a non-sulfated anionic glycosaminoglycan that is widely distributed throughout the body. It is a polysaccharide made up of N-acetyl-D-glucosamine and D-glucuronic acid (Fig. 9.1) (Weissmann and Meyer, 1954). The skin contains the most hyaluronic acid, accounting for half of all hyaluronic acid in the body (Reed et al., 1988), and is involved in skin tissue repair. hyaluronic acid plays a vital role in maintaining the integrity of the epidermis. It is critical to the re-epithelialization process. Being an integral part of the extracellular matrix of basal keratinocytes, it has an involvement in keratinocyte proliferation and migration and the free radical scavenging process (Papakonstantinou et al., 2012). In the dermis, hyaluronic acid controls osmotic pressure, ion flow, and water balance and stabilizes skin through electrostatic interactions. Hyaluronic acid is an important moisturizer and has a high water retention capacity, for human skin. Studies have shown that hyaluronic acid decreases skin wrinkles and enhances skin moisture, smoothness, and suppleness (Göllner et al., 2017; Oe et al., 2017). Root vegetables (such as sweet potatoes, potatoes, carrots, and lotus roots) Almonds and soy products are the common dietary sources of hyaluronic acid.

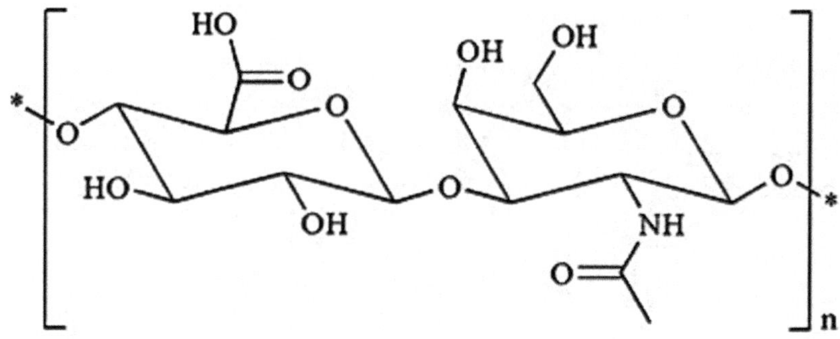

FIGURE 9.1 Structure of hyaluronic acid.

9.3.2 *COENZYME Q10*

The coenzyme Q family, often known as ubiquinone, is a group of enzymes that can be found in all human cells. Coenzyme Q10 (CoQ10) is a lipid-soluble antioxidant that is vital for cellular energy generation, but its endogenous synthesis in human skin declines with age and exposure to external stressors (Sanoobar et al., 2015). It has 10 isoprenoid units attached to benzoquinone rings in its chemical structure (Fig. 9.2). By providing protection against endogenously generated ROS and photoproduction, CoQ10 minimizes fine lines and wrinkles, promotes skin smoothness, and reduces other indications of skin aging (Knott et al., 2015). This leads to the notion that nutraceutical supplementation of CoQ10 will be beneficial. The impact of dietary CoQ10 alone (50/150 mg/day for 12 weeks) on skin parameters was investigated in research, and it was discovered that CoQ10 considerably decreased periorbital wrinkles and microrelief lines, as well as enhanced skin smoothness. It was reported that a larger dose of CoQ10 resulted in significant wrinkle reduction in various areas of the face (upper radial lip lines, corner of the mouth lines, and nasolabial folds) (Zmitek et al., 2020).

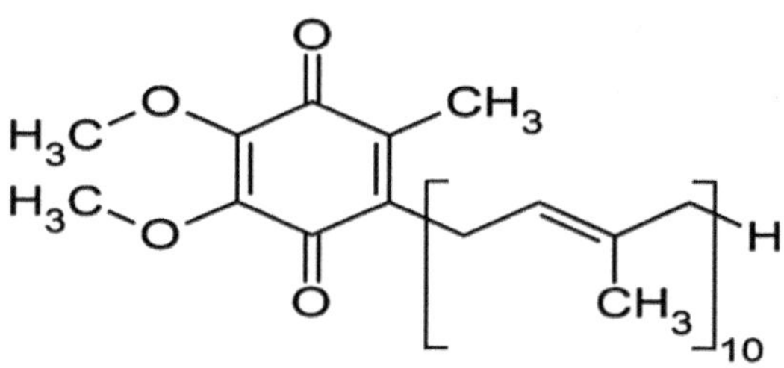

FIGURE 9.2 Structure of CoQ10.

9.3.3 OMEGA-3 FATTY ACIDS

Omega-3 fatty acids are polyunsaturated fatty acids (PUFAs) in which at the third carbon atom from the end of the carbon chain, there is a double bond (C=C). Omega-3 fatty acids are extremely beneficial to health. Exposure to UV-B radiation results in a various adverse effect on the skin (like photosensitivity, immunosuppression, sunburn, and photocarcinogenesis). Although sunscreens applied to the skin can give some protection, their effectiveness is restricted by improper application techniques (Rhodes et al., 1998).

In this, a systemic photoprotective constituent is needed that can protect the skin against harmful UV radiation. According to some research studies, dietary omega-3 PUFAs diminish sunburn sensitivity in humans (Orengo et al., 1992; Rhodes et al., 1994) and also diminish photoimmunosuppression and photocarcinogenesis in mice (Moison et al., 2001). Several studies have shown that omega-3 PUFA metabolites have anti-inflammatory properties in inflammatory skin illnesses and conditions such as contact hypersensitivity, psoriasis, atopic dermatitis, and UV radiation (Moison et al., 2001). Chia seeds, Brussels sprout, algal oil, hemp seed, walnuts, are some plant foods that contain omega-3 fatty acids.

9.3.4 ARGININE

The amino acid arginine (also known as l-arginine) is essential for protein synthesis. It contains an α-amino group, an α-carboxylic acid group, and a three-carbon aliphatic straight chain with a side chain of a guanidino group (Fig. 9.3).

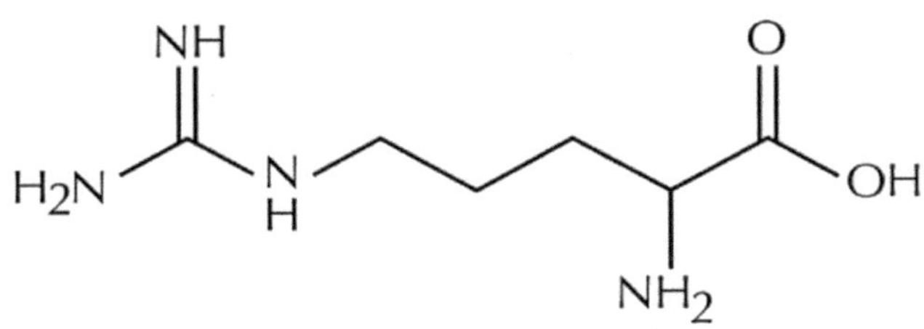

FIGURE 9.3 Structure of arginine.

9.3.5 LYCOPINE

A symmetrical tetraterpene, lycopene, is made up of eight isoprene molecules. Since it is fully composed of hydrogen and carbon, it is categorized as a carotene and belongs to the carotenoid family (Fukushima et al., 2021) Figure 9.4.

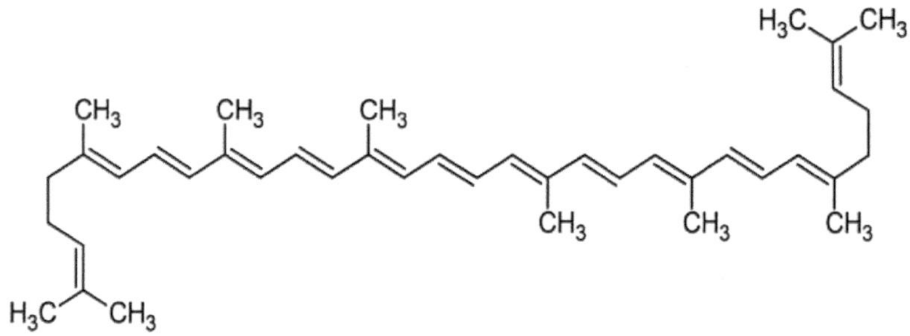

FIGURE 9.4 Structure of lycopine.

Lycopene is a nutritional antioxidant that has been associated with skin photodamage prevention. The antioxidant property of lycopene is well known. It has the ability to prevent oxidation of DNA, proteins, and lipids (Caseiro et al., 2020). It has anti-inflammatory activity also. It has been reported that in hairless SKH-1 mice and human keratinocytes, lycopene treatment inhibited UVB-induced cell proliferation while increasing apoptosis via lowering CDK2 and CDK4 levels (Chen et al., 2018).

9.3.6 *LUTEIN*

Lutein is a type of xanthophyll that is one of 600 carotenoids found in nature. As it is not synthesized by the human body, it must be obtained through diet. Human skin contains lutein as a result of dietary consumption. Researchers have reported that, in animal experiments, lutein has also been found to help protect against skin damage induced by UV radiation exposure. Lutein consumption aids in the prevention of skin hyperplasia, which is induced by UV radiation (González et al., 2003). In a study, it has been reported that oral supplementation of lutein significantly lightens skin tone and improves skin conditions due to its antioxidant ability and photoprotective activity (Juturu et al., 2016). One of the studies on mice reported that following oral supplementation, lutein accumulation in mice skin was found to reduce ROS formation after UV radiation exposure (Lee et al., 2004). As a result, lutein influences the reaction of the skin to UV radiation and may help to safeguard the skin against the detrimental effects of sun exposure.

9.3.7 MINERAL

9.3.7.1 ZINC

Zinc is a chemical element with the symbol Zn. Zinc is a vital mineral and antioxidant that serves a range of purposes in our bodies. It can be found naturally in many foods or taken as a supplement. It has a significant impact on the stimulation of metallothionein (a protein with antioxidant properties) synthesis in the structure of the superoxide dismutase enzyme, which is a responsible catalyst for the conversion of two superoxide ion radicals to H_2 and O_2 peroxide, converting a highly reactive species (superoxide ion) into a less aggressive form (H_2 peroxide). By inhibiting ions (such as copper and iron) it preserves sulfhydryl groups in cell membrane proteins (Koury et al., 2003). Zinc helps to maintain the integrity of cellular membranes by reducing membrane lipid peroxidation (Arlette et al., 1983). Zinc is effective for the treatment of acne and due to its anti-inflammatory properties, it is effective in relieving the discomfort from skin rashes, blisters, sunburn, etc. (Bhowmik et al., 2010).

9.3.7.2 SILICON

Silicon is the second most prevalent chemical element on the planet, with the symbol Si. Silicon is a plentiful element in the human body, and it plays a crucial role in maintaining the integrity of the skin by taking part in collagen synthesis and glycosaminoglycans which influence the health of the skin and its annexes (Araújo et al., 2016). Silicon deficiency has been shown to have a deleterious influence on collagen synthesis and deposition in a variety of tissues (Jugdaohsingh et al., 2015). One of the research works reported that nanosilicon-containing dietary supplements are efficient in enhancing skin hydration, elasticity, roughness, and barrier function (Cameli et al., 2021).

9.3.7.3 SELENIUM

Selenium (Se) is a dietary trace mineral that has a variety of uses. Selenium has the ability to protect the skin against cancer caused by UV exposure. It reduces UV-induced oxidative DNA damage and also inhibits the synthesis of immunosuppressive and inflammatory cytokines, both which hinder immune responses after UV-radiation exposure (McKenzie, 2000).

9.3.8 VITAMINS

9.3.8.1 VITAMIN C

Vitamin C (ascorbic acid, ascorbate) is a water-soluble carbohydrate with a low molecular weight that is required by the body. Despite the fact that vitamin C is a crucial antioxidant, humans cannot synthesize it, therefore they must get it from their diet. Vitamin C is associated with the skin barrier formation and collagen synthesis in the dermis as well as protects the skin against oxidative damage, wrinkle formation, and cell signal pathways of cell differentiation and growth, all of which are linked to the occurrence and progression of various skin diseases (Ponec et al., 1997). In research, it was revealed that taking vitamin C orally boosts the skin's radical-scavenging activity. The action is immediate, and it is amplified by increased vitamin C dosage (Lauer et al., 2013).

9.3.8.2 VITAMIN E

Four tocopherols and four tocotrienols are among the eight fat-soluble components that make up vitamin E. Vitamin E is a potent lipid-soluble antioxidant that protects tissues against lipid peroxidation and, as a result, oxidative damage. Atopic dermatitis and psoriasis can be prevented and treated with supplementation (Berardesca et al., 2021).

9.3.8.3 VITAMIN A

A healthy body requires Vitamin A, a fat-soluble vitamin. The chemical constituents that make up this group include retinol, retinal, retinoic acid, and a variety of provitamin A carotenoids. Two categories of vitamin-A are present in the human diet, namely, (1) preformed vitamin A (retinyl ester, retinol); and (2) provitamin A carotenoids. Vitamin A inadequacy has been related to a high risk of skin infection and inflammation (Roche et al., 2021).

9.3.9 POLYPHENOLS

9.3.9.1 GREEN TEA POLYPHENOL

Green tea is an abundant source of plant polyphenols. Catechins (including epigallocatechin gallate, epigallocatechin, epicatechin gallate, and epicatechin gallate) are the polyphenols found in green tea. Polyphenols protect the skin by acting as antioxidants, chemopreventatives, and immunomodulators. Green tea, which is high in polyphenolic antioxidants, is becoming more popular as a skin care supplement. A study reported that green tea catechin consumption causes the absorption of catechin metabolites into human skin, which is related to a reduction in UV radiation-induced 12-HETE, which might help protect against the inflammation caused by UV radiation or sunburn and perhaps longer-term UV radiation-mediated skin damage (Rhodes et al., 2013).

9.3.9.2 RESVEREATROL

The stilbenoid, resveratrol (3,5,4′-trihydroxystilbene) is a stilbene derivative (Fig. 9.5) produced by several plants. Resveratrol is best known for its powerful antioxidant properties. Other than scavenging ROS or free radicals, resveratrol encourages the production of other naturally occurring enzymatic antioxidants (like catalase, superoxide dismutase, and hemoxygenase-1) within the cell (Patricia et al., 2013). It has antiproliferative, anti-inflammatory, antibacterial, and antiangiogenic, activities also and it has been observed that resveratrol increases fibroblast growth, resulting in an increase in collagen III concentration (Ratz and Arct, 2019). Researchers have reported that in mice oral consumption of resveratrol before UVB exposure reduced epidermal thickening by about 55% compared to only UVB-treated mice and it appeared to provide a marginal level of protection against wrinkle formation (Kim et al., 2019).

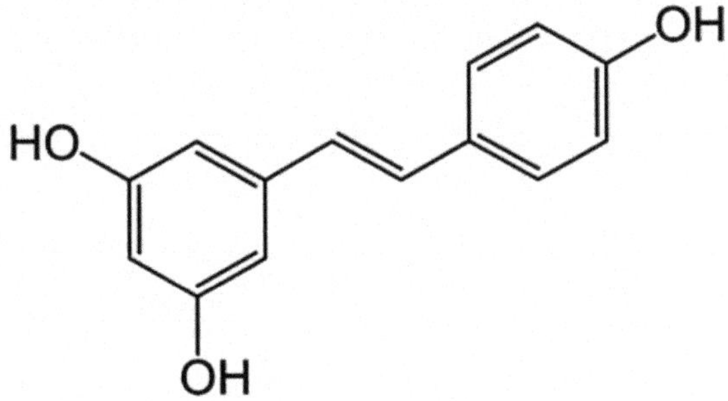

FIGURE 9.5 Structure of resveratrol.

Some of the nutraceutical supplements available in the market are shown in Table 9.2.

TABLE 9.2 Nutraceutical Supplements Available in Market for Skin Care.

Product	Active constituents	Use
What's Up Wellness Beauty Gummies	Biotin, vitamins A to E, folic acid, zinc, aloe vera, sea buckthorn, and grapeseed	Keep skin hydrated and maintain its youthful glow; Helps in softening and moisturizing the skin; Zinc prevents acne
HealthKart HK Vitals Vitamin E Capsules	Tocopherol with evening primrose oil	Antioxidant Support, Controls skin wrinkling, roughness, and dehydration
OZiva Plant-Based Collagen Builder	Vitamin C, silica, and biotin	Antiaging
VLCC Wellscience Green Tea Extract	Polyphenols	Reduce oxidative stress and damage caused by free radicals
Nutrija-Green Tea Extract	Polyphenols, catechin, and epigallocatechin gallate	Powerful source of antioxidants
OZiva Plant-Based Bettr. Zinc+ (With Zinc, Magnesium, Chamomile & Ginger Root)	Magnesium, zinc, and folic acid	Anti-inflammation and acne control
HealthyHey Nutrition Selenium Plus Natural Vitamin E	Selenium and vitamin E	Antioxidant supplement: Prevents lipid peroxidation and protects cells against free radicals
Neuherbs Plant-Based Skin Collagen Booster	Hyaluronic acid, vitamin C, ellagic acid, Carotenoids, Silica	Antiaging and skin repair
Nutriosys hyaluronic acid	Hyaluronic acid	Hydrate dry aged skin
OZiva Plant-Based OMEGA 369	Vegan omega oil concentrate with flaxseed and backseed oil	Promotes skin health

9.4 CONCLUSION

Because good nutrition affects every organ, including the skin, the availability of products related to "Beauty from Within" has expanded in recent years. Nutraceuticals have been discovered to have the ability to lessen the signs of skin aging as well as a variety of skin disorders. It's a combination of nutrients and medications that gives the skin more nourishment. To summarize, obtaining solid scientific data for skin nutraceuticals and demonstrating a cause-and-effect relationship between their components and their skin–beneficial impacts requires a high-quality and rigorous investigation.

KEYWORDS

- **nutraceuticals**
- **herbal**
- **skin care**
- **polyphenols**
- **antiinflammation**

REFERENCES

Araujo, L. A.; Addor, F.; Campos, P. M. Use of Silicon for Skin and Hair Care: An Approach of Chemical Forms Available and Efficacy. *An. Bras. Dermatol.* **2016,** *91* (3), 331–335.

Arlette, J. P. Zinc and the Skin. *Pediatr. Clin. North Am.* **1983,** *30* (3), 583–596.

Berardesca, E.; Cameli, N. Vitamin E Supplementation in Inflammatory Skin Diseases. *Dermatol. Therap.* **2021,** *34* (6), e15160.

Bhowmik, D.; Chiranjib, K.; Kumar, S. A Potential Medicinal Importance of Zinc in Human Health and Chronic. *Int. J. Pharm.* **2010,** *1* (1), 05–11.

Bolke, L.; Schlippe, G.; Gerß, J.; Voss, W. A Collagen Supplement Improves Skin Hydration, Elasticity, Roughness, and Density: Results of a Randomized, Placebo-Controlled, Blind Study. *Nutrients* **2019,** *11* (10), 2494–2500.

Cameli, N.; Silvestri, M.; Mariano, M.; Berardesca, E. Effects of Food Supplements and a Topical Solution Containing Nanosilicon on Skin Hydration, Barrier Function, and Elasticity. *J. Cosmet. Dermatol.* **2021,** *20*, 32–35.

Caseiro, M.; Ascenso, A.; Costa, A.; Creagh-Flynn, J.; Johnson, M.; Simões, S. Lycopene in Human Health. *LWT* **2020,** *127*, 109323.

Chen, P.Xu, S.; Qu, J. Lycopene Protects Keratinocytes Against UVB Radiation-Induced Carcinogenesis via Negative Regulation of FOXO3a Through the mTORC2/AKT Signaling Pathway. *J. Cell Biochem.* **2018,** *119*, 366–377.

Fukushima, Y.; Taguchi, C.; Kishimoto, Y.; Kondo, K. Japanese Carotenoid Database with α- and β-carotene, β-cryptoxanthin, Lutein, Zeaxanthin, Lycopene, and Fucoxanthin and Intake in Adult Women. *Int. J. Vitam. Nutr. Res.* **2021,** *7*, 1–12.

Göllner, I.; Voss, W.; von Hehn, U.; Kammerer, S. Ingestion of an Oral Hyaluronan Solution Improves Skin Hydration, Wrinkle Reduction, Elasticity, and Skin Roughness: Results of a Clinical Study. *J. Evid. Based Complement. Altern. Med.* **2017,** *22* (4), 816–823.

González, S. Astner, S.; Wu, A.; Goukassian, D.; Pathak, M. Dietary Lutein/ Zeaxanthin Decreases Ultraviolet B-Induced Epidermal Hyperproliferation and Acute Inflammation in Hairless Mice. *J. Invest. Dermatol.* **2003,** *121*, 399–405.

Inan, S. The Potential Role of Nutraceuticals in Inflammation and Oxidative Stress. In: Hueda, M. C., Ed.; *Nutraceuticals—Past, Present and Future*; IntechOpen: London, 2019; pp 1–12.

Jones, A. L. The Influence of Shape and Colour Cue Classes on Facial Health Perception. *Evol. Hum. Behav.* **2018,** *39* (1), 19–29.

Jugdaohsingh, R.; Watson, A. I.; Pedro, L. D.; Powell, J. J. The Decrease In Silicon Concentration of the Connective Tissues with Age in Rats Is a Marker of Connective Tissue Turnover. *Bone* **2015,** *75*, 40–48.

Juturu, V.; Bowman, J. P.; Deshpande, J. Overall Skin Tone and Skin-Lightening-Improving Effects with Oral Supplementation of Lutein and Zeaxanthin Isomers: A Double-Blind, Placebo-Controlled Clinical Trial. *Clin. Cosmet. Investig. Dermatol.* **2016,** *9*, 325–332.

Kim, J.; Oh, J.; Averilla, J. N.; Kim, H. J.; Kim, J. S.; Kim, J. S. Grape Peel Extract and Resveratrol Inhibit Wrinkle Formation in Mice Model Through Activation of Nrf2/HO-1 Signaling Pathway. *J. Food Sci.* **2019,** *84* (6), 1600–1608.

Knott, A.; Achterberg, V.; Smuda, C.; Mielke, H.; Sperling, G.; Dunckelmann, K.; Vogelsang, A.; Krüger, A.; Schwengler, H.; Behtash, M.; Kristof, S.; Diekmann, H.; Eisenberg, T.; Berroth, A.; Hildebrand, J.; Siegner, R.; Winnefeld, M.; Teuber, F.; Fey, S.; Möbius, J.; Blatt, T. Topical Treatment with Coenzyme Q10-Containing Formulas Improves Skin's Q10 Level and Provides Antioxidative Effects. *BioFactors* **2015,** *41* (6), 383–390.

Koury JC, Donangelo CM. Zinc, Oxidative Stress and Physical Activity. *Revista de Nutrição* **2003,** *16*, 433–441.

Lauer, A. C.; Groth, N.; Haag, S. F.; Darvin, M. E.; Lademann, J.; Meinke, M. C. Dose-Dependent Vitamin C Uptake and Radical Scavenging Activity in Human Skin Measured with In Vivo Electron Paramagnetic Resonance Spectroscopy. *Skin Pharmacol. Phys.* **2013,** *26* (3), 147–154.

Lee, E. H.; Faulhaber, D.; Hanson, K. M.; Ding, W.; Peters, S.; Kodali, S.; Granstein, R. D. Dietary Lutein Reduces Ultraviolet Radiation-Induced Inflammation and Immunosuppression. *J. Invest. Dermatol.* **2004,** *122* (2), 510–517.

McKenzie, R. C. Selenium, Ultraviolet Radiation and the Skin. *Clin. Exp. Dermatol.* **2000,** *25* (8), 631–636.

Meyer, T.; Stockfl, E.; Chrstophers, E. Immune Response Profiles in Human Skin. *Br. J. Dermatol.* **2007,** *157*, 1–7.

Moison, R. M.; Beijersbergen G. M.; Henegouwen, V. Dietary Eicosapentaenoic Acid Prevents Systemic Immunosuppression in Mice Induced by UVB. *Radiat. Res.* **2001,** *156*, 36–44.

Niki, E. Lipid Oxidation in the Skin. *Free Radic. Res.* **2015,** *49* (7), 827–834.

Oe, M.; Sakai, S.; Yoshida, H.; Okado, N.; Kaneda, H.; Masuda, Y.; Urushibata, O. Oral Hyaluronan Relieves Wrinkles: A Double-Blinded, Placebo-Controlled Study Over a 12-Week Period. Clinical. *Clin. Cosmet. Investig. Dermatol.* **2017,** *10*, 267–273.

Orengo, I. F.; Black H. S.; Wolf, J. E. Influence of Fish Oil Supplementation on the Minimal Erythema Dose in Humans. *Arch. Dermatol. Res.* **1992,** *284*, 219–221.

Papakonstantinou, E.; Roth, M.; Karakiulakis, G. Hyaluronic Acid: A Key Molecule in Skin Aging. *Dermato Endocrinol.* **2012,** *4* (3), 253–258.

Park, K. Role of Micronutrients in Skin Health and Function. *BiomolTher* **2015,** *23* (3), 200–207.

Patricia Farris, M. D.; Jean Krutmann, M. D.; Yuan-Hong Li, M. D.; David McDaniel, M. D.; Krolj, Y. Resveratrol: A Unique Antioxidant Offering a Multi-Mechanistic Approach for Treating Aging Skin. *J. Drugs Dermatol.* **2013,** *12* (12), 1389–1394.

Ponec, M.; Weerheim, A.; Kempenaar, J.; Mulder, A.; Gooris, G. S.; Bouwstra, J. et al. The Formation of Competent Barrier Lipids in Reconstructed Human Epidermis Requires the Presence of Vitamin C. *J. Invest. Dermatol.* **1997,** *109*, 348–355.

Ratz-Łyko, A.; andArct, J. Resveratrol as anActive Ingredient for Cosmetic and Dermatological Applications: A Review. *J. Cosmet. Laser Ther.* **2019,** *21* (2), 84–90.

Reed, R. K.; Lilja, K.; Laurent, T. C. Hyaluronan in the Rat with Special Reference to the Skin. *Acta Physiol. Scand.* **1988,** *134* (3), 405–411.

Rhodes, L. E.; Darby, G.; Massey, K. A.; Clarke, K. A.; Dew, T. P.; Farrar, M. D.; Nicolaou, A. Oral Green Tea Catechin Metabolites Are Incorporated Into Human Skin and Protect Against UV Radiation-Induced Cutaneous Inflammation in Association with Reduced Production of Pro-Inflammatory Eicosanoid 12-Hydroxyeicosatetraenoic Acid. *Br. J. Nutr.* **2013,** *110* (5), 891–900.

Rhodes, L. E. Topical and Systemic Approaches for Protection Against Solar Radiation-Induced Skin Damage. *Clin. Dermatol.* **1998,** *16*, 75–82.

Rhodes, L. E.; Farrell, S.; Jackson M. J.; Friedmann, P. S. Dietary Fish-Oil Supplementation in Humans Reduces UVB-Erythemal Sensitivity but Increases Epidermal Lipid Peroxidation. *J. Investig. Dermatol.* **1994,** *103*, 151–154.

Roche, F. C.; Harris-Tryon, T. A. Illuminating the Role of Vitamin A in Skin Innate Immunity and the Skin Microbiome: A Narrative Review. *Nutrients* **2021,** *13* (2), 302.

Sanoobar, M.; Eghtesadi, S.; Azimi, A.; Khalili, M.; Khodadadi, B.; Jazayeri, S.; Aryaeian, N. Coenzyme Q10 Supplementation Ameliorates Inflammatory Markers in Patients with Multiple Sclerosis: A Double Blind, Placebo, Controlled Randomized Clinical Trial. *Nutr. Neurosci.* **2015,** *18* (4), 169–176.

Spiro, A.; Lockyer, S. Nutraceuticals and Skin Appearance: Is There Any Evidence to Support This Growing Trend? *Nutr. Bull.* **2018,** *43* (1), 10–45.

Taniguchi, M.; Arai, N.; Kohno, K.; Ushio, S.; Fukuda, S. Anti-Oxidative and Anti-Aging Activities of 2-O-α-glucopyranosyl-L-Ascorbic Acid on Human Dermal Fibroblasts. *Eur. J. Pharmacol.* **2012,** *674* (3), 126–131.

Weissmann, B.; Meyer, K. The Structure of Hyalobiuronic Acid and of Hyaluronic Acid from Umbilical Cord. *J. Am. Chem. Soc.* **1954,** *76*, 1753–1757.

Zmitek, K.; Zmitek, J.; RoglButina, M.; Pogačnik, T. Effects of a Combination of Water-Soluble CoenzymeQ10 and Collagen on Skin Parameters and Condition: Results of a Randomised, Placebo-Controlled, Double-Blind Study. *Nutrients* **2020,** *12* (3), 618–625.

CHAPTER 10

NUTRACEUTICAL HERBS IN OBESITY AND DIABETES

ARUN KUMAR SINGH[1], REKHA MEHTA[1], ANITA SINGH[2], and RAJESHWAR KAMAL KANT ARYA[2]

[1]*Amrapali Institute of Pharmacy and Sciences, Haldwani, Uttarakhand, India*

[2]*Department of Pharmaceutical Sciences, Sir J. C. Bose Technical Campus Bhimtal, Kumaun University, Nainital, India*

ABSTRACT

Nutraceuticals are products derived from food or a part of it that have some health benefits along with or without nutritional values. Nutraceuticals mainly improve the health status of individuals by modulating body functions. In recent years, nutraceuticals have gained the importance of being used as an alternative or supplemental medication in combination with pharmaceuticals for the prevention and treatment of a wide range of diseases. Generally, nutraceuticals are nonspecific biological therapies that include proteins, natural herbs and botanicals, vitamins, antioxidants, minerals, amino acids, and fatty acids, which prevent and control disease-related symptoms and promote wellness. Lifestyle diseases are mainly due to the way we lead our lives daily. More work and no play make many of us prone to various diseases, physical ailments, and emotional problems resulting from such diseases. The most common lifestyle diseases around the world today are obesity and diabetes. Lifestyle diseases account for most diseases such as arteriosclerosis, heart disease, stroke, hypertension, nephritis, cancer, chronic obstructive pulmonary disease (COPD), allergies, hearing problems,

Herbals As Nutraceuticals: Their Role in Healthcare.
Raj K. Keservani, Rajeshwar Kamal Kant Arya, and Rajesh K. Kesharwani (Eds.)

depression, anxiety disorders, Alzheimer's disease, and chronic backache. So, nutraceuticals have attracted considerable interest due to their potential nutritional, safety, and therapeutic effects.

10.1 INTRODUCTION

Stephen Defelice first devised the term "nutraceutical" in 1989, and it may be defined as the food or a part of a food commonly used for its medicinal or health-improving purpose, along with the prevention and cure of disease' (Defelice, 1995). The nutraceuticals include pure nutrients as well as dietary supplements like minerals, different vitamins (Singh et al., 2017a), and amino acids (Singh et al., 2017b), along with genetically engineered designed nutrients, and specific types of isolated Phyto-compounds such as flavonoids, glycosides, phytosterols, polyphenols, and tannins, that exhibited the promising therapeutic health benefits besides the traditional nutritional values that also found in foods (Smith-Warner et al., 2000). Most nutraceuticals (Keservani et al., 2010a, 2010b, 2017, 2020) have many medicinal values covering disorders related to hepatic diseases, neurodegenerative diseases, depression, digestion, cold and cough, prevention of cancer, cardiovascular diseases, diabetes, and obesity (Khan et al., 2003). Nutraceuticals are generally classified as Potential and Established nutraceuticals. Establishing any herb as a potent nutraceutical requires a lot of information and scientific evidence regarding the therapeutic or medicinal potential with clinical approval. Pharmacokinetic interferences result in a change in active drug metabolism. For example, cytochrome-P450 (CYP-450) is an enzymatic environment that oxidizes a variety of drugs through several metabolic processes which may be altered or minimized by inducers or inhibitors (Dasgupta, 2003; MacDonald et al., 2009).

The major advantages of naturally occurring nutraceuticals over synthetic pharmaceuticals include very few or no side effects, easily available, cost-effective, and multiple therapeutic effects. It also increases overall health maintenance attributed to its medicinal value for human beings. Recently, much attention has been received for nutraceuticals because of their potential nutritional value, safety, and therapeutic benefits (Golla, 2018). This chapter aims to understand the role of nutraceutical herbs in the therapeutic management and therapy of diabetes mellitus (DM) and obesity.

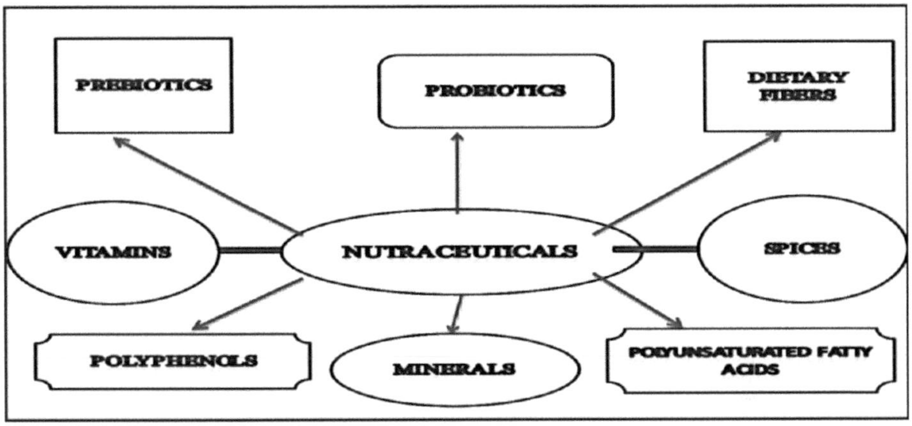

FIGURE 10.1 Herbal nutraceuticals components.

10.2 NUTRACEUTICALS IN DIABETES

Diabetes mellitus is a complex chronic metabolic disorder associated with hyperglycemia because of the disturbance in the process of insulin production/release (insulin insufficiency) insulin resistivity, or both (Kaur et al., 2015). Commonly, diabetes type 2 (insulin resistance) is a global disease affecting about 95% of the population and is expected to afflict 4.4% of the global population by 2030. Generally, Type 2 diabetes is correlated with obesity because, in obesity, the adipose tissue secretes the adiponectin hormone that is responsible for insulin resistance (Kochhar and Pathak, 2018; King et al., 2006)

The major causes of type 2 diabetes are obesity, a bad lifestyle, increasing age, and a bad diet. Gestational diabetes is caused by a family history of overweight or polycystic ovarian syndrome (PCOS). People with diabetes are more likely to have cardiovascular disease, diabetic retinopathy, glaucoma, cataract, and kidney failure. Many synthetic pharmaceutical compounds such as sulfonylureas, thiazolidinediones, glucagon-like peptide (GLP)-1 agonists, biguanides, etc. are commercially available on the market to control for the treatment of diabetes, but yet now complete cure has not been reported (Kaveeshwar and Cornwall, 2018).

Furthermore, the clinically available antidiabetic drugs elicit undesirable side effects like severe hypoglycemic effects, lactic acidosis, sometimes peripheral edema, and abdominal discomfort (Lorenzati et al., 2012).

Therefore, the investigations of new antidiabetic drugs with more therapeutic value and lesser adverse effects have been continued.

TABLE 10.1 Marketed Preparation of Nutraceuticals (Kaur et al., 2015).

Nutraceuticals	Preparation of nutraceuticals available in market	Pre-clinical studies	Clinical studies
Curcumin (Curcuminoids obtained from *Curcuma longa*)	Mervia	Wistar and Swiss albino mice were treated with 1% of pure *C. longa* as well as 5% of curcumin. Duration: 3 months	Curcumin administered in human subjects. Duration: 4 months Dose: 3.6 g–8.0 g/day
Fenugreek	Seeds and their powder	Weanling rats fed with fenugreek seeds. Duration: 3 months	The fenugreek powdered seed was given to diabetic patients. Duration: 24 weeks Dose: 25 mg, daily
Vitamins chromium	Chromium-poly-nicotinate (niacin or vitamin B3)		An open 10-month clinical trial was conducted in type 2 diabetic patients with vitamin B3 and chromium preparation.
Vitamins Magnesium	Tablets		This preparation was given to diabetic patients, and an increment in insulin secretion and lowering in fasting blood glucose level was found.
Vitamin E (α-tocopherol)			The effect of vitamin E on blood glucose level was accessed. Dose:100–1600 mg/kg per day Duration: 2–4 months
Amla	Amla and its formulation	*Emblica officinals* extract was ingested in diabetic rats, and a lowering in neuropathic pain was observed.	*E. officinalis* was administered to diabetic as well as Normal patients daily and it was found that it improves high-density lipoprotein (HDL) and cholesterol. Dose: 2–3 g, daily.

Nutraceuticals or herbs are important for investigating new remedies related to human health problems. Traditionally, many herbs or natural plant actives/nutraceuticals have been prescribed for preventing and treating diabetes since ancient times. The antidiabetic or hypoglycemic effects shown by many plant-based components have been reported by different scientists/researchers, and their published reports are confirmed by in vitro studies and animal models (Patel et al., 2010).

10.3 ROLE OF NUTRACEUTICALS IN DIABETES

Nutraceuticals do not play any direct role in the treatment, but they are commonly used to improve health wellness, inhibit malignant processes, and regulate symptoms. Nutraceuticals can be further divided into three groups, which are as follows

1. The substances of known nutritional values, such as vitamins of different categories, minerals, different types of amino acids, and fatty acids—these nutrients are required for the normal physiology of the body
2. Herbs or isolated natural compounds and extracts—Herbals
3. Dietary supplements such as pyruvate, chondroitin sulfate, and steroidal hormone precursors.

Nutraceutical antidiabetic herbs—A variety of herbal plant materials, as well as nutraceuticals, are commonly utilized for treating diabetes. Their active compounds, their mechanism of action, and related in vivo studies are summarized in Table 10.2.

Several types of nutraceuticals and dietary supplements have been used clinically in recent years, and it has been proven to benefit diabetes, mainly in type 2 and their complications related to targeting the pathogenesis (Rafieian–Kopaei et al., 2013). Such compounds generally contain α-lipoic acid, omega-3 fatty acids, berberine, natural dietary fibers, phytoestrogens (Isoflavones), antioxidants (vitamins C and E), vitamin D, flavonoids (Keservani and Sharma, 2014), conjugated linoleic acid, and minerals such as chromium and magnesium (Davì et al., 2010; Derosa et al., 2014).

10.4 ROLE OF NUTRACEUTICALS IN OBESITY:

Lifestyle diseases are complex diseases, that result from our improper daily routine, less or no physical work with unhealthy food habits make people

TABLE 10.2 Herbal Nutraceutical for the Treatment of DM.

Herbal sources	Part of the plant used	Phytochemicals	In vivo antidiabetic studies		References
			Streptozotocin-induced diabetic rats model	**Alloxan-induced diabetic rats model**	
Peepal (*Ficus religiosa*, family: Moraceae)	Bark, leaf, seeds, fruits, and root	Alkaloids, steroids, flavonoids, glycosides, saponins, catalase, and peroxidase are nonenzymatic and enzymatic constituents in vitamin C	Aqueous extracts of bark Dose: 50–100 mg/kg		Makhija et al. (2010); Pandit et al. (2010)
Blackberry (*Syzygium cumini* or *Eugenia jambolana* Family- Myrtaceae)	Bark, leaf, seeds, stem, and roots	Alkaloids, tannins, flavonoids, saponins, polyphenols, sterols, ellagic acid, salicylic acid, and fibers		Aqueous extract of *S. cumini* seeds Dose: 2.5 g and 5 g/kg, BW Alcoholic extract of *E. jambolana* seeds Dose: 100 mg/kg, BW	Prince et al. (1998); Sharma et al. (2003)
Fenugreek (*Trigonella foenum-graecum*, family: Fabaceae)	Leaf and seeds	Zingirone, steroids, gingerol, eugenol, vanillin, and saponins	Ethanolic extract of seed Dose: 60–70 mg/kg, BW		El-Soud et al. (2007); Al-Daghri et al. (2012)
Bitter melon (*Momordica charantia*, family: Cucurbitaceae),	Seeds, leaves, and pulp	Phenolic compounds, triterpenoids, alkaloids, proteins, steroids, inorganic compounds, resins, saponins, and charantin	The aqueous extract reduced the blood glucose level in mice, which were previously glucose treated by oral as well as i.p. route. The fresh unripe whole fruits in powder or aqueous extract form (20 mg/kg, BW) significantly lowered blood glucose by 48% under fasting conditions		Joseph and Jini (2013); Alam et al. (2015); Jarouliya and Keservani (2019b)
Onion (*Allium cepa*, family: Amaryllidaceae)	Bulb	Alkaloids, terpenes, flavonoids, resins steroids, and cardiac glycosides		Onion Juice Dose: 1 mL/100 g BW	Gazuwa et al. (2013); El-Demerdash et al. (2015)

TABLE 10.2 *(Continued)*

Herbal sources	Part of the plant used	Phytochemicals	In vivo antidiabetic studies		References
			Streptozotocin-induced diabetic rats model	Alloxan-induced diabetic rats model	
Holy basil (*Ocimum sanctum*; family: Lamiaceae)	Leaves	Volatile oil, circimaritin, cirsilineol, apigenin, rosmeric acid, isothymusin, and campesterol	Ethanolic extract of leaves Dose:200 mg/kg, BW		Vats et al. (2004)
Gum arabic (*Acacia nilotica*, family: Fabaceae)	Pods, barks, and leaves	Alkaloids, gallic acid, tannins, and saponins		Aqueous extracts of leaves	Mukundi et al. (2015)
Aloe vera (*Barbadensis mill*, family: Asphodelaceae).	Gel obtained from leaves	Anthraquinones, glycosides, lipids, vitamins, gibberellins, pseudo-prototino-saponin AIII, prototino-saponins AIII, and sterols	Anthraquinones extracted from the pulp Dose: 300 mg/kg, BW		Abo-Youssef and Messiha (2013)
Mulberry (*Morinda citrifolia*, family: Moraceae) and white mulberry (*Morus alba*)	Leaves and fruits	Caffeic acid, rutin, astragalin, isoquercitrin, and kaempferol	Isolated flavonoids and terpenoids were found effective in type 2 diabetes. The protein is extracted from the leaves. Dose:35 mg/kg, BW		Lee et al. (2012); Shoaib et al. (2013); Jeszka Skowron et al. (2014)
Guava (*Psidium guajava*, family: Myrtaceae)	Leaves, bark, buds, flowers, twigs, fruits, and roots	Flavonoids, phenolic, tannic Oxalic and malic acid, amylase, phenyl propyl and butenyl acetate, tannins, resins, and calcium oxalate	Methanolic extracts from leaves Dose:250 mg/kg, BW	Methanolic extracts from leaves Dose:250 mg/kg, BW	Gutiérrez et al. (2008); Sengupta et al. (2011)
Radish (*Raphanus sativus*, family: Brassicaceae)	Leaves and roots	Anthocyanidin, anthocyanin, acetone, acetic acid, trifluoracetic acid, phenols, and sodium acetate	Aqueous extract of the root Dose:300 mg/kg, BW		Hanlon PR, Shukla S
Coriander (*Coriandrum sativum*, family:Umbelliferae)	Leaves, seeds and roots	Tannins, flavonoids, steroids, saponins, and amino acids	Methanolic extract exerted a significant dose-dependent hypoglycemic effect Dose: 200 and 400 mg/kg, BW		Uma B, Mazhar J

TABLE 10.2 *(Continued)*

Herbal sources	Part of the plant used	Phytochemicals	In vivo antidiabetic studies		References
			Streptozotocin-induced diabetic rats model	Alloxan-induced diabetic rats model	
Cumin (*S. cumini* or *Cuminum cyminum*, family: Umbelliferae)	Seeds	Tannins, triterpenoids, flavonoids, saponins, anthraquinones, phytosterol, and steroids	Cumin powder decreased blood glucose levels and glucosuria as well. Dose:1.25% Duration: 8 weeks		Kumar et al. (2009); Willatgamuwa et al. (1998)
Cinnamon (*Cinnamomum cassia*, family: Lauraceae)	Stems and seeds	Methyl-hydroxy chalcone, tannins, flavonoids, glycosides, terpenoids, coumarins, and anthraquinones	Methanol extracts Dose: 3 g/kg, BW		Shihabudeen et al. (2011), Khaki et al. (2014)
Garlic (*Allium sativun*, family: Amaryllidaceae)	Bulb	Terpenoids, phenolic compounds, alkaloids, steroids, saponins, carbohydrates, tannins, and flavonoids	Garlic oil Dose: 50 mg/kg, BW	Vitamin and minerals in the extract of garlic juice Dose:1 mL/100 g, BW	Divya et al. (2017), El-Demerdash et al. (2005); Ohaeri et al. (2001)
Ginger (*Zingiber officinale*, family Zingiberaceae)	Rhizomes	Phenols, terpenoids, tannins, flavonoids, and saponins	Ethanolic extract of ginger–garlic Dose:5 00 mg/kg, BW		Hassan et al. (2017); Islam et al. (2005)
Neem (*Azadirachta indica*, family: Meliaceae)	Leaf, flowers, stem, seeds, bark, and root	Isoazadirolide, nimbandiol, nimbolide, nimbinene, nimbaflavone, and quercetin	Ether extracts of seed Dose: 2 g/kg, BW		Subapriya and Nagini (2005); Gupta et al. (2004)
Gymnema (*Gymnema slyvestre*, family: Asclepiadaceae)	Leaves	Coumarins, flavonoids, saponins, steroids, terpenoids, alkaloids, and tannins	Ethanol extract of leaves		Gopinath et al. (2012); Kang et al. (2013)
Cardamom (*Elettaria cardamomum*, family: Zingiberaceae)	Leaf, fruits, and seeds	Flavonoids, alkaloids, saponins, tannins, coumarins, terpenoids, quinone, glycoside, phenol, steroids, and anthocyanin		The mixture of ginger flavonoids, cinnamon, and cardamom Dose: 250 mg/kg BW	Vishwakarma et al. (2011); El-Yamani (2014)

TABLE 10.2 *(Continued)*

Herbal sources	Part of the plant used	Phytochemicals	In vivo antidiabetic studies		References
			Streptozotocin-induced diabetic rats model	Alloxan-induced diabetic rats model	
Sesame (*Sesamum indicum*, family: Pedaliaceae)	Seed and leaves	Triterpenes, flavonoids, polyphenols, and protein	Ethanolic extract of seeds Dose: 500 mg/kg, BW		Park et al. (2010); Bhuvaneswari and Krishnakumari (2012)
Black pepper (*Piper nigrum*, family:Piperaceae)	Seeds	Flavonoids, lignans, alkaloids, steroids, terpene, and phenolic compounds	The aqueous extract significantly affects antioxidant pathways in diabetes rats. Dose: 300 mg/kg BW		Rauscher et al. (2000); Ahmad et al. (2012)
Curry leaves (*Murraya koenigii/ Rutaceae*) diabetic rats	Leaves	Carbohydrates, alkaloids, phytosterols, alcohol, flavonoids, saponins, tannins, glycosides, and carbohydrates		Aqueous extract Dose:300 mg/kg, BW Dried powdered curry leaves Dose: 35 mg/kg	Arulselvan and Subramanian (2007); Kesari et al. (2007); Yadav et al. (2002)
Ginseng (*Panax quinquefolius*, family: Araliaceae)	Berries, Leaves and flowers	Polysaccharides, polyacetylenes, saponins, nitrogen-containing compounds, triterpene, and phenolic compounds	Aqueous extracts also improved renal damage in diabetic rats Dose: 100 mg/kg		Angelova et al. (2008); Kim et al. (2007)
Stinging nettle (*Urtica pilulifera*, family: Urticaceae)	Flowers, Leaves, and stem	Flavonoids, phenylpropane, histamine, acetylcholine (ACh), caffeic, chlorogenic, and fatty acids	Acetic extracts Dose:100 mg/kg, BW		Kavalalı et al. (2003); Farzami et al. (2003)

prone to develop various diseases such as physical disorders, and serious psychological disorders. Obesity (Jarouliya and Keservani, 2019a) is one of them and is considered a modern lifestyle disease, one of the fastest-growing pathologies in recent years. As per the World Health Organization (WHO) obesity or overweight may be defined as excessive deposition of fat in the body due to an increased intake of energy rather than its consumption, which can be harmful to health, and also leads to serious health problems such as arteriosclerosis, cardiac disease, stroke, high blood pressure, nephritis, tumor, chronic obstructive pulmonary disease (COPD), allergies, hearing impairment, depression, anxiety disorders, Alzheimer's disease, chronic backache (Bischoff et al., 2017). In 2016, WHO also reported a huge population of adults (about 1.9 billion) falls into the overweight category, and about 600 million fall under the obese category (Obesity and Overweight, 2021).

Obesity has become one of the most serious health problems in modern society as it greatly enhances the risk of numerous health diseases, from diabetes and hypertension to insomnia, osteoarthritis, back pain in the lower region, and gallbladder disease. Obesity is also closely associated with cancer disease; coronary artery disease leads to diminished life expectancy. While body mass index (BMI) is generally used to calculate the amount of body fat and the extent of obesity, the overall risk is also affected by the distribution of fat in the body (https://www.aace.com/sites/all/files/Obesity Guidelines Algorithm).

TABLE 10.3 The Categories of Obesity from the AACE*.

Diagnosis	Body mass index (BMI)	Clinical observation
Overweight person	≥25 to 29.9	Mo complications
Obesity of stage 0	Greater than or equal to 30	No complications
Obesity of stage 1	Greater than or equal to 25	One or more mild-to-moderate complications
Obesity of stage 2	Greater than or equal to 25	One or more severe complications

*AACE, American Association of Clinical Endocrinologists.

There are so many options to treat obesity, which include dietary control, physical exercise, changing lifestyle, clinically approved weight-loss medications, and surgeries for weight loss (Chandrasekaran et al., 2012). Natural or herbal supplement (Nutraceuticals) plays an important role in the effective management of weight of individuals (Sharma et al., 2020; Rane et al., 2020a, 2020b, 2020c).

Naturally occurring supplements present in the market primarily help the patient fight the battle against obesity. Various natural plant resources such as herbs, fruits (Keservani et al., 2016b), vegetables, functional fatty acids such as polyunsaturated and conjugated fatty acids, and other naturally available dietary compounds have been used as antiobesity products (Sun et al., 2016).

10.5 ANTIOBESITY: MECHANISMS OF NATURAL DIETARY OR NUTRACEUTICALS AND INCREASE IN ENERGY EXPENDITURE

Energy expenditure is the amount of energy required to carry out normal physical functions like breathing (Respiration), blood circulation, digestion of foods, and physical exercise. Energy is measured in terms of calories, so the total energy expenditure in daily routine equals the number of calories burned daily.

Energy consumption is classified into three categories:

a. Physical activity
b. Obligatory energy expenditure
c. Adaptive thermogenesis

Currently, available antiobesity formulations regulate body weight by increasing obligatory energy expenditure (Kajimura et al., 2014).

10.6 ANTIOBESITY PRODUCTS FROM NATURAL RESOURCES (PARK ET AL., 2011)

The antiobesity products currently available on the market are classified into three categories:

1. Food ingredients: Products prepared from fruits like citrus and berries, some grains such as soybean, vegetables, or beverage drinks (tea leaves) are safer and acceptable to consumers.
2. Herbal ingredients: The herbal remedy contains different herbs, like turmeric and mulberry leaves are generally utilized to treat obese patients.
3. Functional supplements: the probiotic and calcium supplements also showed antiobesity effects.

TABLE 10.4 Nutraceutical Herbs and Their Components with Potential Antiobesity Effects.

Natural herbs	Active constituents	Formulation and does	Experiments	Activity/inference	References
Shiikuwasa (*Citrus depressa*)	Flavonoids	Methanolic extract Dose: 1.0–1.5%	Animal: Male ICR mice Ingested with high-fat content (HFC) (40%, w/w) Duration: one month.	10.0% Reduction in body weight (BW), and 50.9% reduction in organ weight	Lee et al. (2011)
Blueberry (*Vaccinium ashei*) and mulberry (*Morus australis*)	Anthocyanins	Juice	Animal: Male C57BL/6 mice Ingested with HFC (45%, w/w) Duration: 3 months	7.30–9.81% reduction in BW, and reduction in WAT adipocyte sizes.	Wu et al. (2013)
Soybean (*Glycine max*)	Protein isolated	Isolated protein (30%)	Animal: Male Sprague Dawley rats Ingested with HFC (25%, w/w) Duration: 180 days	18.9% reduction in BW, 26.6% in body fat, and higher thermogenin expression	Torre-Villalvazo, et al. (2008)
Coffee (*Coffea arabica*)	Caffeoyl, quinic acids	Aqueous extract Dose: 0.5–1.0%	Animal: Male C57BL/6J mice Ingested with HFC (30%, w/w) Duration: 2–15 weeks	Significant reduction in BW 14.3, 16.2%, and in organ WAT weight and also decreased liver fat	Murase et al. (2011)
Green tea (*Camellia sinensis*)	Polysaccharides, caffeine, and catechins	Aqueous extract Dose: 400–800 mg/kg	Animal: Male Sprague Dawley rats Ingested with HFC (10%w/w) along with egg yolk powder (10%, w/w) Duration: 6 weeks	11.3–16.9% reduction in BW, body fat index, and WAT adipocyte sizes were also reduced.	Xu et al. (2015)
Lotus leaf with taurine (*Nelumbo nucifera*)	Alkaloids, flavonoids, triterpenoids, polyphenols, steroids, glycosides, and taurine	Aqueous extract Dose: 400 mg/kg	Animal: Male Sprague Dawley rats Ingested with HFC (20% w/w) and taurine Duration: 6 weeks	Reduction in BW, and WAT adipocyte sizes and number were reduced.	Du et al. (2010)

TABLE 10.4 *(Continued)*

Natural herbs	Active constituents	Formulation and does	Experiments	Activity/inference	References
Ginger (*Z. officinale*)	Gingerol, paradol, and shogoal	Powder (5%)	Animal: Male albino rats Fed with HFC (30%, w/w) Duration: 5 weeks	Significant reductions in BW (38.6%) and pancreatic lipase activity (45.7%) were noticed	Mahmoud and Elnour (2013)
Black wattle (*Acacia mollissima*)	Robinetinidol and fisetinidol	Aqueous extract (5%)	Animal: Diabetic KKAy mice Ingested with HFC (60%, w/w) Duration: 7 weeks	Reduction in BW (23.2%) and higher expression of energy expenditure-related genes were determined.	Ikarashi et al. (2011)
Chili pepper (*Capsicum annum*)	Capsaicin	Does: 10 mg/kg	Animal: Male Sprague Dawley rats Ingested with HFC (45%, w/w) Duration: 9 weeks	Reduction in BW (8%) ; was observed along with a reduction in WAT weight and adipocyte sizes	Joo et al. (2010)
Coptis Root (*Rhizoma coptidis*)	Berberine	Ethanolic extracts Dose: 200 mg/kg	Animals: Male C57BL/6J mice HFD (16.2%, w/w) Duration: 1.5 months	Reduction in BW and organ WAT weight were observed.	Xie et al. (2011)
Turmeric (*C. longa*)	Curcumin	Ethanolic extract (50%)	Animal: Male Sprague Dawley rats HFD (60%, w/w) Duration: 3 months	A reduction in BW gain was observed by 15.9% and a 31.3% reduction in organ WAT weight was found	Kim et al. (2016)
White mulberry (*Morus alba*)	Rutin, resveratrol anthocyanin, and 1-deoxynojirimycin	Leaves extract Dose:133–333 mg/kg Fruit extract Dose:67–167 mg/kg	Animals: Male C57BL/6 mice HFD (45%) Duration: 3 months	Overall reduction in BW by 53.5%, in organ WAT weight was observed	Lim et al. (2013)

10.7 CONCLUSIONS

Nutraceuticals are functional food supplements obtained from natural sources and have been studied for different lifestyle diseases such as DM and obesity for centuries. The absence of a cure and avoidable side effects of synthetic medicines put efforts towards preventive and effective antidiabetic nutraceuticals. The different plants and their extracts or isolated compounds, such as *Gymnema sylvestre*, *Momordica charantia*, *Nigella sativa*, *Allium sativum*, and *C. cassia* have an effective antidiabetic mechanism, and their efficacy in DM was proven by good-quality clinical trials and previous in vivo animal studies. The Antidiabetic effects of these nutraceutical herbs are because of increased insulin secretion, decreased insulin resistance, and reduced intestinal absorption of glucose.

Obesity is a complex, chronic disorder primarily due to certain factors, which include dietary, lifestyle, genetic, and environmental. The different scientists reported numerous in vivo animal studies along with clinical trials regarding the efficacy of weight loss.

The different mechanisms by which the antiobesity agent acts are by stimulating thermogenesis, lowering lipid formation, increasing lipid breakdown of lipolysis, suppressing the appetite, and decreasing lipid absorption. So the use of naturally occurring antiobesity preparations can be a supportive medicine for weight loss in overweight people. Natural materials such as nutraceuticals provide antiobesity and other health benefits, such as antidiabetic and antihyperlipidemic activities. Therefore, the availability of many natural resources will be valuable for developing novel antiobesity products.

Hence, physicians can use these nutraceuticals as supporting agents, along with currently available antiobesity and hypoglycaemic compounds for the effective management of type 2 diabetic and obese patients.

KEYWORDS

- **obesity**
- **overweight**
- **arteriosclerosis**
- **diabetes mellitus**
- **antidiabetic herbs**
- **adaptive thermogenesis**

REFERENCES

Abo-Youssef, A. M.; Messiha, B. A. Beneficial Effects of Aloe Vera in Treatment of Diabetes: Comparative In Vivo and In Vitro Studies. *Bull. Faculty Pharm. Cairo Univ.* **2013,** *51*, 7–11.

Ahmad, N.; Fazal, H.; Abbasi, B. H.; Farooq, S.; Ali, M.; Khan, M. A. Biological Role of *Piper nigrum* L. (Black Pepper): A Review. *Asian Pac. J. Trop. Biomed.* **2012,** 2: S1945–S1953.

Ahmed, M. F.; Kazim, S. M.; Ghori, S. S.; Mehjabeen, S. S.; Ahmed, S. R.; Ali, S. M. et al. Anti-Diabetic Activity of *Vinca rosea* Extracts in Alloxaninduced Diabetic Rats. *Int. J. Endocrinol.* **2010,** 1–6. DOI: 10.1155/2010/841090.

Al-Daghri, N. M.; Alokail, M. S.; Alkharfy, K. M.; Mohammed, A. K.; Abd Alrahman, S. H.; Yakout, S. M. et al. Fenugreek Extract as an Inducer of Cellular Death via Autophagy in Human T Lymphoma Jurkat Cells. *BMC J. Altern. Complement.* **2012,** *12*, 202.

Alam, M. A.; Uddin, R.; Subhan, N.; Rahman, M. M.; Jain, P.; Reza, H. M. Beneficial Role of Bitter Melon Supplementation in Obesity and Related Complications in Metabolic Syndrome. *J. Lipids* **2015,** 1–18.

Angelova, N.; Kong, H. W.; Heijden, R.; Yang, S. Y.; Choi, Y. H.; Kim, H. K. et al. Recent Methodology in the Phytochemical Analysis of Ginseng. *Phytochem. Anal.* **2008,** *19*, 2–16.

Arulselvan, P.; Subramanian, S. P. Beneficial Effects of *Murraya koenigii* Leaves on Anti-Oxidant Defense System and Ultrastructural Changes of Pancreatic β-Cells in Experimental Diabetes in Rats. *Chem. Biol. Interact.* **2007,** *165*, 155–164.

Bhuvaneswari, P.; Krishnakumari, S. Antihyperglycemic Potential of *Sesamum indicum* (Linn) Seeds in Streptozotocin Induced Diabetic Rats. *Int. J. Pharm. Sci.* **2012,** *4*, 527–531.

Bischoff, S. C. et al. Towards a Multidisciplinary Approach to Understand and Manage Obesity and Related Diseases. *Clin. Nutr.* **2017,** *36* (4), 917–938.

Chandrasekaran, C. V.; Vijayalakshmi M A, Prakash, K.; Bansal, V. S.; Meenakshi, J.; Amit, A. Review Article: Herbal Approach for Obesity Management. *Am. J. Plant Sci.* **2012,** *3*, 1003–1014.

Dasgupta, A. Review of Abnormal Laboratory Test Results and Toxic Effects Due to Use of Herbal medicines. *Am. J. Clin. Pathol.* **2003,** *120*, 127–137.

Davì, G.; Santilli, F.; Patrono, C. Nutraceuticals in Diabetes and Metabolic Syndrome. *Cardiovasc. Ther.* **2010,** *28* (4), 216–226.

Defelice, S. L. The Nutraceutical Revolution—Its Impact on Food–Industry R–and–D. *Trends Food Sci. Tech.* **1995,** *6* (2), 59–61.

Derosa, G.; Limas, C. P.; Macías, P. C. et al. Dietary and Nutraceutical Approach to Type 2 Diabetes. *Arch. Med. Sci.* **2014,** *10* (2), 336–344.

Divya, B. J.; Suman, B.; Venkataswamy, M.; Thyagaraju, K. A Study on Phytochemicals, Functional Groups and Mineral Composition of *Allium sativum* (Garlic) Cloves. *Int. J. Curr. Pharm. Res.* **2017,** *9*, 42–45.

Du, H.; You, J. S.; Zhao, X.; Park, J. Y.; Kim, S. H.; Chang, K. J. Anti-Obesity and Hypolipidemic Effects of Lotus Leaf Hot Water Extract with Taurine Supplementation in Rats Fed a High Fat Diet. *J. Biomed. Sci.* **2010,** *17*.

El-Demerdash, F. M.; Yousef, M. I.; El-Naga, N. A. Biochemical Study on the Hypoglycemic Effects of Onion and Garlic in Alloxaninduced Diabetic Rats. *Food Chem. Toxicol.* **2005,** *43*, 57–63.

El-Demerdash, F. M.; Yousef, M. I.; El-Naga, N. A. Biochemical Study on the Hypoglycemic Effects of Onion and Garlic in Alloxan Induced Diabetic Rats. *Food Chem Toxicol.* **2005,** *43*, 57–63.

El-Soud, N. H.; Khalil, M. Y.; Hussein, J. S.; Oraby, F. S.; Farrag, A. H. Anti-Diabetic Effects of Fenugreek Alkaloid Extract in Streptozotocin-Induced Hyperglycemic Rats. *J. Appl. Sci. Res.* **2007,** *3*, 1073–1083.

El-Yamani, M. A. Cinnamon, Cardamom and Ginger Impacts as Evaluated on Hyperglycemic Rats. *Res. J. Specific Educ.* **2011,** *20*, 665–678.

Farzami, B.; Ahmadvand, D.; Vardasbi, S.; Majin, F. J.; Khaghani, S. H. Induction of Insulin Secretion by a Component of Urtica Dioica Leave Extract in Perfused Islets of Langerhans and Its In Vivo Effects in Normal and Streptozotocin Diabetic Rats. *J. Ethnopharmacol.* **2003,** *89*, 47–53.

Gajalakshmi, S.; Vijayalakshmi, S.; Devi, R. V. Pharmacological Activities of *Catharanthus roseus*: A Perspective Review. *Int. J. Pharma. Bio. Sci.* **2013,** *4*, 431–439.

Gazuwa, S. Y.; Makanjuola, E. R.; Jaryum, K. H.; Kutshik, J. R.; Mafulul, S. G. The Phytochemical Composition of *Allium cepa*/*Allium sativum* and the Effect of Their Aqueous Extracts (Cooked and Raw Forms) on the Lipid Profile and Other Hepatic Biochemical Parameters in Female Albino Wistar Rats. *Asian J. Exp. Biol. Sci.* **2013,** *4*, 406–410.

Golla, U. Emergence of Nutraceuticals as the Alternative Medications for Pharmaceuticals. *Int. J. Complement. Altern. Med.* **2018,** *11* (3), 155–158. DOI: 10.15406/ijcam.2018.11.00388.

Gopinath, S.; Rakesh, C.; Murthy, T.; Dayananda, K. Preliminary Phytochemical Evaluation of Leaf Extracts of *Gymnema sylvestre*, *Phyllanthus amarus*, *Phyllanthus reticulatus* of Siddarabetta, Tumkur district, Karnatka. *J. Pharmacogn. Phytochem.* **2012,** *4*, 109–111.

Gupta, S.; Kataria, M.; Gupta, P. K.; Murganandan, S.; Yashroy, R. C. Protective Role of Extracts of Neem Seeds in Diabetes Caused by Streptozotocin in Rats. *J. Ethnopharmacol.* **2004,** *90*, 185–189.

Gutiérrez, R. M.; Mitchell, S.; Solis, R. V. *Psidium guajava*: A Review of Its Traditional Uses, Phytochemistry and Pharmacology. *J. Ethnopharmacol.* **2008,** *117*, 1–27.

Hanlon, P. R.; Barnes, D. M. Phytochemical Composition and Biological Activity of 8 Varieties of Radish (*Raphanus sativus* L.) Sprouts and Mature Taproots. *J. Food Sci.* **2011,** *76*, 185–192.

Hassan, N. A.; Karunakaran, R.; Sankar, U.; Aye, K. M. Antiinflammatory Effect of *Zingiber officinale* on Sprague Dawley Rats. *Asian J. Pharm. Clin. Res.* **2017,** *10*, 353–355.

https://www.aace.com/sites/all/files/ObesityGuidelinesAlgorithmslidesFINAL2016.pdf.

Ikarashi, N.; Toda, T.; Okaniwa, T.; Ito, K.; Ochiai, W.; Sugiyama, K. Anti-Obesity and Anti-Diabetic Effects of Acacia Polyphenol in Obese Diabetic KKAy Mice Fed High-Fat Diet. *Evid. Based Complement. Altern. Med.* **2011,** *2011*, 952031. DOI: 10.1093/ecam/nep241

Islam, M. S.; Choi, H. Comparative Effects of Dietary Ginger (*Zingiber officinale*) and Garlic (*Allium sativum*) Investigated in a Type 2 Diabetes Model of Rats. *J. Med. Food* **2008,** *11*, 152–159.

Jarouliya, U.; Keservani, R. K. Pathways Leading to Child Obesity. In: *Global Perspectives on Childhood Obesity: Current Status, Consequences and Prevention*; Bagchi, D.; Ed.; Academic Press, Elsevier, 2019a; Chapter 12, pp 137–146.

Jarouliya, U.; Keservani, R. K. Protein Function as Cell Surface and Nuclear Receptor in Human Diseases. In: *Medicinal Chemistry with Pharmaceutical Product Development*; Bharti, S.; Mahapatra, D. K.; Eds.; Apple Academic Press, CRC Press, Taylor & Francis Group, 2019b, Chapter 1, pp 1–32.

Jeszka Skowron, M.; Flaczyk, E.; Jeszka, J.; Krejpcio, Z.; Krol, E.; Buchowski, M. S. Mulberry Leaf Extract Intake Reduces Hyperglycaemia in Streptozotocin (STZ)-Induced Diabetic Rats Fed High-Fat Diet. *J. Funct. Foods* **2014,** *8*, 9–17.

Joo, J. I.; Kim, D. H.; Choi, J. W.; Yun, J. W. Proteomic Analysis for Anti-Obesity Potential of Capsaicin on White Adipose Tissue in Rats Fed with a High Fat Diet. *J. Proteome Res.* **2010,** *9* (6), 2977–2987. DOI: 10.1021/pr901175w

Joseph, B.; Jini, D. Antidiabetic Effects of *Momordica charantia* (Bitter Melon) and Its Medicinal Potency. *Asian Pac. J. Trop. Dis.* **2013,** *3*, 93–102.

Kajimura, S.; Saito, M. A New Era in Brown Adipose Tissue Biology: Molecular Control of Brown Fat Development and Energy Homeostasis. *Annu. Rev. Physiol.* **2014,** *76*, 225–249.

Kang, M. H.; Lee, M. S.; Choi, M. K.; Min, K. S.; Shibamoto, T. Hypoglycemic Activity of Gymnema Sylvestre Extracts on Oxidative Stress and Anti-Oxidant Status in Diabetic Rats. *J. Agric. Food Chem.* **2012,** *60*, 2517–2524.

Kaur, G.; Mukundan, S.; Wani V.; Kumar, M. S. Nutraceuticals in the Management and Prevention of Metabolic Syndrome. *Austin J. Pharmacol. Ther.* **2015,** *3* (1), 1063.

Kavalalı, G.; Tuncel, H.; Göksel, S.; Hatemi, H. H. Hypoglycemic Activity of *Urtica pilulifera* in Streptozotocin-Diabetic Rats. *J. Ethnopharmacol.* **2003,** *84*, 241–245.

Kaveeshwar, S. A.; Cornwall, J. The Current State of Diabetes Mellitus in India. *Aust. Med. J.* **2014,** *7* (1), 45–48.

Kesari, A. N.; Kesari, S.; Singh, S. K.; Gupta, R. K.; Watal, G. Studies on the Glycemic and Lipidemic Effect of *Murraya koenigii* in Experimental Animals. *J. Ethnopharmacol.* **2007,** *112*, 305–311.

Keservani, R. K.; Sharma, A. K. Flavonoids: Emerging Trends and Potential Health Benefits. *J. Chin. Pharm. Sci.* **2014,** *23* (12), 815.

Keservani, R. K.; Kesharwani, R. K.; Vyas, N.; Jain, S.; Raghuvanshi, R.; Sharma, A. K. Nutraceutical and Functional Food as Future Food: A Review. *Der Pharmacia Lettre* **2010a,** *2* (1), 106–116.

Keservani, R.K.; Kesharwani, R.K.; Sharma, A.K.; Vyas, N.; Chadoker, A. Nutritional Supplements: An Overview. *Int. J. Curr. Pharm. Rev. Res.* **2010b,** *1* (1), 59–75.

Keservani, R. K.; Sharma, A. K.; Kesharwani, R. K. Nutraceutical and Functional Foods for Cardiovascular Health. In: *Food Process Engineering*; Apple Academic Press, CRC Press, 2016a; pp 257–278, 291–312.

Keservani, R. K.; Sharma, A. K.; Kesharwani, R. K. Medicinal Effect of Nutraceutical Fruits for the Cognition and Brain Health. *Scientifica* **2016b,** Article ID 3109254, 10 pages. DOI: 10.1155/2016/3109254.

Keservani, R. K.; Sharma, A. K.; Kesharwani, R. K. An Overview and Therapeutic Applications of Nutraceutical and Functional Foods. Recent Adv. Drug Deliv. Technol. **2017,** 160–201.

Keservani, R. K.; Sharma, A. K.; Kesharwani, R. K. (Eds.). *Nutraceuticals and Dietary Supplements: Applications in Health Improvement and Disease Management*; CRC Press, 2020.

Khaki, A.; Khaki, A. A.; Hajhosseini, L.; Golzar, F. S.; Ainehchi, N. The Anti-Oxidant Effects of Ginger and Cinnamon on Spermatogenesis Dysfunction of Diabetes Rats. *Afr. J. Tradit. Complement. Altern. Med.* **2014,** *11*, 1–8.

Khan, V.; Najmi, A. K.; Akhtar, M.; Aqil, M.; Mujeeb, M.; Pillai, K. K. A Pharmacological Appraisal of Medicinal Plants with Anti-Diabetic Potential. *J. Pharm. Bioallied Sci.* **2012,** *4*, 27–42.

Kim, H. Y.; Kang, K. S.; Yamabe, N.; Nagai, R.; Yokozawa, T. Protective Effect of Heat-Processed American Ginseng Against Diabetic Renal Damage in Rats. *J. Agric. Food Chem.* **2007,** *55*, 8491–8497.

Kim, J. H.; Kim, O. K.; Yoon, H. G. et al. Anti-Obesity Effect of Extract from Fermented *Curcuma longa* L. Through Regulation of Adipogenesis and Lipolysis Pathway in High-Fat Diet-Induced Obese Rats. *Food Nutr. Res.* **2016,** *60*, 30428. DOI: 10.3402/fnr.v60.30428.

King, H.; Aubert, R. E.; Herman, W. H. Global Burden of Diabetes, 1995– 2025: Prevalence, Numerical Estimates, and Projections. *Diab. Care* **1998,** *21* (9), 1414–1431.

Kochhar, A.; Pathak, N. Nutraceutical Based Approach to Combat Diabetes Mellitus. *Curr. Res. Diab. Obes. J.* **2018,** *8* (3), 555739. DOI: 10.19080/CRDOJ.2018.08.555739.

Kumar, A.; Ilavarasan, R.; Jayachandran, T.; Decaraman, M.; Aravindhan, P.; Padmanabhan, N. et al. Phytochemicals Investigation on a Tropical Plant, *Syzygium cumini* from Kattuppalayam, Erode District, Tamil Nadu, South India. *Pak. J. Nutr.* **2009,** *8*, 83–85.

Lee, S. Y.; Park, S. L.; Hwang, J. T.; Yi, S. H.; Nam, Y. D.; Lim, S. I. Anti-Diabetic Effect of *Morinda citrifolia* (Noni) Fermented by Cheonggukjang in KK-Ay Diabetic Mice. *J. Evid. Based Complement. Altern. Med.* **2012,** 2256–2264. http://dx.doi.org/10.1155/ 2012/ 163280.

Lee, Y. S.; Cha, B. Y.; Saito, K.; Choi, S.S, Wang, X. X.; Choi, B K, Yonezawa, T, Teruya, T, Nagai, K.; Woo, J. T. Effects of a *Citrus depressa* Hayata (Shiikuwasa) Extract on Obesity in High-Fat Diet-Induced Obese Mice. *Phytomedicine* **2011,** *18*, 648–654.

Lim, H. H.; Lee, S. O.; Kim, S. Y.; Yang, S. J.; Lim, Y. Anti-Inflammatory and Anti-Obesity Effects of Mulberry Leaf and Fruit Extract on High Fat Diet-Induced Obesity. *Exp. Biol. Med. (Maywood)*. **2013,** *238* (10), 1160–1169. DOI: 10.1177/1535370213498982.

Lorenzati, B.; Zucco, C.; Miglietta, S, Lamberti, F.; Bruno, G. Oral Hypoglycemic Drugs: Pathophysiological Basis of Their Mechanism of Action. *Pharmaceuticals* **2010,** *3*, 3005–3020.

MacDonald, L.; Foster, B. C.; Akhtar, H. Food and Therapeutic Product Interactions—A Therapeutic Perspective. *J. Pharm. Pharm. Sci.* **2009,** *12*, 367–377.

Mahmoud, R.; Elnour, W. Comparative Evaluation of the Efficacy of Ginger and Orlistat on Obesity Management, Pancreatic Lipase and Liver Peroxisomal Catalase Enzyme in Male Albino Rats. *Eur. Rev. Med. Pharmacol. Sci.* **2013,** *17*, 75–83.

Makhija, I. K.; Sharma, I. P.; Khamar, D. Phytochemistry and Pharmacological Properties of *Ficus religiosa*: An Overview. *Ann. Biol. Res.* **2010,** *1*, 171–180.

Mazhar, J.; Mazumder, A. Evaluation of Anti-Diabetic Activity of Methanolic Leaf Extract of *Coriandrum sativum* in Alloxan Induced Diabetic Rats. *Res. J. Pharm. Biol. Chem. Sci.* **2013,** *4*, 500–507.

Mukundi, M. J.; Piero, N. M.; Mwaniki, N. E.; Murugi, N. J.; Daniel, A. S.; Peter, G. K. et al. Anti-Diabetic Effects of Aqueous Leaf Extracts of *Acacia nilotica* in Alloxan Induced Diabetic Mice. *J. Diab. Metab.* **2015,** *6*, 1–6.

Murase, T, Misawa, K.; Minegishi, Y.; Aoki, M.; Ominami, H, Suzuki, Y.; Shibuya, Y.; Hase, T. Coffee Polyphenols Suppress Diet-Induced Body Fat Accumulation by Downregulating srebp-1c and Related Molecules in C57BL/6J Mice. *Am. J. Physiol. Endocrinol. ME Table* **2011,** *300*, E122–E133.

Nammi, S.; Boini, M. K.; Lodagala, S. D.; Behara, R. B. The Juice of Fresh Leaves of *Catharanthus roseus* Linn. Reduces Blood Glucose in Normal and Alloxan Diabetic Rabbits. *BMC Complement. Altern. Med.* **2003,** *3*, 1–4.

Obesity and Overweight. (n.d.). https://www.who.int/news-room/fact-sheets/detail/obesity-and-overweight (accessed 24 Nov 2022).

Ohaeri, O. C. Effect of Garlic Oil on the Levels of Various Enzymes in the Serum and Tissue of Streptozotocin Diabetic Rats. *Biosci. Rep.* **2001,** *21*, 19–24.

Pandit, R.; Phadke, A.; Jagtap, A. Antidiabetic Effect of *Ficus religiosa* Extract in Streptozotocin-Induced Diabetic Rats. *J. Ethnopharmacol.* **2010,** *128*, 462–466.

Park, J. P.; Kim, J. H.; Park, M. K.; Yun, J. W. Potential Agents for Cancer and Obesity Treatment with Herbal Medicines from the Green Garden. *Biotechnol. Bioprocess Eng.* **2011,** *16*, 1065–1076.

Park, S. H.; Ryu, S. N.; Bu, Y.; Kim, H.; Simon, J. E.; Kim, K. S. Anti-Oxidant Components as Potential Neuroprotective Agents in Sesame (*Sesamum indicum*). *Food Rev. Int.* **2010,** *26*, 103–121.

Patel, D. K.; Prasad, S. K.; Kumar, R. An overview on Anti-Diabetic Medicinal Plants Having Insulin Mimetic Property. *Asian Pac. J. Trop. Biomed.* **2012,** *2* (4), 320–330.

Prince, P. S.; Menon, V. P.; Pari, L. Hypoglycaemic Activity of *Syzigium cumini* Seeds: Effect on Lipid Peroxidation in Alloxan Diabetic Rats. *J. Ethnopharmacol.* **1998,** *61*, 1–7.

Rafieian–Kopaei, M.; Nasri, H. Ginger and Diabetic Nephropathy. *J. Renal. Inj. Prev*. **2013,** *2* (1), 9–10.

Rane, B. R.; Patil, A. S.; Keservani, R. K.; Jain, A. S. Novel Approaches in Herbal Formulation. In: *Enhancing the Therapeutic Efficacy of Herbal Formulations through Novel Drug Delivery Systems*; Kesharwani, R. K., Keservani, R. K., Sharma, A. K., Eds.; IGI Global International Publisher: Pennsylvania, 2020a; Chapter 2, pp 43–68.

Rane, B. R.; Bharath, M. S.; Patil, R. R.; Keservani, R. K.; Jain, A. S. Novel Approaches in Nutraceuticals. In: *Enhancing the Therapeutic Efficacy of Herbal Formulations through Novel Drug Delivery Systems*; Kesharwani, R. K., Keservani, R. K., Sharma, A. K., Eds.; IGI Global International Publisher: Pennsylvania, 2020b; Chapter 11, pp 241–266.

Rane, B. R.; Tadavi, S. A.; Keservani, R. K. Naturopathy. In: *Herbal Product Development*; Sharma, A. K., Keservani, R. K., Gautam, S. P., Eds.; Apple Academic Press, CRC Press, Taylor & Francis Group, 2020c; Chapter 12, pp 321–347.

Rauscher, F. M.; Sanders, R. A.; Watkins, J. B. Effects of Piperine on Anti-Oxidant Pathways in Tissues from Normal and Streptozotocin-Induced Diabetic Rats. *J. Biochem. Mol. Toxicol.* **2000,** *14*, 329–334.

Sengupta, R.; Sawant, C. S.; Karmarkar, S. M.; Bhagwat, A. M. Antihyperglycaemic Anti Diabetic Effect of the Leaf Extracts of *Albizzia lebbeck* linn. (benth) and *Psidium guajava* linn. On alloxan and Streptozotocin-Induced Diabetic Mice. *Asian J. Pharm. Clin. Res.* **2011,** *4*, 129–131.

Sharma, S. B.; Nasir, A.; Prabhu, K. M.; Murthy, P. S.; Dev, G. Hypoglycaemic and Hypolipidemic Effect of Ethanolic Extract of Seeds of *Eugenia jambolana* in Alloxan-Induced Diabetic Rabbits. *J. Ethnopharmacol.* **2003,** *85*, 201–206.

Sharma, A. K.; Keservani, R. K.; Gautam Surya, P. *Herbal Product Development*; Apple Academic Press, CRC Press, Taylor & Francis Group, 2020.

Shihabudeen, H. M.; Priscilla, D. H.; Thirumurugan, K. Cinnamon Extract Inhibits α-Glucosidase Activity and Dampens Postprandial Glucose Excursion in Diabetic Rats. *Nutr. Metab.* **2011,** *8*, 1–11.

Shoaib, Z. M.; Muhammad, F.; Javed, I.; Akhtar, M.; Khaliq, T.; Aslam, B. et al. White Mulberry (*Morus alba*): A Brief Phytochemical and Pharmacological Evaluations Account. *Int. J. Environ. Agric. Res.* **2013,** *15*, 612–620.

Shukla, S.; Chatterji, S.; Mehta, S.; Rai, P. K.; Singh, R. K.; Yadav, D. K. et al. Anti-Diabetic Effect of *Raphanus sativus* Root Juice. *Pharm. Biol.* **2011,** *49*, 32–37.

Singh, P.; Kesharwani, R. K.; Keservani, R. K. Antioxidants and Vitamins: Roles in Cellular Function and Metabolism. In: *Sustained Energy for Enhanced Human Functions and Activity*; Academic Press, 2017a; pp 385–407.

Singh, P.; Kesharwani, R. K.; Keservani, R. K. Protein, Carbohydrates, and Fats: Energy Metabolism. In: *Sustained Energy for Enhanced Human Functions and Activity*; Academic Press, 2017b; pp 103–115.

Smith-Warner, S. A.; Elmer, P. J.; Tharp, T. M. et al. Increasing Vegetable and Fruit Intake: Randomized Intervention and Monitoring in an at–Risk Population. *Cancer Epidem. Biomar. Prev.* **2000,** *9* (3), 307–317.

Subapriya, R.; Nagini, S. Medicinal Properties of Neem Leaves: A Review. *Curr. Med. Chem. Anticancer Agents* **2005,** *5*, 149–156.

Sun N N, Wu T Y, Chau C F. Natural Dietary and Herbal Products in Anti-Obesity Treatment. Molecules (Basel, Switzerland), **2016,** 21 (10): 1351. https://doi.org/10.3390/molecules21101351

Torre-Villalvazo, I.; Tovar, A. R.; Ramos-Barragán, V. E.; Cerbón-Cervantes, M. A.; Torres, N. Soy Protein Ameliorates Metabolic Abnormalities in Liver and Adipose Tissue of Rats Fed a High Fat Diet. *J. Nutr.* **2008,** *138*, 462–468.

Uma, B.; Prabhakar, K.; Rajendran, S.; Sarayu, Y. L. Antimicrobial Activity and Phytochemical Analysis of *Coriandrum sativum* Against Infectious Diarrhea. *Ethnobot. Leaflets* **2009,** *13*, 590–594.

Vats, V.; Yadav, S. P.; Grover, J. K. Ethanolic Extract of *Ocimum sanctum* Leaves Partially Attenuates Streptozotocin-Induced Alterations in Glycogen Content and Carbohydrate Metabolism in Rats. *J. Ethnopharmacol.* **2004,** *90*, 155–160.

Vishwakarma, S.; Chandan, K.; Jeba, R. C.; Khushbu, S. Comparative Study of Qualitative Phytochemical Screening and Anti-Oxidant Activity of *Mentha arvensis*, *Elettaria cardamomum* and *Allium porrum*. *Indo Am. J. Pharm. Res.* **2014,** *4*, 2538–2556.

Willatgamuwa, S. A.; Platel, K.; Saraswathi, G.; Srinivasan, K. Antidiabetic Influence of Dietary Cumin Seeds (*Cuminum cyminum*) in Streptozotocin Induced Diabetic Rats. *Nutr. Res.* **1998,** *18*, 131–142.

World Health Organization. *Obesity and Overweight*; Geneva: WHO, 2009.

Wu, T.; Tang, Q.; Gao, Z.; Yu, Z.; Song, H.; Zheng, X.; Chen, W. Blueberry and Mulberry Juice Prevent Obesity Development in C57BL/6 Mice. *PLoS ONE* **2013,** *8*, e77585.

Xie, W.; Gu, D.; Li, J.; Cui, K.; Zhang, Y. Effects and Action Mechanisms of Berberine and *Rhizoma coptidis* on Gut Microbes and Obesity in High-Fat Diet-Fed C57BL/6J Mice. *PLoS One* **2011,** *6* (9), e24520. DOI: 10.1371/journal.pone.0024520

Xu, Y.; Zhang, M.; Wu, T.; Dai, S.; Xu, J.; Zhou, Z. The Anti-Obesity Effect of Green Tea Polysaccharides, Polyphenols and Caffeine in Rats Fed with a High-Fat Diet. *Food Funct.* **2015,** *6*, 296–303.

Yadav, S.; Vats, V.; Dhunnoo, Y.; Grover, J. K. Hypoglycemic and Antihyperglycemic Activity of *Murraya koenigii* Leaves in Diabetic Rats. *J. Ethnopharmacol.* **2002,** *82*, 111–116.

CHAPTER 11

HERBAL SUPPLEMENTS

GOVIND RAJ PAL[1], NIHARIKA MISHRA[1], AMIT SHARMA[1], MEENA KAUSAR[1], RAJESHWAR KAMAL KANT ARYA[1], and SUNIL KUMAR PRAJAPATI[2]

[1]*Department of Pharmaceutical Sciences, Sir J.C. Bose Technical Campus Bhimtal, Kumaun University, Nainital, India*

[2]*Institute Of Pharmacy Bundelkhand University Jhansi, Uttar Pradesh, India*

ABSTRACT

Food supplements are a source of nutrients such as vitamins and minerals. A food supplement provides nutrition to its consumers and performs a physiological role. It can be food or a part of food containing health-promoting ingredients from plants, animals, minerals, vitamins, prebiotics, or probiotics. Herbal supplements have been utilized for centuries. The ancestors believed that herbal products had healing properties, such as liquid extracts, tea, tablet, capsules, bath salts, oils, and ointments. Various nutritional and physiological supplements are available on the market in various dosages. The main objective of using food supplements is to provide an adequate quantity of nutrition in the diet, and supplements can provide a measurable amount of essential nutrients that are required by the body. Supplements are added to the diet to improve both physiological and nutritional benefits. In the current chapter, we have discussed the in vitro, in vivo, and clinical studies with their possible mechanisms of action in various herbal supplements available in the market for treating various disorders, for example, weight gain, weight loss, sexual disorders, immunomodulators, and galactagogue.

Herbals As Nutraceuticals: Their Role in Healthcare.
Raj K. Keservani, Rajeshwar Kamal Kant Arya, and Rajesh K. Kesharwani (Eds.)

11.1 INTRODUCTION

The US Food and Drug Administration (USFDA) defines supplements as food products, and the Dietary Supplement Health and Education Act (DSHEA) states supplements as "the formulation of products that are administered orally for supplementing the diet" (Bailey, 2020). Safety testing doesn't require in the case of herbs and vitamins. They are autoregulated by the manufacturers; therefore, no proof is needed to demonstrate their efficacy. Supplements are considered safe until proven unsafe" (Cohen, 2014). FDA does not regulate supplements (Furhad and Bokhari, 2022). Under the DSHEA, substances derived from plants such as oil, roots, seeds, berries, and flowers are called herbal supplements (Herbal Supplements, nd.). The ancestors believed that herbal products have healing properties, such as liquid extracts, tea, tablet, capsules, bath salts, oils, and ointments (Sharma et al., 2020; Rane et al., 2020a, 2020b, 2020c; Furhad and Bokhari, 2022). Food supplements are a source of nutrients in concentrated form (i.e., Vitamins and minerals). Various nutritional and physiological supplements are available on the market in various dosage forms (European Food Safety Authority). A food supplement provides nutrition to its consumers and performs a physiological role. It can be food or a part of food containing health-promoting ingredients from plants, animals, minerals, vitamins, prebiotics, or probiotics. Food supplements are used to provide an adequate quantity of nutrition in the diet. With the help of this, we can get a measurable amount of essential nutrients that are required by the body, for example, stress (tension) also increases the body's need for various nutrients such as chamomile and lavender. People with regular physical work and exercise, such as trainee athletes, need to consume more mineral vitamins and antioxidants. When not consumed in the proper amount, they may be prone to or suspected of degenerative diseases. Supplements are added to the diet to improve both physiological and nutritional benefits. Many researchers said that instead of taking supplements, one should consume or eat a healthy, rich diet, but it is impossible in this working schedule to do both. Fitness freaks primarily consume supplements to build their muscles and shape their bodies. Supplements can also be expensive compared to whole food and an unprocessed diet.

In contrast to supplements, vegetables, whole grains, and legumes are much more affordable and are also rich in nutrients (Polak et al., 2015). Various types of supplements are available in the market for multiple purposes, which include weight loss, weight gain, aphrodisiacs, immunomodulators, and galactagogues. Here, we discussed some commonly available supplements used for treating various ailments.

11.2 SUPPLEMENTS FOR WEIGHT GAIN

There are several supplements now available in the market for weight gain. They help in improving the appetite, improving digestion, and also increasing body mass. Many herbs are available for faster weight gain and have low or negligible side effects. Some reports suggest *Cannabis sativa* helps in gaining weight. However, the role of (*C. sativa*) is controversial, clinical studies show that *C. sativa* improves appetite in HIV patients and increases patient weight (Kirkham et al., 2009), but *C. sativa* also helps in reducing weight by blocking the CB2 receptor, which regulates the body weight in rats (Ignatowska-Jankowska et al., 2011). Some polyherbal herbal supplements are available in the market for weight gain, generally polyherbal formulations, for example, Weight Gain Formula (brand name), which includes *Withania somnifera*, *Terminalia chebula*, *Terminalia berllirica*, *Phyllanthus emblica*, *Foeniculum vulgare*, *Piper longum*, and *Cuminum cyminum* (Weight Gain Formula). The herbal supplements work as weight-gaining supplements by improving digestion and appetite. Some of which are helpful in weight gain are given in Table 11.1:

TABLE 11.1 Herbs Supplements for Weight Gain.

Herbs for weight gain	Mechanism
C. sativa	Increase appetite (Kirkham et al., 2009; Richard and Bayer, 2001)
Cnicus benedictus L.	Improve digestion and increase appetite (Ahmadimoghaddam, 2020)
Gentian roots	Increases the gastric sensations, appetite as well as bile flow (Mirzaee et al., 2017)
Glycyrrhiza glabra	Soothe digestive problems, treats upper digestive problems, and they improve appetite (Bethapudi et al., 2022)
Taraxacum officinale	They increase appetite and they are high content of beneficial nutrients and are recommended especially for pregnant women (Hempen and Fischer, 2009; Romm et al., 2010)
W. somnifera	With healthy and nutritious diet heals dyspepsia and thus increases appetite (Chauhan, 2019)

11.3 SUPPLEMENTS FOR WEIGHT LOSS

Obesity is the most significant global problem, affecting approximately 1.9 billion people internationally (Ahirwar and Mondal, 2019). There are various supplements available in the market that help in weight loss. The supplements mainly diminish the desire for food; some may help burn the deposited fat by enhancing the metabolism. Some of the supplements are discussed here.

11.3.1 GLUCOMANNAN

Glucomannan is a fiber obtained from *Amorphophallus konjac*. Several countries use it as a food additive and supplement (Devaraj, 2019). Glucomannan is a highly swellable fiber (Keithley et al., 2012). The colonic microbiota ferments it. It absorbs water in the colon and forms a dense layer over food; hence the retention time of food is enhanced in the body, eventually giving a sense of satisfaction. It also lowers cholesterol and triglyceride levels by excreting it into faces (Devaraj, 2019). A randomized clinical study also revealed the beneficial effect of glucomannan supplements on weight loss. The study revealed that glucomannan supplements help in reducing weight and cholesterol levels (Kaats et al., 2015).

11.3.2 GREEN TEA

Green tea is obtained from the leaves of *Camellia sinensis* (Grove and Lambert, 2010). *C. sinensis* is supposed to work either by restraining the desire for food, elevating the LDL and fat metabolism, or burning calories (Henning et al., 2018; Watanabe et al., 2020), but there are no scientific reasons for this. High doses may cause vomiting, bloating, gas, diarrhea, dizziness, insomnia, and agitation (Dostal et al., 2015). The main chemicals for weight loss are catechins and caffeine, inhibiting the catechol O-methyltransferase and phosphodiesterase, enhancing fat oxidation, and regulating lipid breakdown (Westerterp-Plantenga, 2010). However, a clinical study shows that green tea can help maintain weight to a certain extent in obese subjects with caffeine, showing promising results (Kovacs et al., 2004). Another clinical study on obese Thai subjects shows that *C. sinensis* control weight by enhancing energy consumption and lipid oxidation (Auvichayapat et al., 2008). *C. sinensis* decreases protein and lipids' absorption and induces adenosine monophosphate (AMP)-activated protein kinase (AMPK), suppressing gluconeogenesis. (Yang et al., 2016).

11.3.3 GREEN COFFEE

Green coffee is commonly recommended for weight loss, and various studies support its benefits. Clinical studies show green coffee extract helps in weight loss (Buchanan and Beckett, 2013). A 22-week-long double-blind clinical study on 16 obese subjects shows that the GCA (brand name), green coffee market formulation, is very effective in reducing weight (Vinson et al., 2012). Another clinical trial was performed on 64 females with obesity that showed the green coffee extract had the potential to reduce weight, and it also reduced the cholesterol, leptin, and free fatty acid levels in blood (Haidari et al., 2017).

11.3.4 GUAR GUM

Guar gum is extracted from the seeds of *Cyamopsis tetragonoloba* and belongs to the Leguminosae family (Prem *et al.,* 2005). The gum reduces hunger and blood glucose in obese subjects (Krotkiewski et al., 1984). The gum reduces obesity by burning calories and inhibiting fat absorption in the body. Researchers are more interested in guar gum for weight loss but clinical studies show it is not effective in reducing body weight (Pittler et al., 2001)

11.3.5 CITRUS AURANTIUM

Citrus aurantium, commonly known as Seville orange or Bitter orange, is native to Africa and tropical Asia. It is also found growing in the Mediterranean, California, and Florida. It is used for losing weight (Penzak et al., 2001). *C. aurantium* fruit contains synephrine, a stimulant similar to ephedrine (Stohs et al., 2012; Bouchard et al., 2005). It is believed to work by uplifting the number of calories burned (Stookey et al., 2012). *C. aurantium* supplement is recommended as an alternative for weight loss after the FDA has banned ephedra (Rossato et al., 2011), and various in vitro, animal, and clinical studies show *C. aurantium* is safe (Stohs et al., 2011). Clinical studies show *that C. aurantium* with caffeine and *Hypericum perforatum* helps lose weight while the subjects were on a controlled diet and regular exercise (Preuss et al., 2002). *C. aurantium* helps in reducing weight by enhancing lipid metabolism and thermogenesis in animals and humans (Peixoto et al., 2012)

11.4 SUPPLEMENTS FOR TREATING SEXUAL DYSFUNCTION

Various herbal compounds show beneficial effects on sexual disorders such as erectile dysfunction (ED) or low sexual desire and premature ejaculation (PE). These include Ginkgo, Red ginseng, and yohimbine. The herbal supplements help relax the smooth muscles, improve blood flow in reproductive organs, improve hormone levels, and enhance nitric oxide (NO) levels. Several animal and clinical studies confirm the potential of herbal drugs for treating sexual disorders, for example, ED, loss of libido, and PE. Korean ginseng berry extract also shows similar results and relaxes the contracted muscle (Cho et al., 2013). An ex vivo and in vivo study suggests that *Panax ginseng* helps relax the contracted muscle by increasing testosterone, NO levels, and cGMP levels (Lee et al., 2016). Another herb *Allium tuberosum* shows relaxing effects on the corpus cavernosum smooth muscle and stimulates sexual desire and performance in rats (Tang et al., 2017). The combination of Chinese chive and *Lepidium meyenii* root extract shows a synergic effect, enhances the NO and cGMP concentration in reproductive organs, and shows effectiveness in ED (Zhang et al., 2019). A double-blind trial also confirms the potential of *L. meyenii* over the placebo. It enhances sexual desire in subjects (Gonzales et al., 2002). The methanolic extract of *Corchorus depressus* also exhibits sexual desire-enhancing potential in an animal study (Kataria et al., 2013). *Pausinystalia yohimbe* is an important plant used for treating various sexual disorders. It contains an important alkaloid (Bylund and Lavicky, 2007), and studies show it is not effective when used alone, but in combination with other supplements, it is effective in ED (Satriyo Arif Wibowo et al., 2021). Some of them are listed below in Table 11.2 with their possible mode of action.

11.5 HERBS FOR BOOSTING THE IMMUNE SYSTEM

Nutritional deficiencies are a major concern in India. People of all ages deal with mild to severe nutritional deficiencies due to improper diet results in a weakened immune system. Using immunomodulators derived from herbs and their products helps boost immunity and decreases the risk of body infection, even though it can be helpful in COVID-19 (Khanal et al., 2022). Various report suggests that *G. glabra* (licorice) has immunomodulatory effects. It is an economical option for enhancing immunity (Bisht et al.,2022). The experiment revealed that the addition of licorice to the diet given to Japanese quail enhanced performance, immunity, and antioxidant capacity and declined intestinal pathogens (Reda et al., 2021). The aqueous

TABLE 11.2 Herbal Supplements for Sexual Dysfunction.

Herb name	Mode of action	Disorder
Ginkgo biloba	Improves blood flow to sexual organs by dilating the blood vessels	Effective in ED (Sohn et al., 2015)
Epimedium	The mode of action is not clear but is believed to alter the body's hormonal level and improve sexual performance	Sexual dysfunction such as ED and loss of libido, and painful intercourse (Shindel et al., 2010)
P. ginseng	The mode of action is not clear. It is believed to work by influencing various hormones responsible for achieving an erection. Enhances the NO level and thus improves blood flow	Effective in ED (Yee and Chan, 2021)
P. yohimbe	It induces receptors in the body that triggers the release of epinephrine and norepinephrine which are responsible to restrict the release of blood vessel-constricting hormones in the penis	Used to treat ED (Hozeifa et al., 2012)
Tribulus terrestris	Increases testosterone levels and builds muscle and strength	Improve sexual performance (Singh et al., 2012), Improve sexual desire in women (Akhtari et al., (2014).)
Muira puama (*Ptychopetalum olacoides*):	It contains sitosterol, campesterol, lupol, and camphor and reduces the sensitivity of the male organ Also, works as a nervine tonic	With *G. biloba,* it improves sexual desire in men (Waynberg and Brewer, 2000; Lim, 2017
Ashwagandha (*W. somnifera*) root extract	It enhances the salivary dehydroepiandrosterone sulfate and testosterone level	Effective in female sexual disorder (Dongre et al., 2015) But not effective in psychogenic ED (Mamidi and Thakar, 2011)
Eurycoma longifolia Jack (Malaysian ginseng)	It enhances testosterone levels	Enhances sexual desire and it is also effective in ED (Tambi et al., 2012, Leitão et al., 2021)
A. tuberosum	Relaxing the contracted muscle	Enhance sexual desire and effective in PE (Zhang et al., 2019;
L. meyenii	Enhances the level of NO and cGMP in	Enhance sexual desire and effective in ED (Gonzales et al., 2002)

extract of *G. glabra* and zinc combination enhances the immunomodulatory effect on animals (Mitra Mazumder et al., 2012). The *Glycyrrhiza uralensis*, or Chinese licorice is a plant of traditional relevance, it shows antitumor activity by enhancing the antitumor cytokines and immunostimulant activity in an animal model (Ayeka et al., 2017).

Ganoderma lucidum is known as a medicinal mushroom and also shows immunomodulatory properties. *G. lucidum* shows immunomodulatory properties in in vitro and in vivo models (Wang and Lin, 2019). *G. lucidum* and Agaricus *brasiliensis* mycelia supplement suppresses cancer progression. It enhances CD4+ and CD8+ cells and decreases CD19+ cell CD19+ cells (Rubel et al., 2018). A study on nine different types of mushrooms shows that they synergistically exhibit an immunomodulatory effect on lipopolysaccharide-induced and noninduced macrophages (Mallard et al., 2019). Recombinant *G. lucidum* induces the hematopoietic stem cells in cyclophosphamide-induced neutropenia. This suggests it can be helpful in neutropenia caused by chemotherapy (Lei et al., 2020). A clinical study on 40 women with breast cancer receiving chemotherapy showed that *G. lucidum* enhances the level of interferons (IFN)-γ and decreases the level of tumor necrosis factor (TNF)-α and interleukin (IL)-8 in serum and showed immunomodulatory effects (Ali et al., 2018)

Astragalus membranaceus, a traditional Chinese herbal supplement, also show immunomodulatory potential. It enhances the release of IL-1β and TNF-α in macrophages and also enhances messenger RNA levels (Qin et al., 2012). The aqueous extract and powder supplement exhibit immunomodulatory properties in vivo (Li et al., 2019). The polysaccharides extracted from *A. membranaceus* showed an immunomodulatory effect on broiler chicks (Farag and Alagawany, 2019). The polysaccharides improve humoral cellular immunity, enhance the effectiveness of recombinant hepatitis B surface antigens, and produce more antibodies (Du et al., 2011).

*Origanum vulgare also shows immunomodulatory properties. The hydroalcoholic extract of O. vulgare possesses antioxidant potential. It suppresses the production of TNF-α and IL-12 and enhances the production of TGF-β, and shows antimycobacterial innate immunity in vitro (*De Santis et al., 2019).

Azadirachta indica is traditionally used in India for the treatment of various diseases, and studies show *A. indica* bark oil improves immunity in mice it stimulates the immune system non specifically that potentiate cell-mediated immune (CMI) mechanisms to elicit an enhanced response to subsequent mitogenic or antigenic challenge (Upadhyay et al., 1992). aqueous extract of the stem bark of *A. indica* improves immunity in a dose-dependent manner by enhancing the production of migration inhibition factor by lymphocytes. (Van der Nat et al., 1987). it enhances immunity by inhibiting the activation of the CD4+ T-cell caused by staphylococcus enterotoxin B in an in vitro study (Olwenyi et al., 2021). an animal study shows the leaf extract of *A. indica* show an immunomodulatory effect by decreasing proinflammatory molecules and affecting the expression of CD4+, CD8+, CD25+, CD62L, IL-10, and IL-17 cells (Supriyanto et al., 2021). The polysaccharides improve humoral cellular immunity, enhance the effectiveness of recombinant hepatitis B surface antigens, and produce more antibodies (Du et al., 2011). Some other important herbal supplements are listed in Table 11.3.

TABLE 11.3 Herbal Supplements for Boosting the Immune System.

Name of herb	Beneficial features
P. ginseng	It acts as a great immune booster and improves strength. Supports endurance, and also (Kang and Min, 2012)
A. embranaceus	It increases the number of active immune cells as well as their activity. It principally boosts the already existing immune system and preventing pathogenic agents from attacking the body (Qin, 2012; Ny, 2021)
Aloe vera	Improves metabolism, vitalizes the immune system, inhibits cancerous growth, reduces oxidative stress, and treats hyperglycemia (Rahmani et al., 2015; Sardashti et al., 2020; Alinejad-Mofrad, S. et al., 2015)
Ocimum basilicum	a rich antioxidant profile as it is loaded with vitamins and also exhibits immunity, antiaging properties, and antitumor properties (Eftekhar et al., 2019)
Galium aparine	Support the lymphatic and immune systems and boost liver and kidney functions (Ilina et al., 2020; Tobyn et al., 2011)
Chamaemelum nobile (Chamomile*)*	It is found to be rich in antioxidants. It vitalizes the immune system and protects the skin from free radical damage (Srivastava, 2010)
Cordyceps	Cordyceps has marked immune-modulatory functions while enhancing overall immunity by raising the body's counts of lymphocytes and natural killer cells (Lin and Li, 2011)
G. lucidum	Show immunomodulatory properties (Wang and Lin, 2019).
Terminalia ferdinandiana	It contains high vitamin C concentration, which is essential for the immune system. It is rich in minerals, ellagic acid, and antioxidant-rich polyphenols Mohanty and Cock, 2012)
A. indica	protect against microbes, malaria, bacteria, viruses, and insects. Immunomodulatory effects of neem oil (Upadhyay et al., 1992).
O. vulgare	Antioxidant, anti-inflammatory, and cancer-suppressing properties. immunomodulators in fish (Alagawany et al., 2020)
G. glabra	The Enhanced immunity (Reda et al., 2021)

11.6 HERBS INCREASE BREAST MILK

Mother's milk is a boon for children. It enhances the immunity of the child gives nutrition and supports the development of the child. Nowadays, insufficient milk production is a common problem, approximately 5% of mothers have a problem in producing milk (agalactia) and about 15% suffer from insufficient supply of their breast milk (hypogalactias), globally (Bekoe et al.., 2018). Several factors impeded milk production in women (Committee Opinion, 2013). In Developing countries, poor nutrition, contaminated drinking water, and low availability of the vaccine lower immunity in children but breastfeeding play a significant role in preventing children from infection and death. Reports suggest breastfeeding for 6 months can prevent 1.3 million (13%) deaths yearly in children under 5 years (Bekoe et al.., 2018). Several plants are believed to promote breastfeeding and boost milk production in lactating mothers (Bazzano et al., 2016). These herbal supplements are called lactogogue or galactogogue (Ali et al., 2020). Herbal supplements are rich in iron and vitamins, build breast tissue and increase breast milk supply. Some herbal supplements are discussed below

Zingiber officinale is traditionally used by lactating mothers in South Asia to improve milk volume. Clinical studies show that 500 mg of *Z. officinale* capsule enhances the milk volume in lactating mothers (Paritakul P, Ruangrongmorakot K, Laosooksathit W, Suksamarnwong M, Puapornpong P. The Effect of Ginger on Breast Milk Volume in the Early Postpartum Period: A Randomized, Double-Blind Controlled Trial. Breastfeed Med. 2016). Clinical trial reports show an interesting point about *Z. officinale:* it improves milk production in women who gave a vaginal birth, not in the cesarian section (Dilokthornsakul et al., 2022). Another clinical study shows the *Z. officinale* supplement with fenugreek and turmeric improves milk production and secretion (Bumrungpert et al., 2018)

Medicago sativa (Alfalfa) is traditionally used by women as galactagogue (Othman et al., 2014); it increases milk production and supply and is very safe for use in women. The dietary Supplement of *M. sativa* improved milk production in cows (Chingala et al., 2013).

C. benedictus is commonly known as the blessed thistle found in North America; it is used for improving milk production in women. It improves the milk flow by improving the blood supply to the breast and stimulating milk production (Westfall, 2003).

Silybum marianum or milk thistle Silymarin increases milk production in cows. Sixty-three days of clinical studies on 50 healthy lactating mothers show an improvement in milk production, and also it was found to be safe and tolerable (Di Pierro et al., 2008). Silitidil (milk thistle extract) improved prolactin levels in female rats, which helps in enhancing milk production. (Capasso, 2014)

F. vulgare is a very old herb that has been used for increasing milk production for centuries (Penagos et al., 2014). A clinical study on 66 women shows herbal tea containing *F. vulgare* enhances milk production in breastfeeding women (Turkyılmaz et al., 2010).

There are some other supplements which are given in Table 11.4.

TABLE 11.4 Herbal Supplements for Increasing Breast Milk.

Supplement	Use
M. sativa	It is rich in vitamins and minerals daidzein, genistein), vitamin K, and the amino acid L-canavanine; it enhances milk production and improves reproductivity in females (Mandle and Chaudhari, 2020)
C. benedictus with fenugreek	Improve breast milk secretion (Bazzano et al., 2016) Marketed as Blessed thistle tea (Nature Way company)
Z. officinale	It increases the breast milk supply of lactating mothers (Paritakul P. *et al*., 2016)
Galega officinalis (Goat's rue)	Build up breast tissue and produce more milk in the breasts. Yet, the fresh goat's rue plant is dangerous to consume. (Penagos et al., 2014)
Trigonella foenum graecum	It improves milk production in goats (Al-Shaikh et al., 1999)
F. vulgare	Boost breast milk production in lactating mothers. Fennel can be incorporated in teas or food as a flavoring agent. (Turkyılmaz et al., 2010)
Garlic	Increases milk supply but changes breast milk flavor (Gangwani, 2021
Silybum marianum	Associated with breastfeeding for centuries. (McGuire, 2018)
Urtica dioica	It enhances the breast milk supply of lactating mothers (Khanal et al., 2017)
R. communis	Crashed leaves are massaged over the breast and stimulate mammary glands that improve milk secretion and flow (Ladda and Bhimashankar, 2014)

11.7 SUPPLEMENTS USED IN ARTHRITIS

Arthritis is a long-term (chronic) inflammatory disorder of joints. The pain in joints occurs due to the deficiency of synovial fluid; if it occurs in children, it is classified as osteoarthritis (OA) and if it occurs in adults classified under rheumatoid arthritis. Some important herbal supplements are discussed below:

Ganoderma lucidum and San-Miao-San have anti-inflammatory potential and are traditionally used to treat inflammatory diseases such as arthralgia syndrome, and animal studies prove this. The supplementation enhances the level of CD4+, CD25+, Foxp3+, Treg and IL-10+ Breg cells and suppresses the level of cytokines responsible for inflammation (Cai et al., 2016). In cell line studies, the *G. lucidum* inhibits the nuclear factor-kappa B (NF-κB) transcription that inhibits the proliferation of synovial fibroblasts IL-1β LPS-induced IL-8 and MCP-1 production (Ho et al., 2007). The combination possesses an antioxidant potential useful in Rheumatoid Arthritis (RA) (Li et al., 2007. A randomized clinical study for 24 days on 32 subjects with RA shows that the combination has analgesic potential but without significant change in antioxidant and anti-inflammatory properties, but it suppresses the level of IL-18 ex vivo (Li et al., 2007),

Camellia sinensis or green tea also show beneficial effects on OA. The hydroalcoholic extract of *C. sinensis* at a dose of 400 mg/kg exhibited anti-oxidant potential and suppressed oxidative stress, and improved the physical disabilities in collagen-induced arthritis in mice (Tanwar et al., 2017), the telomerase suppressing function of *C. sinensis* on is responsible for the antiarthritic properties (Jalali and Saadat, 2017). Animal studies show that the polyphenol found in *C. sinensis* suppresses the IL-17 level and enhances the immunoregulatory cytokine IL-10 and anti-inflammatory cytokine in the arthritic model (Kim et al., 2008). The *C. sinensis* extract contains a high amount of catechins that show good anti-inflammatory response than black tea extract, which contains theaflavins and thearubigins (Ramadan et al., 2017). A clinical study on 50 subjects shows that *C. sinensis* extract ameliorated physical function and showed an analgesic effect on OA patients. (Hashempur et al., 2018).

Z. officinale. In a 6-week clinical study, 61 OA subjects were randomly treated with ginger extract and a placebo. The study shows that ginger extract suppresses OA symptoms with mild gastrointestinal side effects (Altman and Marcussen, 2001).

Curcuma longa (turmeric) contains curcumin that possesses anti-inflammatory activity and is used to treat RA, OA, and other arthritis conditions. (Henrotin et al., 2019). *C. longa* shows better anti-inflammatory, antioxidant and antiarthritic properties than *Z. officinalis* (Ramadan et al., 2011)

TABLE 11.5 Herbal Supplements for Arthritis.

Herb	Properties
A. vera	Anti-inflammatory activities. Orally it is used to relieve OA pain. Topically inhibits inflammation(Cowan, 2010; Bałan et al., 2014)
Boswellia	Boswellic acid has an anti-inflammatory effect that could treat RA, OA, and gout. A dose up to 1 g appears safe, but a high dose causes liver damage (Yu et al., 2020)
Uncaria tomentosa	It reduces swelling in joints. Arthritis suppresses TNF (necrotic tumor factor (Piscoya et al., 2001)
Eucalyptus	Contains tannins used to treat arthritis pain. Eucalyptus aroma therapy may reduce the symptoms of RA (Jun et al. 2013)
Z. officinale	Anti-inflammatory action because of the presence of Zingerone. (Al-Nahain et al., 2014)
C. sinensis	Effective against inflammation that occurs with RA and OA (Katturajan et al., 2021)
Tripterygium wilfordil	Manage inflammation and excess immune activity. It is commonly taken by mouth as a dietary supplement. (Goldbach-Mansky et al., 2009)
C. longa	Show anti-inflammatory activity and is used to treat RA, OA, and others (Henrotin et al., 2019)
Willow bark tea or tablet form	Used for pain and inflammation. It may help to relieve joint pain related to OA and RA (Biegert et al., 2004)
Ganoderma lucidum	Show analgesic and anti-inflammatory response in arthritis (Chen et al., 2020)

11.8 SIDE EFFECTS

Herbal supplements are considered safe for use, and to date, USFDA has not implemented any guidelines regarding the use of herbal supplements. Herbal supplements are thought to be safe until adverse reactions are reported (Cohen, 2014). Sometimes the Supplements used for weight gain may result in muscle cramps and several other adverse side effects. The inadequate dose can lead to heart, kidney, and liver problems. It may also result in dehydration and diarrhea-causing weight gain powder is rich in vitamin C and iron, which results in diarrhea and stomach problems. Few people also experience side effects with green coffee. The presence of caffeine in green coffee may show symptoms of headaches; stomach upset, nervousness, insomnia, and abnormal heart rhythms (Griffin, n.d.)

11.9 CONCLUSION

Herbal products contain various micro- and macronutrients essential for humans, and these nutrients are called supplements. For a long time, herbal supplements have been used for nutrient deficiency and deficiency-related ailments that include sexual disorders, immunity, and insufficiency in producing milk, and supplements are also helpful in reducing body weight or improving body weight. Several studies have been conducted to understand the potential of herbal supplements and their safety. However, to date, proper safety profiles are not available, and no proper guidelines of USFDA and other regulatory bodies have been established. From the above discussion, we can conclude that herbal supplements are more advantageous for nutrient deficiency and deficiency-related disorders. More animal and clinical trials should be conducted to confirm the efficacy and safety of herbal supplements so that everybody can benefit from natural products or herbal supplements.

CONFLICT OF INTEREST

The authors have no conflict of interest

KEYWORDS

- **supplements**
- **weight gain**
- **aphrodisiac**
- **immunobooster**
- **erectile dysfunction**

REFERENCES

Ahirwar, R.; Mondal, P. R. Prevalence of Obesity in India: A Systematic Review. *Diabetes Metab Syndr*. **2019,** *13* (1), 318–321. DOI: 10.1016/j.dsx.2018.08.032

Ahmadimoghaddam, D.; Sadeghian, R.; Ranjbar, A.; Izadidastenaei, Z.; Mohammadi, S. Antinociceptive Activity of *Cnicus benedictus* L. Leaf Extract: A Mechanistic Evaluation. *Res. Pharm. Sci.* **2020,** *15*, 463–472. https://doi.org/10.4103/1735–5362.297849

Akhtari, E.; Raisi, F.; Keshavarz, M. et al. *Tribulus terrestris* for Treatment of Sexual Dysfunction in Women: Randomized Double-Blind Placebo—Controlled Study. *Daru* **2014,** *22*, 40. DOI: 10.1186/2008-2231-22-40

Alagawany, M.; Farag, M. R.; Salah, A. S.; Mahmoud, M. A. The Role of Oregano Herb and Its Derivatives as Immunomodulators in Fish. *Rev. Aquacult.* **2020,** *12* (4), 2481–2492. DOI: https://doi.org/10.1111/raq.12453

Ali, N. A. M.; Saeed, H. A.; Othman, R. T. Immunostimulatory and Anti-Inflammatory Effect of Ganoderma Lucidum on Breast Cancer Patients. *Asian Pac. J. Cancer Biol.* **2018,** *3* (2), 51–57. DOI: 10.31557/apjcb.2018.3.2.51–571

Alinejad-Mofrad, S.; Foadoddini, M.; Saadatjoo, S. A.; Shayesteh, M. Improvement of Glucose and Lipid Profile Status with *Aloe vera* in Pre-Diabetic Subjects: A Randomized Controlled-Trial. *J. Diab. Metab. Disord.* **2015,** *14*, 22. DOI: 10.1186/s40200-015-0137-2

Al-Shaikh, M. A.; Al-Mufarrej, S. I.; Mogawer, H. H. Effect of Fenugreek Seeds (*Trigonella foenumgraecum* L) on Lactational Performance of Dairy Goat. *J. Appl. Anim. Res.* **1999,** *16* (2), 177–183. DOI: 10.1080/09712119.1999.9706279

Altman, R. D.; Marcussen, K. C. Effects of a Ginger Extract on Knee Pain in Patients with Osteoarthritis. *Azrthritis Rheum.* **2001,** *44* (11), 2531–2538. DOI: 10.1002/1529–0131 (200111)44:11<2531::aid-art433>3.0.co;2-j

Auvichayapat, P.; Prapochanung, M.; Tunkamnerdthai, O. et al. Effectiveness of Green Tea on Weight Reduction in Obese Thais: A Randomized, Controlled Trial. *Physiol. Behav.* **2008,** *93* (3), 486–491. DOI: 10.1016/j.physbeh.2007.10.009

Ayeka, P. A.; Bian, Y.; Githaiga, P. M.; Zhao, Y. The Immunomodulatory Activities of Licorice Polysaccharides (*Glycyrrhiza uralensis* Fisch.) in CT *26* Tumor-Bearing Mice. *BMC. Complement. Altern. Med.* **2017,** *17* (1), 536. DOI: 10.1186/s12906-017-2030-7

Bailey, R. L. Current Regulatory Guidelines and Resources to Support Research of Dietary Supplements in the United States. *Crit. Rev. Food Sci. Nutr.* **2020,** *60* (2), 298–309.DOI: 10.1080/10408398.2018.1524364

Bazzano, A. N.; Hofer, R.; Thibeau, S.; Gillispie, V.; Jacobs, M.; Theall, K. P. A Review of Herbal and Pharmaceutical Galactagogues for Breast-Feeding. *Ochsner. J.* **2016,** *16*, 511–524.

Bekoe, E. O.; Kitcher, C.; Gyima, N. A. M. et al. *Medicinal Plants Used as Galactagogues*; IntechOpen, 2018. DOI: 10.5772/intechopen.82199

Bethapudi, B.; Murugan, S. K.; Nithyanantham, M.; Singh, V. K.; Agarwal, A.; Mundkinajeddu, D. Chapter 24—Gut Health Benefits of Licorice and Its Flavonoids as Dietary Supplements. In: *Nutrition and Functional Foods in Boosting Digestion, Metabolism and Immune Health*; Bagchi, D., Ohia, S. E., Eds.; Academic Press, 2022; pp 377–417. DOI: 10.1016/B978-0-12-821232-5.00008-2

Bisht, D.; Rashid, M.; Arya, R. K. K. et al. Revisiting Liquorice (*Glycyrrhiza glabra* L.) as Anti-Inflammatory, Antivirals and Immunomodulators: Potential Pharmacological Applications with Mechanistic Insight. *Phytomed. Plus.* **2022,** *2* (1), 100206. DOI: 10.1016/j.phyplu.2021.100206

Bouchard, N. C.; Howland, M. A.; Greller, H. A.; Hoffman, R. S.; Nelson, L. S. Ischemic Stroke Associated with Use of an Ephedra-Free Dietary Supplement Containing Synephrine. *Mayo Clin. Proc.* **2005,** *80* (4), 541–545. DOI: 10.4065/80.4.541

Buchanan, R.; Beckett, R. D. Green Coffee for Pharmacological Weight Loss. *J. Evid. Based Complement. Altern. Med.* **2013,** *18* (4), 309–313. DOI: 10.1177/2156587213496818

Bumrungpert, A.; Somboonpanyakul, P.; Pavadhgul, P.; Thaninthranon, S. Effects of Fenugreek, Ginger, and Turmeric Supplementation on Human Milk Volume and Nutrient Content in Breastfeeding Mothers: A Randomized Double-Blind Controlled Trial. *Breastfeeding Med.* **2018,** *13* (10), 645–650. DOI: 10.1089/bfm.2018.0159

Bylund, D. B.; Lavicky, M. Yohimbine. In: *XPharm: The Comprehensive Pharmacology Reference*; Enna, S. J., Bylund, D. B., Eds.; Elsevier, 2007; pp 1–5. DOI: 10.1016/B978-008055232-3.62871-5

Cai, Z.; Wong, C. K.; Dong, J. et al. Anti-Inflammatory Activities of *Ganoderma lucidum* (Lingzhi) and San-Miao-San Supplements in MRL/LPR Mice for the Treatment of Systemic Lupus Erythematosus. *Chin. Med.* **2016,** *11*, 23. DOI: 10.1186/s13020-016-0093-x

Capasso, R. Effect of Silitidil, a Standardized Extract of Milk Thistle, on the Serum Prolactin Levels in Female Rats. *Nat. Prod. Commun.* **2014,** *9* (7), 1934578X1400900715. DOI: 10.1177/1934578X1400900715

Chauhan, D. M. *How Ashwagandha Is Helpful in Weight Gain?* Planet Ayurveda. Published August 1, 2019. https://www.planetayurveda.com/library/how-ashwagandha-is-helpful-in-weight-gain/

Cho, K.; Park, C.; Kim, C. et al. Effects of Korean Ginseng Berry Extract (GB0710) on Penile Erection: Evidence from In Vitro and In Vivo Studies. *Asian J. Androl.* **2013,** *15*. DOI: 10.1038/aja.2013.49

Cohen, P. A. Hazards of Hindsight—Monitoring the Safety of Nutritional Supplements. *New Engl. J. Med.* **2014,** *370*, 1277–1280. https://doi.org/10.1056/NEJMp1315559

Committee Opinion No. 570: Breastfeeding in Underserved Women: Increasing Initiation and Continuation of Breastfeeding. *Obstet. Gynecol.* **2013,** *122* (2 Pt 1), 423–427. DOI: 10.1097/01.AOG.0000433008.93971.6a

De Santis, F.; Poerio, N.; Gismondi, A. et al. Hydroalcoholic Extract from *Origanum vulgare* Induces a Combined Antimycobacterial and Anti-Inflammatory Response in Innate Immune Cells. *PLoS One*. **2019,** *14* (3), e0213150. DOI: 10.1371/journal.pone.0213150

Devaraj, R. D.; Reddy, C. K.; Xu, B. Health-Promoting Effects of Konjac Glucomannan and Its Practical Applications: A Critical Review. *Int. J. Biol. Macromol.* **2019,** *126*, 273–281. https://doi.org/10.1016/j.ijbiomac.2018.12.203

Dongre, S.; Langade, D.; Bhattacharyya, S. Efficacy and Safety of Ashwagandha (*Withania somnifera*) Root Extract in Improving Sexual Function in Women: A Pilot Study. *Biomed. Res. Int.* **2015,** *2015,* 284154. https://doi.org/10.1155/2015/284154

Dostal, A. M.; Samavat, H.; Bedell, S. et al. The Safety of Green Tea Extract Supplementation in Postmenopausal Women at Risk for Breast Cancer: Results of the Minnesota Green Tea Trial. *Food Chem. Toxicol.* **2015,** *83*, 26–35. DOI: 10.1016/j.fct.2015.05.019

Drugs and Lactation Database (LactMed). Bethesda, National Library of Medicine (US), 2006. Alfalfa. (Updated 21 Jun 2021). https://www.ncbi.nlm.nih.gov/books/NBK501830/

Du, X.; Chen, X.; Zhao, B. et al. Astragalus Polysaccharides Enhance the Humoral and Cellular Immune Responses of Hepatitis B Surface Antigen Vaccination Through Inhibiting the Expression of Transforming Growth Factor β and the Frequency of Regulatory T Cells. *FEMS Immunol. Med. Microbiol.* **2011,** *63* (2), 228–235. DOI: 10.1111/j.1574-695X.2011.00845.x

Eftekhar, N.; Moghimi, A.; Mohammadian Roshan, N.; Saadat, S.; Boskabady, M. H. Immunomodulatory and Anti-Inflammatory Effects of Hydro-Ethanolic Extract of *Ocimum basilicum* Leaves and Its Effect on Lung Pathological Changes in an Ovalbumin-Induced Rat Model of Asthma. *BMC Complement. Altern. Med.* **2019,** *19* (1), 349. DOI: 10.1186/s12906-019-2765-4

Farag, M. R.; Alagawany, M. The Role of *Astragalus membranaceus* as Immunomodulator in Poultry. *World's Poultry Sci. J.* **2019,** *75* (1), 43–54. DOI: 10.1017/S0043933918000739.

Food supplements | EFSA. https://www.efsa.europa.eu/en/topics/topic/food-supplements

Furhad, S.; Bokhari, A. A. Herbal Supplements. In: *StatPearls*; StatPearls Publishing: Treasure Island, 2022.

Goldbach-Mansky, R.; Wilson, M.; Fleischmann, R. et al. Comparison of *Tripterygium wilfordii* Hook F Versus Sulfasalazine in the Treatment of Rheumatoid Arthritis. *Ann. Intern. Med.* **2009,** *151* (4), 229–240. DOI: 10.7326/0003-4819-151-4-200908180-00005

Gonzales, G. F.; Córdova, A.; Vega, K. et al. Effect of *Lepidium meyenii* (MACA) on Sexual Desire and Its Absent Relationship with Serum Testosterone Levels in Adult Healthy Men. *Andrologia* **2002,** *34* (6), 367–372. DOI: 10.1046/j.1439-0272.2002.00519.x

Griffin, R. M. Green Coffee: Uses and Risks. WebMD. https://www.webmd.com/vitamins-and-supplements/green-coffee-uses-and-risks

Grove, K. A.; Lambert, J. D. Laboratory, Epidemiological, and Human Intervention Studies Show That Tea (*Camellia sinensis*) May Be Useful in the Prevention of Obesity. *J. Nutr.* **2010,** *140* (3), 446–453. DOI: 10.3945/jn.109.115972

Haidari, F.; Samadi, M.; Mohammadshahi, M.; Jalali, M. T.; Engali, K. A. Energy Restriction Combined with Green Coffee Bean Extract Affects Serum Adipocytokines and the Body Composition in Obese Women. *Asia Pac. J. Clin. Nutr.* **2017,** *26* (6), 1048–1054. DOI: 10.3316/informit.106852848201239

Hashempur, M. H.; Sadrneshin, S.; Mosavat, S. H.; Ashraf, A. Green Tea (*Camellia sinensis*) for Patients with Knee Osteoarthritis: A Randomized Open-Label Active-Controlled Clinical Trial. *Clin. Nutr.* **2018,** *37* (1), 85–90. DOI: 10.1016/j.clnu.2016.12.004

Hempen, C. H.; Fischer, T. IV—Herbs That Cool Heat. In: *A Materia Medica for Chinese Medicine*; Hempen, C. H., Fischer, T., Eds.; Churchill Livingstone, 2009; pp 110–263. DOI: 10.1016/B978-0-443-10094-9.00007-8

Henning, S. M.; Yang, J.; Hsu, M. et al. Decaffeinated Green and Black Tea Polyphenols Decrease Weight Gain and Alter Microbiome Populations and Function in Diet-Induced Obese Mice. *Eur. J. Nutr.* **2018,** *57* (8), 2759–2769. DOI: 10.1007/s00394-017-1542-8

Herbal Supplements (WWW Document), n.d. Cleveland Clinic. https://my.clevelandclinic.org/health/drugs/15829-herbal-supplements

Hozeifa, M. H.; Eldin IEMT, Elhadi, M. M. A.; Elwahab, H. M. A.; Nizar, S. E. The Relaxant Effect of Methanolic Extract of Yohimbe Bark (*Pausinystalia yohimbe*) on Rabbit Corpus Cavernosum Smooth Muscles. *Health* **2012,** *4* (9), 656–659. DOI: 10.4236/health.2012.49103

http://blog.arthritis.org/living-with-arthritis/health-benefits-of-ginger/

http://blog.arthritis.org/living-with-arthritis/health-benefits-of-ginger/

https://nccih.nih.gov/health/bromelain

https://www.arthritis.org/living-with-arthritis/treatments/natural/supplements-herbs/guide/turmeric.php

https://www.arthritis-health.com/blog/add-ginger-help-arthritis-pain

https://www.ncbi.nlm.nih.gov/pmc/articles/PMC5003001/

Ignatowska-Jankowska, B.; Jankowski, M. M.; Swiergiel, A. H. Cannabidiol Decreases Body Weight Gain in Rats: Involvement of CB2 Receptors. *Neurosci. Lett.* **2011,** *490* (1), 82–84. DOI: 10.1016/j.neulet.2010.12.031

Ilina, T.; Skowrońska, W.; Kashpur, N. et al. Immunomodulatory Activity and Phytochemical Profile of Infusions from Cleavers Herb. *Molecules* **2020,** *25* (16), 3721. DOI: 10.3390/molecules25163721

Irkin, R.; Korukluoglu, M. Control of *Aspergillus niger* with Garlic, Onion and Leek Extracts, 2007. Accessed 24 Dec 2020, SOURCE.

Jalali, S. Z.; Saadat, F. *Camellia Sinensis* Modulates Telomerase in Collagen-Induced Arthritis. *Zahedan J. Res. Med. Sci.* **2017,** *19* (2), e7210. DOI: 10.5812/zjrms.7210

Kaats, G.; Bagchi, D.; Preuss, H. Konjac Glucomannan Dietary Supplementation Causes Significant Fat Loss in Compliant Overweight Adults. *J. Am. Coll. Nutr*. 2015, 1–7. DOI: 10.1080/07315724.2015.1009194

Kang, S.; Min, H. Ginseng, the "Immunity Boost": The Effects of *Panax ginseng* on Immune System. https://www.ncbi.nlm.nih.gov/pmc/articles/PMC3659612/ (accessed 30 Sept 2022).

Kataria, S.; Kaur, D.; Rao, S. K.; Khajuria, R. K. In Vitro and In Vivo Aphrodisiac Properties of *Corchorus depressus* Linn. on Rabbit Corpus Cavernosum Smooth Muscle Relaxation and Sexual Behavior of Normal Male Rats. *J. Ethnopharmacol*. **2013,** *148* (1), 210–217. DOI: 10.1016/j.jep.2013.04.011

Katturajan, R.; Sabina, E. P. Joint Inflammation: Insights of Osteoarthritis, Gouty and Rheumatoid Arthritis and its Prevalence, Mechanism, Medications and Remedies. *Indian J. Pharm. Sci*. **2021,** *83* (5), 886–898. DOI: 10.36468/pharmaceutical-sciences.840

Keithley, J. K.; Swanson, B.; Mikolaitis, S. L. et al. Safety and Efficacy of Glucomannan for Weight Loss in Overweight and Moderately Obese Adults. *J. Obes*. **2013,** 2013:610908. DOI: 10.1155/2013/610908

Khanal, P.; Duyu, T.; Patil, B. M. et al. Network Pharmacology of AYUSH Recommended Immune-Boosting Medicinal Plants Against COVID-19. *J. Ayurveda Integr. Med*. **2022,** *13* (1), 100374. DOI: 10.1016/j.jaim.2020.11.004

Khanal, D.; Tiwari, I.; Bastola, R.; Upreti, C. R. Beneficial Effects of Stinging Nettle Supplementation on Milk Production. *Nepalese Vet. J*. **2017,** *34*, 60–68. DOI: 10.3126/nvj.v34i0.22904

Kim, H. R.; Rajaiah, R.; Wu, Q. L. et al. Green Tea Protects Rats against Autoimmune Arthritis by Modulating Disease-Related Immune Events. *J. Nutr*. **2008,** *138* (11), 2111–2116. DOI: 10.3945/jn.108.089912.

Kirkham, T. C. Cannabinoids and Appetite: Food Craving and Food Pleasure. *Int. Rev. Psychiatry* **2009,** *21*, 163–171. https://doi.org/10.1080/09540260902782810

Kovacs, E. M. R.; Lejeune, M. P. G. M.; Nijs, I.; Westerterp-Plantenga, M. S. Effects of Green Tea on Weight Maintenance After Body-Weight Loss. *Br. J. Nutr*. **2004,** *91* (3), 431–437. DOI: 10.1079/BJN200410611.

Krotkiewski, M. Effect of Guar Gum on Body-Weight, Hunger Ratings and Metabolism in Obese Subjects. *Br. J. Nutr*. **1984,** *52* (1), 97–105. DOI: 10.1079/bjn19840075

Laxmikant, L. P.; Bhimashankar, K. R. *Ricinus communis* (Castor): An Overview. *Int. J. Res. Pharm. Pharmacotherap*. **2014,** *3* (2), 136–144.

Lee, H. S.; Lee, Y. J.; Chung, Y. H. et al. In Vitro and In Vivo Evaluation of Tissue-Cultured Mountain Ginseng on Penile Erection. *J. Ginseng Res*. **2016,** *40* (4), 334–343. DOI: 10.1016/j.jgr.2015.10.003

Lei, X.; Zhi, C.; Huang, W. et al. Recombinant *Ganoderma lucidum* Immunomodulatory Protein Improves the Treatment for Chemotherapy-Induced Neutropenia. *Front Pharmacol*. **2020**. DOI: 10.3389/fphar.2020.00956

Leitão, A. E.; Vieira, M. C. de, S.; Pelegrini, A.; da Silva, E. L.; Guimarães, A. C. de A. A 6-Month, Double-Blind, Placebo-Controlled, Randomized Trial to Evaluate the Effect of *Eurycoma longifolia* (Tongkat Ali) and Concurrent Training on Erectile Function and Testosterone Levels in Androgen Deficiency of Aging Males (ADAM). *Maturitas*. **2021,** *145*, 78–85. DOI: 10.1016/j.maturitas.2020.12.002

Li, E. K.; Tam, L. S.; Wong, C. K. et al. Safety and Efficacy of *Ganoderma lucidum* (lingzhi) and San Miao San Supplementation in Patients with Rheumatoid Arthritis: A Double-Blind,

Randomized, Placebo-Controlled Pilot Trial. *Arthritis Care Res.* **2007,** *57* (7), 1143–1150. DOI: 10.1002/art.22994

Li, Z. X.; Zhao, G. D.; Xiong, W. et al. Immunomodulatory Effects of a New Whole Ingredients Extract from Astragalus: A Combined Evaluation on Chemistry and Pharmacology. *Chin. Med.* **2019,** *14* (1), 12. DOI: 10.1186/s13020-019-0234-0

Lim, P. H. C. Asian Herbals and Aphrodisiacs Used for Managing, ED. *Transl. Androl. Urol.* **2017,** *6*, 167–175. https://doi.org/10.21037/tau.2017.04.04

Lin, B. q.; Li, S. p. *Cordyceps as an Herbal Drug*; CRC Press/Taylor & Francis, 2011. https://www.ncbi.nlm.nih.gov/books/NBK92758/ (accessed 29 Sept 2022).

Mallard, B.; Leach, D. N.; Wohlmuth, H.; Tiralongo, J. Synergistic Immuno-Modulatory Activity in Human Macrophages of a Medicinal Mushroom Formulation Consisting of Reishi, Shiitake and Maitake. *Plos One.* **2019,** *14* (11), e0224740. DOI: 10.1371/journal.pone.0224740

Mamidi, P.; Thakar, A. B. Efficacy of Ashwagandha (*Withania somnifera* Dunal. Linn.) in the Management of Psychogenic Erectile Dysfunction. Ayu **2011,** *32* (3), 322–328. DOI: 10.4103/0974–8520.93907Mandle, D. R. G.; Chaudhari, D. V. M. Ashvabala (*Medicago sativa* Linn.) Nari Aushadhi—A Review. *J. Ayur. Integr. Med. Sci.* **2020,** *5* (05), 315–321.

McGuire, T. M. Drugs affecting milk supply during lactation. *Aust. Prescr.* **2018**. DOI: 10.18773/austprescr.2018.002

Mitra Mazumder, P.; Pattnayak, S.; Parvani, H.; Sasmal, D.; Rathinavelusamy, P. Evaluation of Immunomodulatory Activity of *Glycyrhiza glabra* L. Roots in Combination with Zing. *Asian Pac. J. Trop. Biomed.* **2012,** *2* (1, Supplement), S15–S20. DOI: 10.1016/S2221-1691(12)60122-1

Mohanty, S.; Cock, I. E. The Chemotherapeutic Potential of *Terminalia ferdinandiana*: Phytochemistry and Bioactivity. *Pharmacogn. Rev.* **2012,** *6* (11), 29–36. DOI: 10.4103/0973-7847.95855

Ny, V.; Houška, M.; Pavela, R.; Tříska, J. Potential Benefits of Incorporating *Astragalus membranaceus* into the Diet of People Undergoing Disease Treatment: An Overview. *J. Funct. Foods.* **2021,** *77*, 104339. DOI: 10.1016/j.jff.2020.104339

Olwenyi, O. A.; Asingura, B.; Naluyima, P. et al. In Vitro Immunomodulatory Activity of *Azadirachta indica* A. Juss. Ethanol: Water Mixture Against HIV Associated Chronic CD4+ T-Cell Activation/Exhaustion. *BMC Complement. Med. Therap.* **2021,** *21* (1), 114. DOI: 10.1186/s12906–021–03288–0

Othman, N.; Lamin, R. A. C.; Othman, C. N. Exploring Behavior on the Herbal Galactagogue Usage among Malay Lactating Mothers in Malaysia. *Procedia—Soc. Behav. Sci.* **2014,** *153*, 199–208. DOI: 10.1016/j.sbspro.2014.10.054

Paritakul, P.; Ruangrongmorakot, K.; Laosooksathit, W.; Suksamarnwong, M.; Puapornpong, P. The Effect of Ginger on Breast Milk Volume in the Early Postpartum Period: A Randomized, Double-Blind Controlled Trial. *Breastfeed Med.* **2016,** *11*, 361–365. DOI: 10.1089/bfm.2016.0073

Paul, S.; Modak, D.; Chattaraj, S. et al. *Aloe vera* Gel Homogenate Shows Anti-Inflammatory Activity Through Lysosomal Membrane Stabilization and Downregulation of TNF-α and Cox-2 Gene Expressions in Inflammatory Arthritic Animals. *Futur. J. Pharm. Sci.* **2021,** *7* (1), 12. DOI: 10.1186/s43094-020-00163-6

Peixoto, J. S.; Comar, J. F.; Moreira, C. T. et al. Effects of *Citrus aurantium* (Bitter Orange) Fruit Extracts and p-Synephrine on Metabolic Fluxes in the Rat Liver. *Molecules* **2012,** *17* (5), 5854–5869. DOI: 10.3390/molecules17055854

Penagos Tabares, F.; Bedoya Jaramillo, J. V.; Ruiz-Cortés, Z. T. Pharmacological Overview of Galactogogues. *Vet. Med. Int.* **2014,** *2014*, 602894. DOI: 10.1155/2014/602894

Penzak, S. R.; Jann, M. W.; Cold, J. A.; Hon, Y. Yi, Desai, H. D.; Gurley, B. J. Seville (sour) Orange Juice: Synephrine Content and Cardiovascular Effects in Normotensive Adults. *J. Clin. Pharmacol.* **2001,** *41* (10), 1059–1063. DOI: 10.1177/00912700122012652

Piscoya, J.; Rodriguez, Z.; Bustamante, S. A.; Okuhama, N. N.; Miller, M. J.; Sandoval, M. Efficacy and Safety of Freeze-Dried Cat's Claw in Osteoarthritis of the Knee: Mechanisms of Action of the Species *Uncaria guianensis*. *Inflamm. Res.* **2001,** *50* (9), 442–448. DOI: 10.1007/PL00000268

Pittler, M. H.; Ernst, E. Guar Gum for Body Weight Reduction: Meta-Analysis of Randomized Trials. *Am. J. Med.* **2001,** *110* (9), 724–730. DOI: 10.1016/s0002-9343 (01)00702-1

Polak, R.; Phillips, E. M.; Campbell, A. Legumes: Health Benefits and Culinary Approaches to Increase Intake. *Clin. Diab.* **2015,** *33* (4), 198–205. DOI: 10.2337/diaclin.33.4.198

Prem, D.; Singh, S.; Gupta, P. P.; Singh, J.; Kadyan, S. P. S. Callus Induction and *de novo* Regeneration from Callus in Guar (*Cyamopsis tetragonoloba*). *Plant Cell Tiss. Organ Cult.* **2005,** *80* (2), 209–214. DOI: 10.1007/s11240-004-0738-9

Preuss, H. G.; DiFerdinando, D.; Bagchi, M.; Bagchi, D. *Citrus aurantium* as a Thermogenic, Weight-Reduction Replacement for Ephedra: An Overview. *J Med.* **2002,** *33* (1–4), 247–264.

Qin, Q.; Niu, J.; Wang, Z.; Xu, W.; Qiao, Z.; Gu, Y. *Astragalus membranaceus* Extract Activates Immune Response in Macrophages via Heparanase. *Molecules* **2012,** *17* (6), 7232–7240. DOI: 10.3390/molecules17067232

Qin, Q.; Niu, J.; Wang, Z.; Xu, W.; Qiao, Z.; Gu, Y. *Astragalus embranaceus* Extract Activates Immune Response in Macrophages via Heparanase. *Molecules (Basel, Switzerland)* **2012,** *17* (6), 7232–7240. https://doi.org/10.3390/molecules17067232

Rahmani, A. H.; Aldebasi, Y. H.; Srikar, S.; Khan, A. A.; Aly, S. M. *Aloe vera*: Potential Candidate in Health Management via Modulation of Biological Activities. *Pharmacogn. Rev.* **2015,** *9*, 120–126. https://doi.org/10.4103/0973-7847.162118

Ramadan, G.; Al-Kahtani, M. A.; El-Sayed, W. M. Anti-Inflammatory and Antioxidant Properties of *Curcuma longa* (Turmeric) Versus *Zingiber officinale* (Ginger) Rhizomes in Rat Adjuvant-Induced Arthritis. *Inflammation* **2011,** *34* (4), 291–301. DOI: 10.1007/s10753-010-9278-0

Ramadan, G.; El-Beih, N. M.; Talaat, R. M.; Abd El-Ghffar, E. A. Anti-Inflammatory Activity of Green Versus Black Tea Aqueous Extract in a Rat Model of Human Rheumatoid Arthritis. *Int. J. Rheum. Dis.* **2017,** *20* (2), 203–213. DOI: 10.1111/1756-185X.12666

Rane, B. R.; Patil, A. S.; Keservani, R. K.; Jain, A. S. Novel Approaches in Herbal Formulation, In: *Enhancing the Therapeutic Efficacy of Herbal Formulations Through Novel Drug Delivery Systems*; Kesharwani, R. K., Keservani, R. K., Sharma, A. K., Eds.; IGI Global International Publisher: Pennsylvania, 2020a; Chapter 2, pp 43–68. ISBN13: 9781799844532.

Rane, B. R.; Bharath, M. S.; Patil, R. R.; Keservani, R. K.; Jain, A. S. Novel Approaches in Nutraceuticals, In: *Enhancing the Therapeutic Efficacy of Herbal Formulations Through Novel Drug Delivery Systems*; Kesharwani, R. K., Keservani, R. K., Sharma, A. K., Eds.; IGI Global International Publisher: Pennsylvania, 2020b; Chapter 11, pp 241–266.

Rane, B. R.; Tadavi, S. A.; Keservani, R. K. Naturopathy. In: *Herbal Product Development*; Sharma, A. K., Keservani, R. K., Prakash Gautam, S., Eds.; Apple Academic Press, CRC Press, Taylor & Francis Group, 2020c; Chapter 12, pp 321–347.

Richard, E.; Bayer, M. D. F. Therapeutic Cannabis (Marijuana) as an Antiemetic and Appetite Stimulant in Persons with Acquired Immunodeficiency Syndrome (AIDS). *J. Cannabis Therap.* **2001,** *1*, 5–16. https://doi.org/10.1300/J175v01n03_02

Romm, A. Chapter 13—Pregnancy: First Trimester. In: *Botanical Medicine for Women's Health*; Romm, A., Hardy, M. L., Mills, S., Eds.; Churchill Livingstone, 2010; pp 347–361. DOI: 10.1016/B978-0-443-07277-2.00015-5

Rossato, L. G.; Costa, V. M.; Limberger, R. P.; Bastos, M. de, L.; Remião, F. Synephrine: From Trace Concentrations to Massive Consumption in Weight-Loss. *Food Chem. Toxicol.* **2011,** *49* (1), 8–16. DOI: 10.1016/j.fct.2010.11.007

Rubel, R.; Santa, H. S. D.; Santos LF dos, Fernandes, L. C.; Figueiredo, B. C.; Soccol, C. R. Immunomodulatory and Antitumoral Properties of *Ganoderma lucidum* and *Agaricus brasiliensis* (Agaricomycetes) Medicinal Mushrooms. IJM **2018,** *20* (4). DOI: 10.1615/IntJMedMushrooms.2018025979

Satriyo Arif Wibowo, D. N.; Soebad, D. M. Yohimbine as a Treatment for Erectile Dysfunction: A Systematic Review and Meta-Analysis. *Turk. J. Urol.* **2021,** *47* (6), 482–488. DOI: 10.5152/tud.2021.21206

Sharma, A. K.; Keservani, R. K.; Gautam Surya, P. *Herbal Product Development*; Apple Academic Press, CRC Press, Taylor & Francis Group, 2020.

Shindel, A. W.; Xin, Z. C.; Lin, G. et al. Erectogenic and Neurotrophic Effects of Icariin, a Purified Extract of Horny Goat Weed (Epimedium spp.) In Vitro and In Vivo. *J. Sex Med.* **2010,** *7* (4 Pt 1), 1518–1528. DOI: 10.1111/j.1743-6109.2009.01699.x

Singh, S.; Nair, V.; Gupta, Y. K. Evaluation of the Aphrodisiac Activity of *Tribulus terrestris* Linn. in Sexually Sluggish Male Albino Rats. *J. Pharmacol. Pharmacother.* **2012,** *3*, 43–47. https://doi.org/10.4103/0976-500X.92512

Sohn, M.; Sikora, R. *Ginkgo biloba* Extract in the Therapy of Erectile Dysfunction. *J. Sex Educ. Therap.* **1991,** *17* (1), 53–61. DOI: 10.1080/01614576.1991.11074006

Stohs, S. J.; Preuss, H. G.; Shara, M. A Review of the Human Clinical Studies Involving *Citrus aurantium* (Bitter Orange) Extract and its Primary Protoalkaloid p-Synephrine. *Int. J. Med. Sci.* **2012,** *9* (7), 527–538. DOI: 10.7150/ijms.4446

Stohs, S. J.; Preuss, H. G.; Shara, M. The Safety of *Citrus aurantium* (Bitter Orange) and Its Primary Protoalkaloid p-Synephrine. *Phytother. Res.* **2011,** *25* (10), 1421–1428. DOI: 10.1002/ptr.3490

Stookey, J. D.; Hamer, J.; Espinoza, G. et al. Orange Juice Limits Postprandial Fat Oxidation After Breakfast in Normal-Weight Adolescents and Adults. *Adv. Nutr.* **2012,** *3* (4), 629S-635S. DOI: 10.3945/an.112.0019901.

Supriyanto, S.; Widjanarko, S. B.; Rifa'i, M.; Yunianta, Y. Immunomodulatory Activity of Methanol Leaf Extract of Neem (*Azadirachta indica* Juss) Against Suppressor and Proinflammatory Molecules. *J. Trop. Life Sci.* **2021,** *11* (3), 309–315. DOI: 10.11594/jtls.11.03.07

Tambi, M. I. B. M.; Imran, M. K.; Henkel, R. R. Standardised Water-Soluble Extract of *Eurycoma longifolia, Tongkat ali,* as Testosterone Booster for Managing Men with Late-Onset Hypogonadism? *Andrologia* **2012,** *44* (s1), 226–230. DOI: 10.1111/j.1439-0272.2011.01168.x

Tang, X.; Olatunji, O. J.; Zhou, Y.; Hou, X. In Vitro and In Vivo Aphrodisiac Properties of the Seed Extract from *Allium tuberosum* on Corpus Cavernosum Smooth Muscle Relaxation and Sexual Behavior Parameters in Male Wistar Rats. *BMC. Complement. Altern. Med.* **2017,** *17* (1), 510. DOI: 10.1186/s12906-017-2008-5

Tanwar, A.; Chawla, R.; Ansari, Md. M. et al. In Vivo Antiarthritic Efficacy of *Camellia sinensis* (L.) in Collagen Induced Arthritis Model. *Biomed. Pharmacother.* **2017,** *87*, 92–101. DOI: 10.1016/j.biopha.2016.12.089

Tobyn, G.; Denham, A.; Whitelegg, M. *Galium Aparine*, Goosegrass, 2011; pp 173–180. https://doi.org/10.1016/B978-0-443-10344-5.00022-7

Turkyılmaz, C.; Onal, E.; Hirfanoglu, I. M. et al. The Effect of Galactagogue Herbal Tea on Breast Milk Production and Short-Term Catch-Up of Birth Weight in the First Week of Life. DOI: 10.1089/acm.2010.0090. https://home.liebertpub.com/acm

Upadhyay, S. N.; Dhawan, S.; Garg, S.; Talwar, G. P. Immunomodulatory Effects of Neem (*Azadirachta indica*) Oil. *Int. J. Immunopharmacol.* **1992,** *14* (7), 1187–1193. DOI: 10.1016/0192-0561 (92)90054-o

Van der Nat, J. M.; Klerx JPAM, van Dijk, H.; De Silva, K. T. D.; Labadie, R. P. Immunomodulatory Activity of an Aqueous Extract of *Azadirachta indica* Stem Bark. *J. Ethnopharmacol.* **1987,** *19* (2), 125–131. DOI: 10.1016/0378-8741 (87)90036-5

Vinson, J. A.; Burnham, B. R.; Nagendran, M. V. Randomized, Double-Blind, Placebo-Controlled, Linear Dose, Crossover Study to Evaluate the Efficacy and Safety of a Green Coffee Bean Extract in Overweight Subjects. *Diab. Metab. Syndr. Obes.* **2012,** *5*, 21–27. DOI: 10.2147/DMSO.S27665

Wang, X.; Lin, Z. Immunomodulating Effect of Ganoderma (Lingzhi) and Possible Mechanism. *Adv. Exp. Med. Biol.* **2019,** *1182*, 1–37. DOI: 10.1007/978-981-32-9421-9_1

Watanabe, M.; Risi, R.; Masi, D. et al. Current Evidence to Propose Different Food Supplements for Weight Loss: A Comprehensive Review. *Nutrients* **2020,** *12* (9), 2873. DOI: 10.3390/nu12092873

Waynberg, J.; Brewer, S. Effects of Herbal vX on Libido and Sexual Activity in Premenopausal and Postmenopausal Women. *Adv. Ther.* **2000,** *17*, 255–262. https://doi.org/10.1007/BF02853164

WEIGHTGAINFORMULA.store.planetayurveda.com.https://store.planetayurveda.com/products/weight-gain-formula

Westerterp-Plantenga, M. S. Green Tea Catechins, Caffeine and Body-Weight Regulation. *Physiol. Behav.* **2010,** *100* (1), 42–46. DOI: 10.1016/j.physbeh.2010.02.005

Westfall, R. E. Galactagogue Herbs: A Qualitative Study and Review. **2003,** *2* (2), 6.

Yang, C. S.; Zhang, J.; Zhang, L.; Huang, J.; Wang, Y. Mechanisms of Body Weight Reduction and Metabolic Syndrome Alleviation by Tea. *Mol. Nutr. Food Res.* **2016,** *60* (1), 160–174. DOI: 10.1002/mnfr.201500428

Yee, A. T. S.; Chan, L. T. Effectiveness of Ginseng in Treating Erectile Dysfunction: A Review Paper. *Open Access Lib. J.* **2021,** *8* (10), 1–17. DOI: 10.4236/oalib.1108043

Yu, G.; Xiang, W.; Zhang, T.; Zeng, L.; Yang, K.; Li, J. Effectiveness of Boswellia and Boswellia Extract for Osteoarthritis Patients: A Systematic Review and Meta-Analysis. *BMC Complement Med Ther.* **2020,** *20*, 225. DOI: 10.1186/s12906-020-02985-6

Zhang, Y.; Zhou, F.; Ge, F. Effects of Combined Extracts of *Lepidium meyenii* and *Allium tuberosum* Rottl. on Erectile Dysfunction. *BMC Complement. Altern. Med.* **2019,** *19* (1), 135. DOI: 10.1186/s12906-019-2542

PART III
MISCELLANEOUS

CHAPTER 12

HERBAL SUPPLEMENTS AND THEIR REGULATION

BHUSHAN R. RANE[1], SHIVANI S. PAWAR[1], PUNAM S. GADEKAR[1], AKASH J. AMKAR[1], JULIE A. PATIL[1], AMOL D. PATIL[1], ASHISH S. JAIN[1], and VIRENDRA KUMAR PATEL[2]

[1]*Shri D. D. Vispute College of Pharmacy & Research Center, Panvel, Raigad, Maharashtra, India*

[2]*Faculty of Pharmacy, RKDF University, Bhopal, Madhya Pradesh, India*

ABSTRACT

A large number of individuals take herbal supplements, generally without their doctors' consent. In many nations, there is a steady rise in the number of people using herbal supplements. The way that diseases are treated, however, has evolved over time as a result of the industrial revolution and scientific science. However, nowadays, in many developing nations, 80% of the population uses herbal medications or herbal supplements in an effort to improve their health and treat common conditions like colds, heart disease, inflammation, diabetes, and illnesses of the central nervous system. Common herbal supplements like saw palmetto, black cohosh, ginseng, echinacea, ginkgo, garlic, kava, milk thistle, and others are discussed along with their uses and applications. Herbal supplements offer a wide range of pharmacologic and therapeutic qualities, but their production purity, safety, and effectiveness are unregulated. As medicinal medications and certain plants are frequently delivered together, the possibility of drug–herb interactions increases, and these interactions may have significant clinical relevance. As there are significant differences in the definitions and regulations governing herbal supplements among regulatory agencies in different nations, it is difficult

Herbals As Nutraceuticals: Their Role in Healthcare.
Raj K. Keservani, Rajeshwar Kamal Kant Arya, and Rajesh K. Kesharwani (Eds.)

to get their approval for marketing authorization and their quality may get impacted. This review highlights a large amount of discussion relevant to learn briefly about several herbal supplements that are typically used to promote human health and the status of herbal medications and supplements across various national legislatures.

12.1 INTRODUCTION

Herbal supplements or herbal medicine is the oldest type of treatment and has been utilized for many years in both developed and developing nations. Primitive humans have always relied on nature for their food, clothing, housing, and medical needs. These people were able to discriminate between healthy herbs that were useful and poisonous or inert herbs (Kunle et al., 2012). Around 50,000 plant species are said to have medical characteristics, according to literature (Barboza et al., 2009). As a result, the basis for contemporary pharmaceuticals such as aspirin, morphine, digitoxin, and quinine was developed through the scientific evaluation of herbal medicine (Wachtel-Galor et al., 2011; Pal et al., 2003). The progressive advancement and transmission of knowledge about plant-based medicines served as the basis for many conventional medical systems all over the world (Kunle et al., 2012). Approximately 80% of the earth's population still uses herbal medicine as their primary healthcare provider, especially in underdeveloped nations (Okigbo et al., 2009). We have gained useful formulas for the identification, preparation, and use of herbal treatments as a result of the usage of traditional medicine over time. To make alternative therapies to conventional medicine a viable option, the same rigorous clinical and scientific methodology must be used to confirm the efficacy and safety of curative goods (Pal et al., 2003).

Plants or botanicals are used to make herbal remedies, herbal products, or phytomedicines, which are used to prevent illness or treat it. Products designed primarily for internal use include herbal supplements (Sharma et al., 2020; Rane et al., 2020a, 2020b, 2020c). Many over-the-counter and prescription medications are made from plant compounds. Because they use the US Food and Drug Administration (FDA)-regulated, pure components, they are different from herbal supplements. The FDA does not, however, control the production of herbal supplements. As a result, preparations can range in consistency and contain either a small or large amount of the plant. Herbal supplements can be purchased in liquid or powder form, but are commonly sold in solid form (capsules, pills, tablets, and lozenges) (Furhad et al., 2022).

For thousands of years, people have employed herbal remedies. According to national rules, these products are currently either promoted as herbal medicines or as nutritional supplements in the European Union (EU). The European Commission (EC) has taken steps to ensure free movement in the internal EU market and to standardize the legal framework for herbal goods across the EU. The new European legislation on conventional herbal medicines in theory allows for the registration of a medical product in every member state. Some obstacles must still be overcome in a number of circumstances, though. Considering legal, technical needs, and commercial viewpoints, this assessment evaluates the current and projected state of herbal goods (Hooyenga et al., 2009).

In total, there were 972,073 exposures to pharmaceutical goods that resulted in adverse events in 2006, according to data from the American Association of Poison Control Centre (AAPCC), with 507 fatalities and 6809 serious outcomes (Bronstein et al., 2007). Of these exposures, 76,364 (7.9%) were attributable to dietary supplements and vitamins, with only 3 fatalities (0.6% of all fatalities) and 42 major outcomes (or 0.6% of all major outcomes). Because of this, these products seldom cause fatal side effects or injuries. Only 0.2% of instances in an assessment of hospital admissions in Hong Kong were attributable to adverse responses to herbal products (Chan et al., 1992). From 1983 to 1991, a London unit received 5536 inquiries about traditional medicine and food supplement exposures, but only 12% of the patients exhibited symptoms. Despite the widespread usage of these products, fatalities and bad outcomes rarely occasionally occur. It is reasonable to say that when used properly, nutritional supplements and herbal medicines are generally rather safe.

Descriptions of people using herbs to treat ailing members of society may be found throughout medical records dating back to the beginning of time. However, we saw the emergence of allopathic medicine concurrent with the start of the industrial revolution. Herbal medicine was also a successful kind of treatment, but it was less liked (Tirtha et al., 1998). In the middle of the 20th century, herbal remedies were abandoned for use in conventional medicine, not necessarily because they were ineffective but rather because they were less financially lucrative than the new synthetic pharmaceuticals (Tyler et al., 1999). Early in the 19th century, scientific methods became increasingly sophisticated and popular, and the use of herbal remedies was derided as quackery. The usage of herbal supplements and curiosity in "natural health" rose during the 1960s as a result of worries about the iatrogenic side effects of traditional medicine and a desire for greater independence. The Office of

Alternative Medicine was established by the National Institutes of Health USA in 1992 as a result of awareness of the growing use of herbal medicines and other non-conventional treatments. The WHO urged developing nations to employ traditional plant medicine to treat ailments that modern systems were unable to treat, which gave herbal medicine a boost globally (Winslow and Kroll et al., 1998).

12.2 HISTORICAL POINT OF VIEW OF HERBAL SUPPLEMENTS

According to records from ancient Babylon, the usage of plants as medicines dates back 60,000 years. Publications on herbal medicine or herbal supplements are at least 5000 years old in Egypt and China, and 2500 years old in Asia Minor and Greece (Qazi and Molvi et al., 2016). There are several herbal medicinal systems, and each one's practices and beliefs are shaped by the area where it initially emerged. Traditional Chinese medicine, which has been practiced for thousands of years, is the national medical system of China. The world's oldest herbal book is still in print today. Around 2000 years ago, the Devine Farmer's Classic of Herbal Remedy was written in China (Wnag et al., 2005). Several herbal pharmacopeias and other monographs on particular herbs are also available through written material on herbs (Wachtel-Galor et al., 2011). Ancient Hindu physicians and saints created the healthcare system known as Ayurveda, which has been practiced in India for over 5000 years. Over 1500 plants and 10,000 formulas are fully described in Ayurveda. Comparing Ayurveda to western medicine, the Indian government has declared it to be a complete healthcare system (Goyal and Sasmal et al., 2012). With over 148 formulas, Kampo medicine is a Japanese herbal remedy that has been used for more than 1500 years (Watanabe and Matsuura et al., 2011). Ayurveda is an approximately 5000-year-old Indian medical system. The Atharvaveda, which dates from 1500 to 1000 BC, is where the Ayurvedic text is found. Thousands of years of oral transmission resulted in the ancient Ayurvedic books being preserved and written down in Sanskrit. However, the primary goal of traditional Ayurveda was to improve health rather than treat illness. In Ayurveda, there are eight different fields of study: Kaya Chikitsa (General Medicine), Goyal M, Sasmal Bhruthya (Paediatrics), Bhutha Vidhya (Psychiatry), Salakya (ENT and Ophthalmology and dentistry), Shalya (Surgery), Agada Tantra (Toxicology), Rasayana (Rejuvenation Therapy) and Vajeekarana (sexual vitality) (Goyal and Sasmal et al., 2012). One of the greatest early Greek philosophers and natural scientists, Theophrastus (about 300 BC), is referred to be the "father of botany." He researched the medicinal properties and quirks of herbs and wrote nearly 227 volumes as a result (Firenzuoli et al., 2007). Natural History, written by Pliny the Elder, has influenced science up to the Middle Ages. There are 37 books in The Natural History (Popa et al., 2010) (130–200 AD) was a physician and pharmacist who worked in Rome. Galen's ideas for making and compounding medications governed the Western world for 1500 years, and his name is still attached to the category of pharmaceuticals known as galenicals that are manufactured mechanically (Rogers et al., 2010).

12.3 COMMONLY USE HERBAL SUPPLEMENTS

12.3.1 SAW PALMETTO

To treat benign prostatic hyperplasia (BPH), a benign enlargement of the prostate, men frequently take herbal supplements containing saw palmetto (*Serenoa repens*). Men suffering from prostate cancer and BPH are both frequent consumers of these supplements (Penugonda and Lindshield et al., 2013). Saw palmetto fruit is used to address issues with the urinary and reproductive systems. Saw palmetto extract is frequently utilized to treat BPH (Kwon et al., 2019). Saw palmetto extracts are distinct from other extracts, vegetable oils, and nut oils in that they are a rich source of the saturated, medium-chain fatty acids laurate (12:0) and myristate (14:0). Saw palmetto extracts are mostly composed of fatty acids of about 90% (Schantz et al., 2008). Recent clinical investigations have found their effectiveness at a dose of 320 mg daily for BPH (Barry et al., 2011).

12.3.2 GARLIC

One of the oldest and most significant herbs that have been utilized in traditional medicine since ancient times is garlic (*Allium sativum* L.; Family: Amaryllidaceae), an aromatic herbaceous plant (El-Saber et al., 2020). Allium species and their active ingredients are said to have capabilities that are validated by epidemiological data from human clinical research to lower the risk of diabetes and cardiovascular disorders, defend against infections by boosting the immune system, and have antimicrobial, antifungal, antiaging, and anticancer properties (Rahman et al., 2001). Numerous phytochemicals have been found in *A. sativum* bulbs, including ajoenes, thiosulfinates (allicin), vinyldithiins, sulfides (diallyl disulfide, diallyl trisulfide), and others that made up 82% of the total sulfur content of garlic (Al-Snafi et al., 2013). S-allyl-cysteine (SAC) has anti-inflammatory, antioxidant, controlled redox, proenergetic, antiapoptotic, and signaling properties, whereas S-allyl-mercapto cysteine (SAMC) exhibits an anticancer effect by inhibiting the cancer cell proliferation (Liu et al., 2015). From alliin, SAC and SAMC are derived. Allicin (allyl thiosulfinate) is attributed to antioxidant activity (Lawson and Hunsaker et al., 2018). For type 2 diabetics, dietary intake is a daily intake of about 0.5–1.5 g of allicin substantially reduced HbA1c levels within 12 weeks (Wang et al., 2017). Significant blood pressure lowering was observed with a dose of 240–960 mg of old garlic extract containing 0.6–2.4 of S-allylcysteine throughout a 12-week period (Bayan et al., 2014).

12.3.3 GINGKO BILOBA

Standardized extract of *G. biloba* leaves (EGb 761) is one of the most popular phytomedicines worldwide. This extract has 24% flavanol glycosides, 6% terpene lactones, and significant levels of additional compounds such as proanthocyanidins and organic acids (DeFeudis and Drieu et al., 2000). Many phytochemical components, including terpenes and trilactones (ginkgolides and bilobalides), flavonoids, and other substances, have been identified as *G. biloba*'s phytochemical constituents (Isah et al., 2015). Recent research has demonstrated that *G. biloba* L. leaf extract is effective in treating Alzheimer's disease, neurodegenerative diseases, cerebral insufficiencies, neurosensory issues (such as tinnitus and vertigo), eye ailments, vascular insufficiencies, age-related memory deficits, and oxidative stress. It has been demonstrated that ginkgolide B may have therapeutic benefits such as antiangiogenesis, antiinflammatory, and antiasthmatic properties (Mohanta et al., 2014).

12.3.4 ECHINACEA

The Asteraceae family's genus *Echinacea* contains a variety of species, three of which are considered as medicinal herbs: *Echinacea angustifolia* (DC. Hell.), *Echinacea pallida* (Nutt.), and *Echinacea purpurea* (L.) Moench (Barnes et al., 2005). From *Echinacea* species, numerous important bioactive substances with pharmacological properties have been identified. The most essential components of *E. purpurea* (L.) Moench include alkyl amides, polysaccharides, glycoproteins, flavonoids, and phenolic compounds, which include derivatives such as caffeic acid, chicoric acid, caftaric acid, chlorogenic acid, and echinacoside (Burlou-Nagy et al., 2022). Echinacea has been the subject of numerous studies on its effects on immune system stimulation or on conditions such as the common cold and other upper respiratory tract diseases. Herpes simplex virus strains that are acyclovir-resistant and acyclovir-susceptible can both be treated with *E. purpurea* (L.) Moench aqueous extract. *Echinacea* species is also used to treat snakebites, heal wound infections, and as a primitive antibiotic. The immunomodulatory activities of the roots of *E. purpurea* are chemically regulated by glycoprotein, alkyl amides, and polysaccharides (Aarland et al., 2017).

12.3.5 BLACK COHOSH

Black cohosh root (*Cimicifuga racemosa*, formerly *Actaea racemosa*) consists of triterpene glycosides and phenolic acids are two widely distributed chemical compounds. The major components are triterpene glycosides. Actein and 23-epi-26-deoxyactein are two of the most prevalent triterpenes, and they are commonly employed as markers for the standardization of black cohosh preparations (Nikolić et al., 2015). Black cohosh is one of the herbal supplements for menopausal symptoms that have undergone the most thorough research (Viereck et al., 2005; Tiran et al., 2006). A number of female problems, such as pain during childbirth, uterine colic, and dysmenorrhea, have traditionally been treated with extracts of the rhizome of *C. racemosa* (Borrelli and Ernst et al., 2008). The rhizome of *C. racemosa* includes phenolic acids (isoferulic acid and fukinolic acid), flavonoids, volatile oils, tannins, and other important bioactive substances. It also contains triterpene glycosides (actein, 23-epi26-deoxyactein (formerly known as 27-deoxyactein), and cimicifugoside) (Nuntanakorn et al., 2006). The serotonergic system may be modulated by black cohosh, which is generally considered not to be estrogenic; this could be comparable to how antidepressants work in their effects (Dietz et al., 2016).

12.3.6 GINSENG

Ginseng, a member of the Araliaceae family and the genus *Panax*, is a popular herbal supplement due to its potent therapeutic benefits (Wang et al., 2019). There are 17 species in the *Panax* genus, but the three most frequently utilized as functional foods and medicines are *P. ginseng* (Asian or Korean ginseng), *Panax quinquefolius* (American ginseng), and *Panax notoginseng* (Chinese ginseng) (Mohanan et al., 2018; Zhang et al., 2020). There are nearly 30 triterpene glycosides, or ginsenosides, that are the bioactive components in ginseng. According to their aglycone moieties, these glycosides are either categorized as the 20(S) protopanaxadiol group (PPD), which includes the ginsenosides Rb1, Rb2, Rg3, Rc, and Rd, or the 20(S) protopanaxatriol (PPT), which contains the ginsenosides Re, Rg1, Rg2, and Rh1 (Mancuso and Santangelo et al., 2017). Ginsenosides Rb1, Rb2, Rc, Re, and Rg1 make up the majority of ginsenosides in fresh ginseng (70–80% of all ginsenosides) (Koh et al., 2015). It has been demonstrated that a standardized *P. ginseng* extract improves cardiac function. Both anticoagulant and antiplatelet actions were seen in *P. ginseng* and *P. quinquefolius* extracts. Numerous preclinical investigations have demonstrated that *P. ginseng* inhibits the growth of cancer cells through a number of methods, including the suppression of angiogenesis and the induction of apoptotic cell death (Sagar et al., 2006; Mancuso and Santangelo et al., 2017).

12.3.7 KAVA

The botanical name for kava is *Piper methysticum*, although it is often referred to as kava kava (Ulbricht et al., 2005). Kava is a member of the Piperaceae family of peppers. It has traditionally been used for its relaxing and medicinal value as a painkiller, a muscle relaxant, and a treatment for anxiety, anxiousness, and insomnia (Bian et al., 2020). The term "kava lactones" refers to a group of lactones that are abundant and found nearly in kava (Rivers et al., 2016). By changing the structural properties of kava components, numerous researchers have strengthened the anti-inflammatory properties (Huck et al., 2019). Inhibition of cell proliferation, activation of apoptosis, and over-expression of cancer suppressor genes are likely the mechanisms by which kava exhibits protective properties against urothelial cell carcinoma (UCC), prostate cancer, and colon cancer (Bian et al., 2020). Kava and kava lactones has also been proven to have neuroprotective properties (Tzeng and Lee et al., 2015)

12.3.8 MILK THISTLE

The milk thistle, often known as wild artichoke (*Silybum marianum* L. Gaernt), is a useful therapeutic herbal supplement. The milk thistle is the part of Asteraceae family. Mothers have utilized the fruits of *S. marinum* to increase milk production. Since the time of ancient physicians and herbalists, *S. marianum* has been used to treat a variety of liver and gallbladder diseases (Abenavoli et al., 2010). Silymarin is an isomeric mixture of specific flavonoid complexes called flavonolignans and makes up 1.5–3% of the dry weight of the fruit. Along with flavonolignans, the milk thistle fruit also contains additional flavonoids, 5,7-dihydroxychromone, dehydroconiferyl alcohol, fixed oils, tocopherol, sterols, carbohydrates, and proteins in its chemical components (Bijak et al., 2017). However, silybin, the primary bioactive component of the extract, which has been validated in numerous studies, has the highest concentration and makes up roughly 50–70% of the extract (Lee et al., 2007).

TABLE 12.1 Some Other Herbal Supplements with Their Used and Scientific Name.

Sr. no	Common name	Scientific name	Application	References
1.	Aloe vera	*Aloe vera*	Applied topically for osteoarthritis, psoriasis, and burns. Additionally, it is taken orally to treat digestive problems including constipation or gastritis.	Sánchez et al. (2020); Hęś et al. (2019)
2.	Chamomile	*Matricaria recutita*	Used topically to treat skin disorders. Furthermore, effective for treating diarrhea, gas, and stomach discomfort	Williams et al. (2021)
3.	Flaxseed	*Linum usitatissimum*	It has been in use for a long time as a laxative and cardiovascular health. Additionally, it is used to treat menopausal symptoms, atherosclerosis, obesity, diabetes mellitus, hyperlipidemia, and premature birth. Both fiber and omega-3 fatty acids are rich in it.	Williams et al. (2021)
4.	Peppermint oil	*Mentha piperita*	Used to treat digestive issues such as indigestion, stomach issues, intestinal disorders, and nausea.	Shams et al. (2015)
5.	Soy	*Glycine max*	Used to address memory issues, high cholesterol, and menopausal symptoms.	Vincent et al. (2007)
6.	St. John's Wort	*Hypericum perforatum*	Commonly used to treat anxiety and depression. Additionally, it is said to be beneficial in treating menopausal symptoms, somatoform disorders, obsessive-compulsive disorder, attention-deficit hyperactivity disorder (ADHD), angioplasty, migraine headaches, premenstrual syndrome, glioma, and polycystic ovary syndrome.	Williams et al. (2021)

TABLE 12.1 *(Continued)*

Sr. no	Common name	Scientific name	Application	References
7.	Tea tree oil	NA	Used topically to treat a variety of disorders, such as dandruff, wounds, athlete's foot, infections, nail fungus, lice, and oral yeast infections.	Pazyar et al. (2013); Hoffmann et al. (2020)
9.	Evening primrose	*Oenothera biennis*	It is used to treat conditions such as atopic dermatitis, chronic fatigue syndrome, dry eye syndrome, dyslexia, premenstrual syndrome, rheumatoid arthritis, obesity, ulcerative colitis, and diabetes mellitus.	Timoszuk et al. (2018); Bayles et al. (2009); Kokke et al. (2008)
10.	Feverfew	*Tanacetum parthenium*	Menstrual cramps and migraine headaches are two conditions it is frequently used to treat Additionally; it has been used to treat menstruation irregularities, arthritis, vertigo, tinnitus, allergies, and psoriasis.	Williams et al. (2021)
11.	Goldenseal	*Hydrastis canadensis*	It is frequently used for colds and the flu and is well known for its healing and antibacterial characteristics. Additionally, it is frequently used to relieve uncomfortable or inflammatory nose lining. Treat gastrointestinal irritability in addition to skin and eye infections.	Furhad et al. (2022)
12.	Hawthorn	*Crataegus monogyna*	It is mostly used to treat heart-related disorders such as angina, heart failure, and excessive blood pressure.	Furhad et al. (2022)

12.4 REGULATIONS FOR HERBAL SUPPLEMENTS

12.4.1 REGULATIONS ACCORDING TO THE FDA

According to the Dietary Supplement Health and Education Act (DSHEA) of 1994, which was updated by the Federal Food, Drug, and Cosmetic Act (the FD&C Act) ((FDA), 2019; Denham et al., 2011; Hathcock et al., 2001; Ross et al., 2000), the US-FDA regulates items advertised as dietary supplements. A product that is designed to supplement the diet and contains one or more of the following dietary elements is considered a dietary supplement or herbal supplement under the FD&C Act. Herbal or Dietary supplements are generally monitored through post-market surveillance and are not authorized at all, in contrast to FDA drug regulation, where safety and efficacy must be established prior to approval. Pre-marketing approval is not necessary for the marketing of dietary or herbal supplements in the United States, but it is the manufacturer's responsibility to make sure that the product is safe for the general public and that the label's instructions are followed. Herbal or Dietary supplements cannot be advertised in the United States as a means of disease diagnosis, treatment, mitigation, or prevention (Thakkar et al., 2020).

12.4.2 REGULATION IN AUSTRALIA

According to the Therapeutic Goods Act (TGA), which was formed in 1989, the majority of natural products, including herbal, vitamin, mineral, and nutritional supplements, are considered as "complementary medicines" in Australia with the goal of promoting improved health and are referred to as "therapeutic goods" (TGA, 2013). The Australian Regulatory Guidance for Complementary Medicine (ARGCM), which offers information on the requirements to be low risk or high risk, provides the regulatory guidance. Currently, not many complementary medicines have gone through the "registered" path; most are only "listed." Products that are "registered" must be examined for their quality, safety, and efficacy. Only 40 complementary supplements or medicines, specifically, are registered (TGA, 2018).

12.4.3 REGULATION IN JAPAN

The majority of modern Japanese legislation, excluding a few exceptions, is based on American legislation. In Japan, foods are classified as either "Foods in General" or "Food with Health Claims" (Tanaka et al., 2004). There are three major categories under "Food with Health Claims" (Shimizu et al., 2003):

i. "Food with Nutrient Function Claims" (FNFC), which are mostly vitamins and minerals,
ii. "Food for Specified Health Uses" (FOSHU), which are other uses, and
iii. "Foods with Function Claims" (FFC).

These groups are governed by several regulatory processes, from the strict FOSHU pathway to the quick FFC pathway (Ohama et al., 2008).

Health Foods may only be promoted as FOSHU by the Consumer Affairs Agency if the claims made on the labeling requirements have been filed for review. For product approval, companies must present proof of the items' safety and efficacy (Martirosyan et al., 2015). Products that are used in Kampo Medicine are referred to as "Medicines" in Japan (Saito et al., 2000). Currently, Japan's national health insurance program covers 148 different types of prescription Kampo medications. In brief, Kampo medicine and chemically manufactured medicine are not differentiated under Japanese legislation (Yu et al., 2006; Thakkar et al., 2020).

12.4.4 REGULATION IN EUROPEAN UNION

In the EU, the two categories "Classified as Medicines" and "Classified as Supplements" are used to classify products. Food supplements that are governed as foods under Directive 2002/46/EC fall under the classification of "Classified as Supplements" (EU) (Gulati and Ottaway et al., 2006; Thakkar et al., 2020). The EU does not have a centralized pre-market authorization process for dietary or herbal supplements. Herbal medical products are "Classified as Medicines" and are governed under Directive 2004/24/EC of the EU (European Commission, 2004) (Fisher and Ward et al., 1994). The European Medicines Agency (EMA) is responsible for the safety, effectiveness, and premarket approval of these herbal products or supplements (Benzi and Ceci et al., 1997; Blumenthal et al., 2000; Calapai et al., 2008).

12.5 CURRENT SCENARIO OF HERBAL SUPPLEMENTS

Since ancient times, many people have used herbal medicine to treat common illnesses like colds, allergies, upset stomachs, and toothaches. This trend is only continuing to increase (Khan et al., 2019). But by the middle of the 20th century, herbal products were no longer used in mainstream medicine. This was due to the fact that they were less monetarily successful than the more modern synthetic medications and because they were less effective (Tyler et al., 1999). Later, with the development of scientific procedures, herbal medicines could be studied and proven to be effective medications. In addition, the 1960s saw a demand for safer and more affordable medications to support "natural health" due to worries about the harmful and iatrogenic consequences of conventional medicine. As a result, the trend worldwide has changed from synthetic to natural medications. Additionally, herbal medicine gained popularity internationally when the WHO encouraged developing nations to employ traditional plant medicine to fulfill requirements that modern systems were unable to address (Miller et al., 1998).

Now, the WHO has acknowledged herbal medicine as an essential part of primary healthcare. Nearly 25% of current medications, according to the WHO, are derived from plants that are also used in traditional medicine. Many other synthetic analogue are made using model chemicals extracted from plants (Khan et al., 2019).

12.5.1 GENERAL OVERVIEW OF HERBAL SUPPLEMENTS IN THE US

In the US, there are 20,000 or more herbal products available, the majority of which lack proof of effectiveness. The US National Centre for Complementary and Integrative Health (NCCIH) has prepared a series of data sheets named "Herbs at a Glance" that cover more than 50 herbs and offer basic herbal knowledge for the general audience (Hassen et al., 2022).

12.5.1.1 CHALLENGES

Users of herbal medicines run the risk of developing severe poisoning and serious health issues since makers of botanical dietary supplements lack knowledge, which is worsened by the general public's lack of risk awareness (Hudson et al., 2018).

12.5.1.2 THE FEDERAL TRADE COMMISSION

Through legal action, advocacy, and outreach, the Federal Trade Commission (FTC) defends consumers and the industry from dishonest and unfair business practices (Hassen et al., 2022).

12.5.2 GENERAL OVERVIEW OF HERBAL SUPPLEMENTS IN THE INDIA

Except for Allopathic, all of India's officially recognized health systems—Ayurveda, Yoga, Unani, Siddha, Homeopathy, and Naturopathy—include a significant portion of herbal medications. 70% of the 1.1 billion people in India still rely on these non-allopathic medical systems. According to the Indian Drugs Act, there is currently no distinct category for herbal medications or nutritional supplements (Vaidya et al., 2007). India is now the world's top exporter of value-added goods (such as essential oils and gum) as well as medical plants, plant pharmaceuticals, and other products. In India, there are around 9000 traditional pharmaceutical manufacturers, the majority of which produce Ayurvedic medicines. Nearly 960 different types of medicinal plants are traded, of which 178 kinds consume more than 100 tons annually. It is estimated that 2000 tons of medicinal plant raw materials are required annually. Of these, more than 1500 different types of herbs are also sold as dietary supplements or traditional ethnic medicines. The market for health products is estimated to be about USD 1 billion and is expanding, while the export value of Indian medicinal plants and related items is around USD 110 billion (Sen et al., 2017).

India is home to 8% of the world's plant resources, at least 44% of which can be turned into pharmaceuticals. However, many conventional medicine patents have been copied as a result of a lack of technical tools, care, and understanding of protecting rights. A number of actions have been taken by the Indian government to protect and rationally develop traditional medical knowledge in view of the universality and seriousness of this situation, including the revision of the patent law, the collection and filing of traditional medical knowledge, the creation of a registration and invention patent system, and the creation of a special investment foundation. Indian traditional medicine patents have been well protected due to these strategies (Yu et al., 2020).

12.6 HERBAL DRUG INTERACTIONS

Herbal-drug interactions might show up pharmacodynamically or pharmacokinetically. In general, pharmacodynamic interactions are less common and entail the interacting agents' direct pharmacologic activities, which are independent of changes in the blood concentrations of the herb or medicine (Ramirez et al., 2022). Pharmacokinetic interactions, which impact both the drug's concentration in the blood and its pharmacologic activity, are the most common form of clinically significant herb-drug interactions (Williams et al., 2021). This kind of interaction involves the passage of the drug from absorption to excretion through the body. When a coadministered medicine and the herbal supplement share the same mechanisms of absorption, distribution, metabolism, or excretion, a pharmacokinetic interaction risk arises (Makino et al., 2002). The medicine's concentration at the site of action could alter as a result of competition between the herbal supplement and drug for a common ADME mechanism (Ramirez et al., 2022).

12.7 CONCLUSION

Worldwide, several disorders are treated and managed with the use of herbal supplements. Some of the natural herbal supplements discussed in this review here have risk–benefit ratios that seem favorable are ginkgo, garlic, echinacea, black cohosh, ginseng, kava kava, milk thistle, and saw palmetto. Despite the fact that they include natural substances, many believe that they are safe and healthy for them. As a result of acute toxicity or chronic/cumulative toxicity, there have been a number of recorded adverse effects of herbals, ranging from minor side effects to clinically severe disorders. It is sometimes argued that the expensive expenditures of long-term research cannot be sustained by the herbal industry, because herbal supplements are usually not patented. However, since one-third to one-half of all pharmaceuticals were originally produced from plants, it is obvious that there is a chance to find efficient treatments in the environment. The existing regulation of herbs does not guarantee the safety of the goods that are on the market, and misleading and unlawful marketing claims are frequent. The appropriate usage of herbs could be significantly improved by making a few little improvements to the regulations governing these goods.

12.8 FUTURE PROSPECTIVE

Many specialists are concerned about the current lack of quality control and standardization of herbal therapeutic supplements. Whenever such supplements are independently examined, the findings reveal that a concerning percentage is either under-dosed or contaminated (for instance, by herbicides, pesticides, or heavy metals). It is evident that this condition is intolerable. It endangers the consumers and eventually works against the industry's interests. These issues should be addressed through regulation of the herbal sector by establishing quality standards and thereafter labeling noncompliant producers as such. While this is going on, consumers and healthcare professionals are unsure of which brands of herbal medications to purchase or suggest.

KEYWORDS

- **herbal supplements**
- **herb–drug interaction**
- **regulations**
- **safety**
- **marketing authorization**
- **disorders**

REFERENCES

Aarland, R. C.; Bañuelos-Hernández, A. E.; Fragoso-Serrano, M.; Sierra-Palacios, E. del C.; Díaz de León-Sánchez, F.; Pérez-Flores, L. J.; Rivera-Cabrera, F.; Mendoza-Espinoza, J. A. Studies on Phytochemical, Antioxidant, Anti-Inflammatory, Hypoglycaemic and Antiproliferative Activities of *Echinacea purpurea* and *Echinacea angustifolia* Extracts. *Pharm. Biol.* **2016,** *55* (1), 649–656. https://doi.org/10.1080/13880209.2016.1265989.

Abenavoli, L.; Capasso, R.; Milic, N.; Capasso, F. Milk Thistle in Liver Diseases: Past, Present, Future. *Phytother. Res.* **2010,** *24* (10), 1423–1432. https://doi.org/10.1002/ptr.3207.

Ahmad Khan, M. S.; Ahmad, I. Herbal Medicine. *New Look Phytomed.* **2019,** 3–13. https://doi.org/10.1016/b978-0-12-814619-4.00001-x.

Al-Snafi, A. Pharmacological Effects of Allium Species Grown in Iraq. An overview. *Int. J. Pharm. Health Care Res.* **2013,** *1*, 132–147.

Barboza, G. E.; Cantero, J. J.; Núñez, C.; Pacciaroni, A.; Ariza Espinar, L. Medicinal Plants: A General Review and a Phytochemical and Ethnopharmacological Screening of the Native Argentine Flora. *Kurtziana* **2009,** *34* (1–2), 7–365.

Barnes, J.; Anderson, L. A.; Gibbons, S.; Phillipson, J. D. Echinacea species (*Echinacea angustifolia* (DC.) Hell., *Echinacea pallida* (Nutt.) Nutt., *Echinacea purpurea* (L.) Moench): A Review of Their Chemistry, Pharmacology and Clinical Properties. *J. Pharm. Pharmacol.* **2005,** *57* (8), 929–954. https://doi.org/10.1211/0022357056127.

Barry, M. J. Effect of Increasing Doses of Saw Palmetto Extract on Lower Urinary Tract Symptoms. *JAMA* **2011,** *306* (12), 1344. https://doi.org/10.1001/jama.2011.1364.

Bayan, L.; Koulivand, P. H.; Gorji, A. Garlic: A Review of Potential Therapeutic Effects. *Avicenna J. Phytomed.* **2014,** *4* (1), 1–14.

Bayles, B.; Usatine, R. Evening Primrose Oil. *Am. Fam. Phys.* **2009,** *80* (12), 1405–1408.

Benzi, G.; Ceci, A. Herbal Medicines in European Regulation. *Pharmacol. Res.* **1997,** *35* (5), 355–362. https://doi.org/10.1006/phrs.1997.0132.

Benzie, I. F. F.; Wachtel-Galor, S. *Herbal Medicine: Biomolecular and Clinical Aspects*, 2nd ed.; CRC Press, Cop: Boca Raton, 2011.

Bian, T.; Corral, P.; Wang, Y.; Botello, J.; Kingston, R.; Daniels, T.; Salloum, R. G.; Johnston, E.; Huo, Z.; Lu, J.; Liu, A. C.; Xing, C. Kava as a Clinical Nutrient: Promises and Challenges. *Nutrients* **2020,** *12* (10). https://doi.org/10.3390/nu12103044.

Bijak, M. Silybin, a Major Bioactive Component of Milk Thistle (*Silybum marianum* L. Gaernt.)—Chemistry, Bioavailability, and Metabolism. *Molecules* **2017,** *22* (11), 1942. https://doi.org/10.3390/molecules22111942.

Blumenthal, M., Goldberg, A., Brinckmann, J. *Herbal Medicine: Expanded Commission E Monographs*; Integrative Medicine Communications, 2000.

Borrelli, F.; Ernst, E. Black Cohosh (*Cimicifuga racemosa*) for Menopausal Symptoms: A Systematic Review of Its Efficacy. *Pharmacol. Res.* **2008,** *58* (1), 8–14. https://doi.org/10.1016/j.phrs.2008.05.008.

Bronstein, A. C.; Spyker, D. A.; Cantilena, L. R.; Green, J.; Rumack, B. H.; Heard, S. E. 2006 Annual Report of the American Association of Poison Control Centers' National Poison Data System (NPDS). *Clin. Toxicol.* **2007,** *45* (8), 815–917. https://doi.org/10.1080/15563650701754763.

Burlou-Nagy, C.; Bănică, F.; Jurca, T.; Vicaş, L. G.; Marian, E.; Muresan, M. E.; Bácskay, I.; Kiss, R.; Fehér, P.; Pallag, A. *Echinacea purpurea* (L.) Moench: Biological and Pharmacological Properties: A Review. *Plants* **2022,** *11* (9), 1244. https://doi.org/10.3390/plants11091244.

Calapai, G. European Legislation on Herbal Medicines: A Look into the Future. *Drug Saf.* **2008,** *31* (5), 428–431. https://doi.org/10.2165/00002018-200831050-00009.

Chan, T. Y.; Chan, A. Y.; Critchley, J. A. Hospital Admissions due to Adverse Reactions to Chinese Herbal Medicines. *J. Trop. Med. Hygiene* **1992,** *95* (4), 296–298.

DeFeudis, F.; Drieu, K. *Ginkgo biloba* Extract (EGb 761) and CNS Functions Basic Studies and Clinical Applications. *Curr. Drug Targets* **2000,** *1* (1), 25–58. https://doi.org/10.2174/1389450003349380.

Denham, B. E. Dietary Supplements—Regulatory Issues and Implications for Public Health. *JAMA* **2011,** *306* (4), 428. https://doi.org/10.1001/jama.2011.982.

Dietz, B. M.; Hajirahimkhan, A.; Dunlap, T. L.; Bolton, J. L. Botanicals and Their Bioactive Phytochemicals for Women's Health. *Pharmacol. Rev.* **2016,** *68* (4), 1026–1073. https://doi.org/10.1124/pr.115.010843.

El-Saber Batiha, G.; Magdy Beshbishy, A.; G. Wasef, L.; Elewa, Y. H. A.; A. Al-Sagan, A.; Abd El-Hack, M. E.; Taha, A. E.; M. Abd-Elhakim, Y.; Prasad Devkota, H. Chemical Constituents and Pharmacological Activities of Garlic (*Allium Sativum* L.): A Review. *Nutrients* **2020,** *12* (3), 872. https://doi.org/10.3390/nu12030872.

FDA. New Dietary Ingredients (NDI) Notification Process, 2019. https://www.fda.gov/food/dietary-supplements/new-dietary-ingredients-ndi-notification-process

Fisher, P.; Ward, A. Medicine in Europe: Complementary Medicine in Europe. *BMJ* **1994,** *309* (6947), 107–111. https://doi.org/10.1136/bmj.309.6947.107.

Furhad, S.; Bokhari, A. A. Herbal Supplements, 2022. https://www.ncbi.nlm.nih.gov/books/NBK536964/.

Goyal, M.; Sasmal, D.; Nagori, B. P. Ayurveda the Ancient Science of Healing: An Insight. *Drug Discov. Res. Pharmacogn.* **2012**. https://doi.org/10.5772/31818.

Gulati, O. P.; Berry Ottaway, P. Legislation Relating to Nutraceuticals in the European Union with a Particular Focus on Botanical-Sourced Products. *Toxicology* **2006,** *221* (1), 75–87. https://doi.org/10.1016/j.tox.2006.01.014.

Hassen, G.; Belete, G.; Carrera, K. G.; Iriowen, R. O.; Araya, H.; Alemu, T.; Solomon, N.; Bam, D. S.; Nicola, S. M.; Araya, M. E.; Debele, T.; Zouetr, M.; Jain, N. Clinical Implications of Herbal Supplements in Conventional Medical Practice: A US Perspective. *Cureus* **2022,** *14* (7). https://doi.org/10.7759/cureus.26893.

Hathcock, J. Dietary Supplements: How They Are Used and Regulated. *J. Nutr.* **2001,** *131* (3), 1114S1117S. https://doi.org/10.1093/jn/131.3.1114s.

Hęś, M.; Dziedzic, K.; Górecka, D.; Jędrusek-Golińska, A.; Gujska, E. Aloe Vera (L.) Webb: Natural Sources of Antioxidants—a Review. *Plant Foods Human Nutr.* **2019,** *74* (3), 255–265. https://doi.org/10.1007/s11130-019-00747-5.

Hoffmann, J.; Gendrisch, F.; Schempp, C. M.; Wölfle, U. New Herbal Biomedicines for the Topical Treatment of Dermatological Disorders. *Biomedicines* **2020,** *8* (2), 27. https://doi.org/10.3390/biomedicines8020027.

Hooyenga, P.; Witkamp, R.; Groen, K. Herbal Products: Marketing Strategies and Legislation. *Int. J. Green Pharm.* **2009,** *3* (4), 270. https://doi.org/10.4103/0973-8258.59730.

Huck, O.; Han, X.; Mulhall, H.; Gumenchuk, I.; Cai, B.; Panek, J.; Iyer, R.; Amar, S. Identification of a Kavain Analog with Efficient Anti-Inflammatory Effects. *Sci. Rep.* **2019,** *9* (1). https://doi.org/10.1038/s41598-019-49383-8.

Hudson, A.; Lopez, E.; Almalki, A. J.; Roe, A. L.; Calderón, A. I. A Review of the Toxicity of Compounds Found in Herbal Dietary Supplements. *Planta Medica* **2018,** *84* (09/10), 613–626. https://doi.org/10.1055/a-0605-3786.

Isah, T. Rethinking *Ginkgo biloba* L.: Medicinal Uses and Conservation. *Pharmacogn. Rev.* **2015,** *9* (18), 140. https://doi.org/10.4103/0973-7847.162137.

Johnson, D. A.; Shams, R.; Oldfield, E. C.; Copare, J. *Peppermint Oil: Clinical Uses in the Treatment of Gastrointestinal Diseases*, 2015.

Kamboj, V. P. Herbal Medicine. *Curr. Sci.* **2000,** *78* (1), 35. http://www.sciepub.com/reference/189774 (accessed 8 Apr 2022).

Koh, E.; Jang, O.-H.; Hwang, K.-H.; An, Y.-N.; Moon, B. Effects of Steaming and Air-Drying on Ginsenoside Composition of Korean Ginseng (*Panax ginseng* C.A. Meyer). *J. Food Process. Preserv.* **2014,** *39* (2), 207–213. https://doi.org/10.1111/jfpp.12412.

Kokke, K. H.; Morris, J. A.; Lawrenson, J. G. Oral Omega-6 Essential Fatty Acid Treatment in Contact Lens Associated Dry Eye. *Contact Lens Anterior Eye* **2008,** *31* (3), 141–146. https://doi.org/10.1016/j.clae.2007.12.001.

Kunle. Standardization of Herbal Medicines—A Review. *Int. J. Biodivers. Conserv.* **2012,** *4* (3). https://doi.org/10.5897/ijbc11.163.

Kwon, Y. Use of Saw Palmetto (*Serenoa repens*) Extract for Benign Prostatic Hyperplasia. *Food Sci. Biotechnol.* **2019,** *28* (6), 1599–1606. https://doi.org/10.1007/s10068-019-00605-9.

Lawson, L.; Hunsaker, S. Allicin Bioavailability and Bioequivalence from Garlic Supplements and Garlic Foods. *Nutrients* **2018,** *10* (7), 812. https://doi.org/10.3390/nu10070812.

Lee, J. I.; Narayan, M.; Barrett, J. S. Analysis and Comparison of Active Constituents in Commercial Standardized Silymarin Extracts by Liquid Chromatography—Electrospray Ionization Mass Spectrometry. *J. Chromatogr. B* **2007,** *845* (1), 95–103. https://doi.org/10.1016/j.jchromb.2006.07.063.

Lee, M.-J.; Tzeng, Y.-M. Neuroprotective Properties of Kavalactones. *Neural Regen. Res.* **2015,** *10* (6), 875. https://doi.org/10.4103/1673-5374.158335.

Liu, Y.; Yan, J.; Han, X.; Hu, W. Garlic-Derived Compound S-Allylmercaptocysteine (SAMC) Is Active against Anaplastic Thyroid Cancer Cell Line 8305C (HPACC). *Technol. Health Care* **2015,** *23* (s1), S89–S93. https://doi.org/10.3233/thc-150936.

Makino, T.; Wakushima, H.; Okamoto, T.; Okukubo, Y.; Deguchi, Y.; Kano, Y. Pharmacokinetic Interactions between Warfarin and Kangen-Karyu, a Chinese Traditional Herbal Medicine, and Their Synergistic Action. *J. Ethnopharmacol.* **2002,** *82* (1), 35–40. https://doi.org/10.1016/s0378-8741(02)00175-7.

Mancuso, C.; Santangelo, R. *Panax ginseng* and *Panax quinquefolius*: From Pharmacology to Toxicology. *Food Chem. Toxicol.* **2017,** *107*, 362–372. https://doi.org/10.1016/j.fct.2017.07.019.

Martirosyan, D. M.; Singh, J. A New Definition of Functional Food by FFC: What Makes a New Definition Unique? *Funct. Foods Health Dis.* **2015,** *5* (6), 209. https://doi.org/10.31989/ffhd.v5i6.183.

Miller, L. G. Herbal Medicinals. *Arch. Intern. Med.* **1998,** *158* (20), 2200. https://doi.org/10.1001/archinte.158.20.2200.

Mohanan, P.; Subramaniyam, S.; Mathiyalagan, R.; Yang, D.-C. Molecular Signaling of Ginsenosides Rb1, Rg1, and Rg3 and Their Mode of Actions. *J. Ginseng Res.* **2018,** *42* (2), 123–132. https://doi.org/10.1016/j.jgr.2017.01.008.

Mohanta, T. K.; Tamboli, Y.; Zubaidha, P. K. Phytochemical and Medicinal Importance of *Ginkgo biloba* L. *Nat. Prod. Res.* **2014,** *28* (10), 746–752. https://doi.org/10.1080/14786419.2013.879303.

Nikolić, D.; Lankin, D. C.; Cisowska, T.; Chen, S.-N.; Pauli, G. F.; van Breemen, R. B. Nitrogen-Containing Constituents of Black Cohosh: Chemistry, Structure Elucidation, and Biological Activities. *Recent Adv. Phytochem.* **2015,** *45*, 31–75. https://doi.org/10.1007/978-3-319-20397-3_2.

Nuntanakorn, P.; Jiang, B.; Einbond, L. S.; Yang, H.; Kronenberg, F.; Weinstein, I. B.; Kennelly, E. J. Polyphenolic Constituents of *Actaea racemosa*. *J. Nat. Prod.* **2006,** *69* (3), 314–318. https://doi.org/10.1021/np0501031.

Ohama, H.; Ikeda, H.; Moriyama, H. Health Foods and Foods with Health Claims in Japan. *Toxicology* **2006,** *221* (1), 95–111. https://doi.org/10.1016/j.tox.2006.01.015.

Okigbo, R.; Anuagasi, C.; Amadi, J. Advances in Selected Medicinal and Aromatic Plants Indigenous to Africa. *J. Med. Plants Res.* **2009,** *3* (2), 86-095.

Pal, S. K.; Shukla, Y. Herbal Medicine: Current Status and the Future. *Asian Pac. J. Cancer Prev* **2003,** *4* (4), 281–288.

Pazyar, N.; Yaghoobi, R.; Bagherani, N.; Kazerouni, A. A Review of Applications of Tea Tree Oil in Dermatology. *Int. J. Dermatol.* **2012,** *52* (7), 784–790. https://doi.org/10.1111/j.1365-4632.2012.05654.x.

Penugonda, K.; Lindshield, B. Fatty Acid and Phytosterol Content of Commercial Saw Palmetto Supplements. *Nutrients* **2013,** *5* (9), 3617–3633. https://doi.org/10.3390/nu5093617.

Phua, D. H.; Zosel, A.; Heard, K. Dietary Supplements and Herbal Medicine Toxicities—When to Anticipate Them and How to Manage Them. *Int. J. Emerg. Med.* **2009,** *2* (2), 69–76. https://doi.org/10.1007/s12245-009-0105-z.

Qazi, M. A; Molvi, K. Herbal Medicine: A Comprehensive Review. *J. Pharm. Res.* **2016,** *8* (2), 1–5.

Rahman, K. Historical Perspective on Garlic and Cardiovascular Disease. *J. Nutr.* **2001,** *131* (3), 977S979S. https://doi.org/10.1093/jn/131.3.977s.

Ramirez, L. I.; Kanwugu, O. N.; Ivantsova, M. N. Impact of Herbal Supplements Nowadays: An Overview. *Chim. Techno Acta* **2022,** *9*, 202292S4. https://doi.org/10.15826/chimtech.2022.9.2.s4.

Rane, B. R.; Patil, A. S.; Keservani, R. K.; Jain, A. S. Novel Approaches in Herbal Formulation, In: *Enhancing the Therapeutic Efficacy of Herbal Formulations through Novel Drug Delivery Systems*; Kesharwani, R. K., Keservani, R. K., Sharma, A. K., Eds.; IGI Global International Publisher: Pennsylvania, 2020a; Chapter 2, pp 43–68.

Rane, B. R.; Bharath, M. S.; Patil, R. R.; Keservani, R. K., Jain, A. S. Novel Approaches in Nutraceuticals, In: *Enhancing the Therapeutic Efficacy of Herbal Formulations Through Novel Drug Delivery Systems*; Kesharwani, R. K., Keservani, R. K., Sharma, A. K., Eds.; IGI Global International Publisher: Pennsylvania, 2020b; Chapter 11, pp 241–266.

Rane, B. R.; Tadavi, S. A.; Keservani, R. K. Naturopathy. In: *Herbal Product Development*; Sharma, A. K., Keservani, R. K., Prakash Gautam, S., Eds.; Apple Academic Press, CRC Press, Taylor & Francis Group, 2020c; Chapter 12, pp 321–347.

Rivers, Z. Kava as a Pharmacotherapy of Anxiety Disorders: Promises and Concerns. *Med. Chem.* **2016,** *6* (2). https://doi.org/10.4172/2161-0444.1000329.

Ross, S. Functional Foods: The Food and Drug Administration Perspective. *Am. J. Clin. Nutr.* **2000,** *71* (6), 1735S1738S. https://doi.org/10.1093/ajcn/71.6.1735s.

Sagar, S. M.; Yance, D.; Wong, R. K. Natural Health Products That Inhibit Angiogenesis: A Potential Source for Investigational New Agents to Treat Cancer—Part 2. *Curr. Oncol.* **2006,** *13* (3), 99–107. https://doi.org/10.3747/co.v13i3.88.

Saito, H. Regulation of Herbal Medicines in JAPAN. *Pharmacol. Res.* **2000,** *41* (5), 515–519. https://doi.org/10.1006/phrs.1999.0645.

Sánchez, M.; González-Burgos, E.; Iglesias, I.; Gómez-Serranillos, M. P. Pharmacological Update Properties of Aloe Vera and Its Major Active Constituents. *Molecules* **2020,** *25* (6), 1324. https://doi.org/10.3390/molecules25061324.

Schantz, M. M.; Bedner, M.; Long, S. E.; Molloy, J. L.; Murphy, K. E.; Porter, B. J.; Putzbach, K.; Rimmer, C. A.; Sander, L. C.; Sharpless, K. E.; Thomas, J. B.; Wise, S. A.; Wood, L. J.; Yen, J. H.; Yarita, T.; NguyenPho, A.; Sorenson, W. R.; Betz, J. M. Development of Saw Palmetto (*Serenoa repens*) Fruit and Extract Standard Reference Materials. *Anal. Bioanal. Chem.* **2008,** *392* (3), 427–438. https://doi.org/10.1007/s00216-008-2297-0.

Sen, S.; Chakraborty, R. Revival, Modernization and Integration of Indian Traditional Herbal Medicine in Clinical Practice: Importance, Challenges and Future. *J. Tradit. Complement. Med.* **2017,** *7* (2), 234–244. https://doi.org/10.1016/j.jtcme.2016.05.006.

Sharma, A. K.; Keservani, R. K.; Gautam Surya, P. *Herbal Product Development*; Apple Academic Press, CRC Press, Taylor & Francis Group, 2020.

Shimizu, T. Health Claims on Functional Foods: The Japanese Regulations and an International Comparison. *Nutr. Res. Rev.* **2003,** *16* (2), 241–252. https://doi.org/10.1079/nrr200363.

TGA. *Listed Complementary Medicines*, Vol. 2019; 2018.

Tanaka, H.; Kaneda, F.; Suguro, R.; Baba, H. Current System for Regulation of Health Foods in Japan. *Jpn Med. Assoc. J.* **2004**. *47* (9), 436-440.

Thakkar, S.; Anklam, E.; Xu, A.; Ulberth, F.; Li, J.; Li, B.; Hugas, M.; Sarma, N.; Crerar, S.; Swift, S.; Hakamatsuka, T.; Curtui, V.; Yan, W.; Geng, X.; Slikker, W.; Tong, W. Regulatory Landscape of Dietary Supplements and Herbal Medicines from a Global Perspective. *Regul. Toxicol. Pharmacol.* **2020,** *114*, 104647. https://doi.org/10.1016/j.yrtph.2020.104647.

Timoszuk, M.; Bielawska, K.; Skrzydlewska, E. Evening Primrose (*Oenothera biennis*) Biological Activity Dependent on Chemical Composition. *Antioxidants* **2018,** *7* (8), 108. https://doi.org/10.3390/antiox7080108.

Tiran, D. Integrated Healthcare: Herbal Remedies for Menopausal Symptoms. *Br. J. Nurs.* **2006,** *15* (12), 645–648. https://doi.org/10.12968/bjon.2006.15.12.21395.

Tirtha, S. S. S. Overview of Ayurveda. In *The Ayurveda Encyclopedia: Natural Secrets to Healing, Prevention and Longevity*;. Khalsa, A. K., Satyaguru, R. P.; 1998; pp 3–11.

Tyler, V. E. Phytomedicines: Back to the Future. *J. Nat. Prod.* **1999,** *62* (11), 1589–1592. https://doi.org/10.1021/np9904049.

Ulbricht, C.; Basch, E.; Boon, H.; Ernst, E.; Hammerness, P.; Sollars, D.; Tsourounis, C.; Woods, J.; Bent, S. Safety Review of Kava (*Piper methysticum*) by the Natural Standard Research Collaboration. *Expert Opin. Drug Saf.* **2005,** *4* (4), 779–794. https://doi.org/10.1517/14740338.4.4.779.

Vaidya, A. D. B.; Devasagayam, T. P. A. Recent Advances in Indian Herbal Drug Research Guest Editor: Thomas Paul Asir Devasagayam Current Status of Herbal Drugs in India: An Overview. *J. Clin. Biochem. Nutr.* **2007,** *41* (1), 1–11. https://doi.org/10.3164/jcbn.2007001

Viereck, V.; Emons, G.; Wuttke, W. Black Cohosh: Just Another Phytoestrogen? *Trends Endocrinol. Metabol.* **2005,** *16* (5), 214–221. https://doi.org/10.1016/j.tem.2005.05.002.

Vincent, A.; Fitzpatrick, L. A. Soy Isoflavones: Are They Useful in Menopause? *Mayo Clin. Proc.* **2000,** *75* (11), 1174–1184. https://doi.org/10.4065/75.11.1174.

Wang, H.; Xu, F.; Wang, X.; Kwon, W.-S.; Yang, D.-C. Molecular Discrimination of Panax ginseng Cultivar K-1 Using Pathogenesis-Related Protein 5 Gene. *J. Ginseng Res.* **2019,** *43* (3), 482–487. https://doi.org/10.1016/j.jgr.2018.07.001.

Wang, J.; Zhang, X.; Lan, H.; Wang, W. Effect of Garlic Supplement in the Management of Type 2 Diabetes Mellitus (T2DM): A Meta-Analysis of Randomized Controlled Trials. *Food Nutr. Res.* **2017,** *61* (1), 1377571. https://doi.org/10.1080/16546628.2017.1377571.

Watanabe, K.; Matsuura, K.; Gao, P.; Hottenbacher, L.; Tokunaga, H.; Nishimura, K.; Imazu, Y.; Reissenweber, H.; Witt, C. M. Traditional Japanese Kampo Medicine: Clinical Research between Modernity and Traditional Medicine—The State of Research and Methodological Suggestions for the Future. *Evid. Based Complement. Altern. Med.* **2011,** *2011*, 1–19. https://doi.org/10.1093/ecam/neq067.

Williams, C.T. Herbal Supplements: Precautions and Safe Use. *Nurs. Clin. North Am.* **2021,** *56* (1), 1–21. https://doi.org/10.1016/j.cnur.2020.10.001.

Winslow, L. C.; Kroll, D. J. Herbs as Medicines. *Arch. Intern. Med.* **1998,** *158* (20), 2192. https://doi.org/10.1001/archinte.158.20.2192.

Wnag S, Li Y. Trdaitiional Chinese medicine. In: *Complementary and Alternative Therapies for Epilepsy*; Devinsky, O., Schachter, S., Pacia, S., Eds.; New York: Demos Medical Publishing, 2005; pp 177–182.

Yu, F.; Takahashi, T.; Moriya, J.; Kawaura, K.; Yamakawa, J.; Kusaka, K.; Itoh, T.; Morimoto, S.; Yamaguchi, N.; Kanda, T. Traditional Chinese Medicine and Kampo: A Review from the Distant Past for the Future. *J. Int. Med. Res.* **2006,** *34* (3), 231–239. https://doi.org/10.1177/147323000603400301.

Yu, S.; Chao, Z.; Xiaodong, L. Traditional Indian Medicine, Part I About Traditional Indian Medicine. *J. Tradit. Chin. Med. Sci.* **2020**. https://doi.org/10.1016/j.jtcms.2020.06.007

Zhang, H.; Abid, S.; Ahn, J. C.; Mathiyalagan, R.; Kim, Y.-J.; Yang, D.-C.; Wang, Y. Characteristics of *Panax ginseng* Cultivars in Korea and China. *Molecules* **2020,** *25* (11), 2635. https://doi.org/10.3390/molecules25112635.

CHAPTER 13

RECENT ADVANCEMENTS IN HERBAL NUTRACEUTICALS

VIMAL KUMAR YADAV[1], AJAY KUMAR SHUKLA[1], MANISH KUMAR[2], MALTI ARYA[3], and RITESH TIWARI[3]

[1]*Institute of Pharmacy, Dr Ram Manohar Lohia Avadh University Ayodhya Uttar Pradesh, India*

[2]*Department of Pharmacy, Madhav University, Pindwara, Rajasthan, India*

[3]*Chandra Shekhar Singh College of Pharmacy, Uttar Pradesh, India*

ABSTRACT

A survey conducted in the United Kingdom, Germany, and France revealed that consumers place a higher value on nutrition for maintaining good health than exercise or inherited factors. This sparked the idea for nutraceutical evidence-based treatments. Nutraceuticals are nutritional supplements that are used to improve health, postpone ageing, fend against disease, and maintain normal bodily functions. Due to its potential for both nutrition and treatment, nutraceuticals are currently receiving a lot of attention. Herbal medicine is becoming more and more necessary. The contemporary, competitive, and stressful environments are a haven for illnesses of all stripes. For the majority of disorders, there are allopathic treatments available, but they are expensive and have a wide range of negative effects. Because of this, a lot of individuals are leaning toward changing their lifestyle and using herbal products. Herbal nutraceuticals are utilized as a potent tool to promote ideal wellbeing, life span, and personal satisfaction by acting against intense and persistent issues brought about by unfortunate nourishment. The tempting

Herbals As Nutraceuticals: Their Role in Healthcare.
Raj K. Keservani, Rajeshwar Kamal Kant Arya, and Rajesh K. Kesharwani (Eds.)

possibility of a "nutraceutical" breakthrough in the pharmaceutical warzone as a substitute for contemporary medicine has been created by the rising consumer awareness of health care. The food area will become research-situated, similar as the drug business, because of this change, introducing another period of medication and wellbeing.

13.1 INTRODUCTION

The words "sustenance" and "pharmaceutics" are the underlying foundations of the expression "nutraceuticals." The word is utilized to depict merchandise that are isolated from home-grown items, dietary enhancements (supplements), certain eating regimens, and handled dinners like cereals, soups, and beverages that are likewise used as medication as well as giving sustenance (Trottier et al., 2010). As opposed to meds, nutraceuticals are intensifiers that frequently don't have patent insurance. Albeit both drug and nutraceutical substances might be utilized to treat or forestall illness, just drug substances are endorsed by the public authority. An item is considered to be a dietary enhancement on the off-chance that it bears or incorporates at least one of the dietary parts recorded underneath—a focus, metabolite, part, concentrate, or combination of the following substances. a mineral, nutrient, amino corrosive, therapeutic plant, or other herbal (Sharma et al., 2020; Rane et al., 2020a, 2020b, 2020c), healthful enhancement utilized by people to help day-to-day utilization. The dietary enhancements known as nutraceuticals are used for ailments other than nourishment (Kathleen et al., 2013). Clients favor dietary enhancements over drugs to help their wellbeing due to the negative results of medications. This caused an unrest in the nutraceuticals business on a world-wide scale. Notwithstanding the fundamental nourishing substance tracked down in food, nutraceuticals offer extra wellbeing benefits. Items might make claims differing relying upon the locale, like supporting the construction or capability of the body, deferring the maturing system, further developing wellbeing, and forestalling illnesses (Tank Dharti et al., 2010).

Any item created from food sources that offers extra wellbeing benefits notwithstanding the wholesome substance present in food varieties is alluded to as a nutraceutical. These food sources incorporate nutrients, minerals, proteins, fats, carbs, and other fundamental supplements. Nutraceuticals are utilitarian food sources that aid the avoidance or potential treatment of infections as well as problems other than pallor. The exemption for frailty is thought about to recognize the words "utilitarian food" and "nutraceutical" on the grounds that most of practical food sources work in some way as antianemics obviously. Thus, what fills in as useful nourishment for one shopper might act as a nutraceutical. Citrus leafy foods dairy items (like milk) are instances of nutraceuticals (e.g., squeezed orange) (Whitman et al., 2001).

Nutraceutical items can be considered vague organic meds that are expected to work on in general wellbeing, oversee side effects, and stop malignant cycles. As per the ancient Greek doctor, Hippocrates, who is viewed as the pioneer behind medication, the idea of basic nutraceuticals is to place an accentuation on counteraction. He once commented, "Let food be your medication." One of the most pivotal fields of exploration is their capability in human nourishment, which has significant repercussions for shoppers, medical care experts, controllers, food makers, and merchants. Because of their true capacity for having wholesome, safe, and remedial effects, nutraceuticals certainly stand out enough to be noticed (Nelson et al., 2005). As per a new industry study, the worldwide market for nutraceuticals is developing and will reach US $250 billion by 2018 (Whitman et al., 2001).

Lipids, nutrients, minerals, proteins, carbs, and other fundamental components can be found in nutraceuticals. Studies have shown that nutraceuticals are viable therapies for various ailments, including cardiovascular circumstances (Sosnowska et al., 2007), diabetes (Ashwlayan et al., 2018), atherosclerosis (Aquila et al., 2019), disease (McClements et al., 2019), and neurological circumstances (Sarris et al., 2019).

13.2 HEALTH BENEFITS OF NUTRACEUTICALS

From the perspective of the consumer, nutritional supplements have the following advantages.

- Promote longer life.
- Assist in preventing specific medical disorders.
- Taking care of oneself might be beneficial psychologically.
- It could be thought of as more "natural" and less likely to have negative side effects than conventional medication.
- May provide meals for special needs populations.

Both traditional and contemporary therapies can be successfully derived from herbal medicine. About 80% of rural populations, as indicated by the World Health Organization (WHO), depend on natural medication for their essential medical care. Regardless of the way that they have not been very much examined, restorative plants are a rich wellspring of a few wellbeing benefits (Israel et al., 2010).

13.3 PHYSIOLOGICAL BENEFITS OF NUTRACEUTICALS

The following are some physiological advantages of nutraceuticals (DeFelice et al., 2002):

- Cardiovascular agents,
- Anticancer agents,
- Anti-diabetics agents,
- Anti-obese agents, and
- Help treat chronic inflammatory disorders.

13.4 CATEGORIES OF NUTRACEUTICALS

1. Substances with referred-to healthy benefits are alluded to as supplements. Nutrients, minerals, amino acids, unsaturated fats, polysaccharides, and so on are a couple of such models (Dahiya et al., 2013).
2. Herbals or botanical products—Products made from herbs or botanicals that are concentrated or extracted. Aloe vera, wheat grass, ginger, garlic, and some examples (Kokate et al., 2008).
3. Dietary supplements—Substances taken from other sources that are used for specialized purposes such as meal replacements, weight loss supplements, and sports nutrition (Singh et al., 2012).

13.5 NUTRACEUTICALS ARE ORDERED BASED ON FOOD SOURCES ACCESSIBLE ON THE LOOKOUT (RAJAM ET AL., 2011)

13.5.1 TRADITIONAL NUTRACEUTICALS

These are organic components that don't alter the diet in any way. They are categorized based on the following:

13.5.1.1 CHEMICAL CONSTITUENTS

a. Nutrients

Nutrients incorporate supplements that play deep rooted healthful parts, like nutrients, minerals, amino acids, and unsaturated fats.

Nutraceutical are being used in therapy for malignant growth, diabetes, waterfalls, osteoporosis, coronary illness, and strokes. Patients suffering from osteoporosis and related sickness can profit from the minerals present in plant, creatures, and dairy sources. Omega 3-PUFAs, which are found in flax seed and salmon, are strong calming specialists that likewise keep up with mental well-being and lower cholesterol aggregation (Anita et al., 2012).

b. **Herbals**

For example, willow bark (Salix nigra) contains salicin, a functioning fixing that has mitigating, pain relieving, antipyretic, astringent, and antiarthritic properties. Parsley (Petroselinum cripsum) is diuretic, carminative, and antipyretic, as well as contains flavonoids (apiol, psoralen) (Keservani and Sharma, 2014; Keservani et al., 2016a, 2016b). Menthol, a functioning fixing in peppermint (Mentha piperita), is utilized to treat colds and seasonal virus. The spice lavender (Lavendula angustifolia) contains tannin, which is valuable for treating asthma, colds, hacks, stress, misery, and hypertension. Proanthocyanadin, tracked down in cranberries (Vaccinium erythrocarpum), has been accounted to be useful in treating malignant growth, ulcers, and urinary plot diseases (Anita et al., 2012).

c. **Phytochemicals**

These are classified in view of the compound name doled out as per their phytochemical attributes. Models incorporate isoprenoids, vegetables (like chickpeas and soybeans), and carotenoids. Berries, natural products, vegetables, and vegetables contain flavonoid polyphenolics, which are solid cell reinforcements, phytoestrogens, and can forestall diabetes, bosom disease, and prostate malignant growth. Dim grapes, raisins, berries, peanuts, and turmeric roots contain non-flavonoid polyphenolics, which have intense mitigating, cancer prevention agent, and anticlotting properties and lower cholesterol. Chlorella and spirulina are extra-dietary enhancements (Silva et al., 2011).

13.5.1.2 PROBIOTIC MICROORGANISMS

The expression "probiotics," and that signifies "forever," alludes to living microorganisms that, when taken in adequate amounts, benefit the host's wellbeing. Microorganisms that help a solid stomach-related framework and sustenance ingestion work to uproot sickness-causing creatures like yeasts, different microscopic organisms, and infections and structure a valuable beneficial interaction with the human gastrointestinal framework. Modified microbiota restrain microorganism, and connect to the gastrointestinal epithelium; rival microorganisms are assets fundamental for endurance, give a serum impact, and opposite specific impacts of contamination on the digestive epithelium, like secretary changes and neutrophil relocation. By creating the specific chemical (B galactosidase) that can hydrolyze the disturbing lactose into its part sugars, probiotics assist with treating lactose narrow mindedness. Prebiotics are intensifiers that, when taken, are not processed by us; they are a moderately late expansion to our vocabulary. All things being equal, they act as a wellspring of supplements for valuable probiotic microorganisms (El Sohaimy et al., 2012).

13.5.1.3 NUTRACEUTICAL ENZYMES

The origins of these enzymes include microbial, plant, and animal sources.

Non-Traditional Nutraceuticals

These are biotechnologically produced fake foods. Bioactive ingredients found in food samples are used to create products for human wellbeing. They are broken down into,

a. Fortified nutraceuticals
b. Recombinant nutraceuticals

a. Fortified nutraceuticals

For instance, squeezed orange invigorated with calcium, cereals with extra nutrients or minerals, and wheat with extra folic corrosive are instances of sustained food varieties from horticultural reproduced or added supplements as well as parts. Furthermore, cholecalciferol-strengthened milk is utilized to treat vitamin D inadequacy. Milk that has been enhanced with prebiotics and probiotics and contains Bifidobacterium lactis HN019, which is utilized to treat extreme pediatric illnesses such looseness of the bowels and respiratory diseases. Kumar made the revelation that bananas with the soybean ferritin enhances iron content in body (Borkar et al., 2015).

b. Recombinant nutraceuticals

As the name suggests, these nutrients that provide energy were created through "recombination" with the use of biotechnology. For instance: bread, booze, cheese, yoghurt, fermented starch, vinegar, etc.

13.6 WHY NUTRACEUTICALS SEEM ATTRACTIVE?

- Humans regularly consume a variety of meals that are high in phenolics.
- Side effects are quite rare.
- Their half-life is comparatively long.
- After consumption, they may be quickly absorbed in the gut.
- They are freely accessible without a prescription and do not require a doctor's appointment.
- When conventional therapies for particular conditions fail, some people resort to these goods (Yates et al., 2017).

13.7 MODE OF ACTION OF NUTRACEUTICALS

It has been said that they work by enhancing the body's supply of vital building blocks. There are two ways to get these necessary building pieces.

a. By decreasing illness symptoms as buffering agents for alleviation, and
b. Immediately enhancing the person's health (Whitman et al., 2001).

13.8 CONCEPT OF NUTRACEUTICALS

It is necessary for the pharmaceutical development process to obtain clinical test findings from animal experiments and research to confirm their therapeutic benefits. However, in the past, there was no way to verify the effectiveness of foods in avoiding illnesses when it came to nutrition. However, as the link between lifestyle-related illnesses and dietary composition has been established scientifically, it has recently become a societal concern. Nutraceutical items are known to have medical advantages, including bringing down the gamble of malignant growth and coronary illness, as well as forestalling or treating hypertension, elevated cholesterol, stoutness, osteoporosis, diabetes, loss of water in body, macular degeneration (which brings about irreversible visual impairment), menopausal side effects, sleeping disorders, memory impairment, stomach-related illnesses, and also migraines. Different nutraceuticals are used as solutions for different circumstances. As one of the techniques for keeping away from such sicknesses, the possibility of nutraceuticals is starting to acquire acknowledgment (Heyland et al., 2001; Elizabeth et al., 2002; Kalra et al., 2003; Sumi et al., 2008, Sengupta et al., 2003; Klein et al., 2000).

13.9 NUTRACEUTICAL GROWTH

Nutraceuticals and practical food sources have become an extravagant area in the overall commercial center, as per gauges. The need to precisely mark and assess the wellbeing effects of nutraceuticals and utilitarian food sources is putting significant global extension limitations on this industry. Improvement of substitute nutraceuticals and practical food varieties with particular and more trustworthy wellbeing and food characteristics will be made conceivable by determination for steady creation of high and low efficiency of dynamic plant parts inside specific natural zones. The utilitarian food and nutraceutical industry in the United States of America (USA) is presently the biggest and expanding the quickest on the planet (Natural Health Products Regulations, 2003).

Many therapeutic plants, spices, and tree species may be found in India, where they have a substantial local market. Functional foods and nutraceuticals are sold under many brand names in India as traditional Indian Ayurvedic medicines. The public can purchase ayurvedic and nutraceutical health products in India without a prescription because there are no strong pharmaceutical laws in place for them. India exports goods to many nations and has a sizable portion of the global market for functional foods and nutraceuticals. However, the USA and Japan are India's top export markets (Hardy et al., 2003).

13.10 REGULATIONS

Assuming a case was made about a nutraceutical item that inferred a restorative advantage, for example, that it was a strengthened food or dietary enhancement, the item would need to stick to the administrative prerequisites for restorative items, including security, viability, and quality testing as well as showcasing approval methods. Dietary enhancements were managed by the FDA as food sources for a long time to guarantee wellbeing, health, and precise marking. The Food Handling and Standard Demonstration was laid out by the Indian government in 2006 with the goal to synchronize and improve on the various principles administering food varieties, dietary enhancements, and nutraceuticals. The Food Handling and Principles Authority should be laid out, as per the law (FSSA) (Pandey et al., 2003).

13.11 APPLICATIONS OF NUTRACEUTICALS

The demands of the modern age, quick lifestyle, longer working days, and mental anxieties have fundamentally modified human ways of life and expanded the predominance of diabetes, corpulence, different malignancies, and vascular issues. Most of nutraceuticals are regular substances with endogenous starting points that invigorate healthy individuals' exercises. Dietary nutraceuticals, for example, soy isoflavones, alpha-lipoic corrosive, and carotenoids referred to a person, can be easily found in food sources. Increasingly more plant parts are becoming available, including some as single substances, similar to the resveratrol tracked down in wine and grapes, and others as complicated combinations of parts, as pycnogenol. Glucosamine and chondroitin, the two glycosaminoglycans (GAGs), are acquired from different animal beings (Rinaldi et al., 2005).

13.12 SAFETY AND EFFICACY

The way nutraceuticals are used influences how safe they are to be used. It is oftentimes more straightforward to demonstrate a nutraceuticals' wellbeing than its viability. Various nutraceuticals have been used as alternative for food and medicine. Without sufficient proof to validate the wellbeing and adequacy of their merchandise, numerous producers make unlawful commitments. Clients hence need affirmation that an item is secure and, in a perfect world, equipped for proceeding as guaranteed. Nutraceuticals ought to be protected above all the other things (http://www.aboutbioscience.org/pdfs).

13.13 HERBAL NUTRACEUTICALS

They are optional or tertiary metabolites that come from plants and are bioactive phytochemicals (Dillard et al., 2000). These protected food-extricate supplements have a large number of clinical characteristics, which permits them to impact the medical services framework and give fantastic insurance against numerous horrible illnesses, including malignant growth. Natural nutraceuticals are sorted into many classes in view of their construction and capability, including polyphenols, cannabinoids, flavonoids, saponins, sitosterols, anthocyanins, carotenoids, isothiocyanates, terpenes, isofavones, and so on (Das et al., 2012).

A developing group of exploration demonstrates that home-grown nutraceuticals might be powerful and safe devices for (i) detoxifying ingested and ecological cancer-causing agents through the enactment of cell reinforcement catalysts, growth silencer proteins, and DNA fix pathways; and (ii) stopping the movement of disease by empowering apoptosis and forestalling EMT, angiogenesis, and metastasis (Wargovich et al., 2010) Furthermore, a developing collection of examination shows that natural nutraceuticals may lessen the development, endurance, and improvement of hypoxic malignant growth cells by obstructing HIF-1 flagging pathways. Natural nutraceuticals work by smothering the pathways that empower HIF-1 creation as well as by making HIF-1 corrupt and diminishing resulting flagging occasions. Home-grown nutraceuticals have likewise been displayed to (i) switch hypoxia-actuated drug opposition, (ii) increment the helpful file of malignant growth therapies against hypoxic cancers, and (iii) diminish the negative results of these cytotoxic prescriptions, notwithstanding their enemy of disease impacts (Ranzato et al., 2014).

13.14 REGULATION OF NUTRACEUTICAL INDUSTRY IN INDIA

A new term in the Indian medical care industry is "nutraceutical." This thrilling word addresses gigantic market possibilities for drug and nutraceutical ventures working both locally and abroad. In the Indian medical services area, nutraceuticals have a bewildering yearly development pace of 25%. As indicated by gauges, the overall market for nutraceuticals may arrive at USD 80 billion toward the finish of 2007. Albeit various definitions for nutraceuticals have been given by different specialists, the term's importance of "food as medication" continues as before. Indeed, even now, it tends to be confounding to recognize nutraceuticals and terms like useful food, dietary enhancements, and fashioner food. The expansion of nutraceutical makers in India is brought about by the absence of administrative principles for these items. Amid savage competition, organizations might put forth item quality trade-offs trying to get by. To stop the exorbitant expansion of nutraceutical organizations, this study suggests fostering a brought together definition for nutraceuticals as well as fitting administrative standards (German et al., 2000; Houston et al., 2005).

13.15 NUTRACEUTICALS AND DISEASES

Nutraceuticals are currently recognized as being useful in treating conditions including Parkinson's and Alzheimer's illness along with stoutness, diabetes, malignant growth, osteoporosis, and other persistent and degenerative issues. Proof proposes that various organic cycles, including the enactment of cancer prevention agent protections, signal transduction pathways, cell endurance-related quality articulation, cell multiplication and separation, and conservation of mitochondrial trustworthiness, are associated with the robotic activities of normal mixtures. These attributes appear to be very significant in the safeguard against the pathologies of many age-related or constant sicknesses. It is vital that the supplements remembered for various dinners, including leafy foods, make the notable wellbeing benefits. These should be taken oftentimes to bring down risk factors including elevated cholesterol, hypertension, and diabetes, as well as other clear wellbeing benefits. Botanicals including St. John's wort, echinacea, ginkgo biloba, saw palmetto, and ginseng are the absolute most popular nutraceutical items now available. Various endeavors produce and sell nutraceuticals, but these items' adverse consequences are as often as possible unreported or unconfirmed (especially when ingested in high sums). The general population ought to be educated, and every customer ought to know about the suitable day to day measurements of these nutraceuticals to have logical figuring out about them (Frydoonfar et al., 2003). We should make a flourishing nutraceutical research local area to transform most of forthcoming nutraceuticals into deep-rooted ones and really give their extraordinary benefits to us all considering the quickly developing interest in the nutraceutical upset (Limer et al., 2004).

13.15.1 CARDIO-VASCULAR DISEASES (CVD)

As indicated by reports, an unfortunate utilization of leafy foods is connected to a high demise rate for the treatment of cardiovascular infection hypertension (hypertension), coronary illness (cardiovascular failure), cerebrovascular sickness (stroke), cardiovascular breakdown, fringe vascular infection, and so on as diseases of the heart and veins (Rissanen et al., 2003). For the counteraction and treatment of CVD, nutraceuticals like cancer prevention agents, dietary fiber, omega-3 polyunsaturated unsaturated fats (- 3 PUFAs), nutrients, and minerals are encouraged notwithstanding active work. Beta carotene, vitamin A, and vitamin E enhancements might increment mortality. The improvement of nutraceuticals to diminish and forestall apoplexy risk in females with thrombophilic quality transformations is plausible (Temple et al., 2003).

13.15.2 OBESITY

The nourishing enhancement's mix of glucomannan, chitosan, fenugreek, G sylvestre, and L-ascorbic acid decisively brought down body weight and energized fat misfortune in hefty individuals. To decide a drawn out viability and hindering effect potential, more examination is required. To forestall and treat stoutness, which has a generally high occurrence around the world, diet and exercise are urgent. Enormous scope investigation into the utilization of nutraceuticals as weight control and stoutness treatments is currently in progress. Formed linoleic corrosive (CLA), capsaicin, momordica charantia (MC), and psyllium fiber are instances of nutraceuticals that might have against heftiness impacts. For body weight reduction, a charming and productive nutraceutical that might raise caloric admission as well as lower energy use is normally liked. Natural energizers have been found effective in advancing body weight decrease incorporate ephedrine, caffeine, chitosan, and green tea (Daly et al., 1993; Boozer et al., 2001; Dulloo et al., 1999; Schiller et al., 2001). Be that as it may, in light of the fact that they could have unfortunate results, their use is questionable. Buckwheat seed proteins work like regular filaments tracked down in diet and help with weight and blockage. Green tea separate expands how much energy utilized, which might assist with weight decrease, yet 5-hydroxytryptophan dials back weight reduction (Bell et al., 2002).

13.15.3 DIABETES

The foundation of gestational diabetes mellitus care is diet treatment. Barely any natural dietary enhancements that are remembered to assist individuals with type-2 diabetes mellitus have been deductively shown to do as such in sufficiently directed randomized examinations, notwithstanding their far reaching utilization. Isoflavones are phytoestrogens; they have been taken by individuals all around the world and are primarily and practically equivalent to human estrogen. The best examination has been finished on soy isoflavones of all the phytoestrogens. Low frequency and demise paces of type-II diabetes, coronary illness, osteoporosis, and a few malignancies are connected to high isoflavone utilization (20-100 mg/day). In those inclined to diabetes, omega-3 unsaturated fats might bring down glucose resistance. The heart might be particularly powerless against their exhaustion in diabetes since long chain n-3 unsaturated fat combination relies upon insulin. Patients with diabetes might profit from n-3 unsaturated fat ethyl esters. Docosahexaenoic corrosive controls insulin obstruction and is fundamental for the improvement of the neurovascular framework. Magnesium supplementation brings down the possibility creating diabetes and upgrades insulin awareness. Chromium picolinate, calcium, and vitamin D are said to upgrade glycemic control and increment insulin awareness in specific diabetics; severe melon and cinnamon concentrates might be utilized to treat and maybe forestall diabetes. Nutraceuticals with huge measurements of mixes, in any case, have been proposed to altogether forestall and most likely be marketed legitimately.(McCarty et al., 2005).

13.15.4 ALZHEIMER'S DISEASE

Medicines including phytochemicals for mental deterioration and sorrow may be progressed to a more noteworthy clinical preliminary level on individuals because of their low poisonousness and high retention. The seriousness and event of neurodegenerative and other age-related problems can be diminished in the future by consuming nutraceuticals on a present moment or long haul premise through diet. The capability of nutraceuticals to forestall age-related ailment overall and mental deterioration specifically by turning around hurtful neurodegenerative and obsessive cycles has drawn in critical public and logical interest lately. As opposed to the single-target activity of most of meds used to treat Alzheimer's sickness, Camila Calfio claims that since it is a multifactorial disease, nutraceutical substances give the advantage of a multitarget approach, labeling a few biochemical areas in the human mind. Zeroing in on various pharmacological impacts, they examined top to bottom critical restorative plants that have proactively been explored for helpful purposes against Alzheimer's sickness. Acetylcholinesterase restraint, amyloid decrepit plaques, oxidation items, provocative pathways, specific cerebrum receptors, and so forth are some of these objectives, and pharmacological activities such as fluctuated as mitigating, memory-improving, nootropic impacts, glutamate excitotoxicity, antidepressants, and cell reinforcements are among them. Also, we discuss the instruments of activity of tau protein totals centered phytopharmaceutical equations and nutraceutical substances. This complete perspective on nutraceuticals makes another road for continuous exploration and clinical preliminaries that are supposed to give a few discoveries upheld by logical information (Calfio et al., 2020).

13.15.5 CANCER

Also, we discuss the systems of activity of tau protein totals centered phytopharmaceutical recipes and nutraceutical substances. This extensive perspective on nutraceuticals makes another road for continuous exploration and clinical preliminaries that are supposed to give a few discoveries upheld by logical information (Calfio et al., 2020).

13.16 ANTIOXIDANT ACTIVITY

"A dietary cell reinforcement is a synthetic in food that extensively limits the hurtful impacts of ROS, RNS or both on typical physiological capability in people," as per the Board on Dietary Cancer prevention agents and Related Mixtures of the Food and Sustenance Sheets (Benzie et al., 2000; Yamini et al., 2005; Zheng et al., 2016; Hiramatsu et al., 2005; Food and Nutrition Board, Institute of Medicine).

The few phases of the cancer-causing process incorporate free extremists. DNA transformations can be welcomed by peroxyl extremists and lipid peroxidation, and these progressions are fundamental for the beginning of the tumorigenic cycle (Zhang et al., 2015). By forestalling DNA harm, cancer prevention agent phytochemicals might control the beginning of carcinogenesis (Vargas-Mendoza et al., 2019). As per their source, cell reinforcements might be isolated into exogenous dietary parts such phenolics, flavonoids, phenolic acids, carotenoids, nutrients, and minerals, as well as endogenous union parts like chemicals and little particles (Ratnam et al., 2006). Because of their cell reinforcement abilities and positive effects on human wellbeing, phytochemicals, plant-determined non-nutritive atoms, and various normally happening builds found in food are of critical interest (Na et al., 2008). Consuming organic products, vegetables, and entire grains frequently brings down the gamble of persistent diseases connected to oxidative harm, as per epidemiological and creature research. Ongoing aggravation has been found to make mitigating impacts, making it a fundamental part that might add to the etiology of numerous constant ailments, including disease and type-2 diabetes (Dahlen et al., 2014).

13.17 DRUG DELIVERY OF NUTRACEUTICALS

Creation and plan of nutraceutical conveyance frameworks The viable improvement of nutraceutical conveyance frameworks requires a comprehension of their characteristics as well as the usage of appropriate materials and assembling processes. Since numerous nutraceuticals are powerless to intensity and high temperatures during the exemplification cycle, which could bring about a deficiency of their bioactivity, picking the legitimate embodiment strategy is really a significant stage. Various hardships could emerge while planning nutraceutical conveyance frameworks that are food-grade, financially savvy, have physical and synthetic security, and are in fact attainable. Moreover, it is extremely more enthusiastically to accomplish a decent bioavailability for such nutraceuticals.

13.18 DELIVERY SYSTEMS DESIGN

13.18.1 EMULSIONS

Emulsions can be fluid in fluid or fluid in strong, as well as oil-in-water O/W or water-in-oil (W/O). The sort and content of the oil, the proportion of the surfactant to the oil, and the co-solvents or co-solutes utilized all influence the drop size and emulsion strength. The essential epitomizing specialists utilized incorporate lipids, low atomic weight surfactants, proteins, and polysaccharides (Livney et al., 2015).

13.18.2 LIPOSOMES

A watery center is encircled by one (unilamellar) or more (multilamellar) concentric phospholipid layers in liposomes, which are bilayered vesicular frameworks made of phospholipids. Lim et al. (2012) and Verma et al. (2019) presented the expression "shut, ceaseless bilayered structures comprising for the most part of lipid and additionally phospholipid atoms" to depict liposomes. The US Food and Medication Organization (USFDA) explicitly expressed that liposomes are not emulsion frameworks or a medication lipid blend to forestall misconception with other lipid-based conveyance strategies (Mozafari et al., 2000). The capacity of liposomes to contain both lipid-dissolvable (in lamellae) and water-solvent (in the fluid center region) bioactive synthetics is their chief benefit in application (U.S. Food and Drug Administration 2018).

A plan method called Oneself Emulsifying Nutraceutical Conveyance Framework (SENDS) takes into consideration the solubilization and ensuing ingestion of substances that are just imperceptibly water dissolvable. It offers a great many benefits for your plans, including:

- ➢ Improved bioavailability
- ➢ Enhanced formulation effectiveness to lower dosage
- ➢ Reduced food impact
- ➢ Protection of ANI during processing or manufacture and in the gastrointestinal environment
- ➢ Applicable for dose in liquid or solid forms
- ➢ Each SENDS is specially created to satisfy your formulation requirements.

13.19 NANOTECHNOLOGY IN NUTRACEUTICALS

The nanoencapsulation procedure offers the opportunity to protect the synthetic structure of nutraceuticals from natural aggravations like pH, light, temperature, free extremists, or oxygen (Cristiano et al., 2021; Gumezescu et al., 2017), supports their bioavailability, empowers designated conveyance to target destinations (Meenambal et al., 2020; Fresta et al., 2020), and grants a controlled arrival of the typified compound (Assadpour et al., 2019). The basic issue for epitomizing nutraceuticals is the limit of nanosystems to manage the arrival of the provided dynamic mixtures, which comprises of a predefined fixation/time discharge profile at the ideal site of activity. To deliver its substance in light of specific improvements like pH, dampness, proteins, and temperature, an ideal conveyance framework should likewise have the option to safeguard the nutraceutical from those equivalent upgrades (Kumari et al., 2010). Furthermore, once the nutraceutical has been stacked into the transporter, highlights rely more upon the physico-synthetic properties of the vesicle than the captured drug, which expands the solvency of the nutraceutical compounds. Also, nanosystems empower the synchronous conveyance of mixtures that are solvent in the two lipids and water, supporting their synergistic activity (Jain et al., 2014; Desfrançois et al., 2018; Gumezescu et al., 2017).

13.20 NANOPARTICLES

The nanoencapsulation procedure offers the opportunity to protect the synthetic structure of nutraceuticals from natural aggravations like pH, light, temperature, free extremists, or oxygen (Cristiano et al., 2021; Rasti et al., 2017), supports their bioavailability, empowers designated conveyance to target destinations (Meenambal et al., 2020; Fresta et al., 2020), and grants a controlled arrival of the typified compound (Assadpour et al., 2019). The basic issue for epitomizing nutraceuticals is the limit of nanosystems to manage the arrival of the provided dynamic mixtures, which comprises of a predefined fixation/time discharge profile at the ideal site of activity. To deliver its substance in light of specific improvements like pH, dampness, proteins, and temperature, an ideal conveyance framework should likewise have the option to safeguard the nutraceutical from those equivalent upgrades (Wei et al., 2020; Yuan et al., 2020). Furthermore, once the nutraceutical has been stacked into the transporter, it highlights rely more upon the physico-synthetic properties of the vesicle than the captured drug, which expands the solvency of the nutraceutical compounds. Also, nanosystems empower the synchronous conveyance of mixtures that are solvent in the two lipids and water, supporting their synergistic activity (Liu et al., 2019).

13.21 LIPOSOMES

An adaptable and biocompatible medication conveyance innovation, liposomes are lipid-based vesicles that are used to exemplify both hydrophilic and hydrophobic prescriptions. Liposomes are utilized to move dynamic fixings, including those from nutraceuticals and other pharmacological exercises. A metabolite of vitamin An is called all trans-retinoic corrosive (ATRA). It is a nutraceutical substance whose anticancer capacity has been widely investigated. It was safeguarded against debasement in a new report by being exemplified in liposomes, with an ensnarement viability of around 82%. Specifically, Cristiano et al. showed that ATRA might be safeguarded from photograph debasement occasions that would hinder its pharmacological adequacy by being exemplified inside liposomes (Cui et al., 2009).

In spite of the fact that liposomes are potential conveyance vehicles for nutraceuticals, they have downsides such as the solidness of the lipid bilayer. Beginning from them, a few examination groups have changed the synthetic cosmetics and assembling cycles of liposomes to deliver novel vesicular frameworks with higher porousness across physiological obstructions and more prominent bilayer adaptability. These examinations led to ethosomes (Cristiano et al., 2020; Houghton et al., 2019), transfersomes (Mancuso et al., 2020), and niosomes. Niosomes specifically are spheroidal designs made by non-ionic surfactants, and they were as of late read up for their capability to move—tocopherol. In the first place, the creators investigated what the organization of niosomes meant for their ability to exemplify and convey nutraceuticals orally. They found that niosomes made of Range 60 and Tween 60 have an exceptionally high epitome viability, containing 99.07% of tocopherol, and that, all the more critically, the presence of cholesterol and dicetyl phosphate in the organization worked on the security of tocopherol. The delivery profile of tocopherol was one more huge result of this examination. Fully intent on duplicating the fate of niosomes in the gastrointestinal framework, in vitro discharge examinations were completed in a reenacted gastric liquid (for 2 h) and in a recreated digestive liquid (for 8 h). The discoveries show that niosomes can deliver conveyed tocopherol in a supported way, with a medication arrival of under half throughout the span of 8 h, demonstrating high dependability of the niosomal detailing in gastro-intestinal liquids and high propensity of tocopherol to hold in nanosystems (Dima et al., 2015; Basiri et al., 2017).

13.22 EVALUATION OF NUTRACEUTICAL

Indeed, even while it tends to be basic for a formulator to make a drug type of bioactive food (tablets, powders, cases, suppositories, and so on), it very well may be very challenging to get a reasonable bioavailability for such nutraceuticals. Low dissolvability, dependability, or potentially porousness of the bioactive part in the GIT habitually compromise bioavailability.

The principal purposes behind curcumin's low oral bioavailability, for example, are its very low blood levels, obliged tissue dispersion, evident fast digestion, and brief flow half-life (Anand et al., 2007). Another representation is the as of late evolved probiotics conveyance approach known as probiotics embodiment innovation (or PET). While making probiotics, factors that support cell suitability, for example, biomaterial decision, dissolvable sort and harmfulness, and innovation decision, are of most extreme significance. Biomaterials can be produced using engineered or regular polymers. Physical and compound properties (substance piece, morphology, mechanical strength, and solidness in GI liquids), poisonousness, and assembling and disinfecting methodology are a portion of the variables that should be considered (Gbassi et al., 2012).

13.23 FUTURE PERSPECTIVE

Due to their ease for modern lifestyles, nutraceuticals are destined to play significant part in future therapeutic breakthroughs and will continue to appeal to the general public. Due to their alleged safety and their promise nutritional and therapeutic effects, they have attracted a lot of attention. Pharmaceutical and nutritional companies are aware of the financial success enjoyed by those providing benefits for the numerous consumers seeking better health as well as the changing trends causing the spread of these value-added products claiming to treat conditions ranging from cancer to heart health. It's important to pay attention, especially when it comes to the doses that work, specifically when carriers are employed. Studies must also focus on ascertaining whether nutritional deficiency and chronic administration of nutraceuticals do, in fact, have a role in health, given the continuous discussion over the usefulness of these products. It is clear that the potential benefits of nutraceuticals have already motivated researchers to find novel chemicals that would ensure long, healthy lives. Major experts think that the field of nutraceuticals is about to enter an exciting new area thanks to enzymes. "Enzymes have performed poorly. They'll be quite popular in the future. Using microorganisms to produce novel food items using fermentation technology also seems promising. It makes sense to use nutraceuticals in sports medicine. Because of its health advantages, tocotrienol seems potential for nutraceuticals. New and diverse health claims will undoubtedly arise as a result of advances in technical methods used to assess the effects of nutraceuticals, such as metabolomics, as well as the emergence of new fields of related study, such as nutrigenomics. These claims must be carefully examined" (Verma et al., 2018; Defrates et al., 2018).

13.24 CONCLUSION

The growth of the nutraceutical sector outpaces that of the food and pharmaceutical industries by a wide margin. They are in demand because they are more affordable and accessible than prescription medications. Long-term study is necessary to understand the function that nutraceuticals play in health and how illnesses are treated. Nutraceuticals are produced and sold by several companies, but their negative effects, particularly when eaten in high numbers, are frequently unreported or unverified. Undoubtedly, the current body of information concerning nutraceuticals is a significant challenge for nutritionists, doctors, food technologists, and food chemists. The preparation of the supplement, its stability under various circumstances, and its safety are the main issues. Due to the need in the Indian and foreign markets, scientists and researchers are creating nutraceutical items. Nutraceuticals are receiving attention at the moment for their potential use in the treatment of conditions including osteoporosis, Parkinson's disease, Alzheimer's disease, cancer, diabetes, obesity, and coronary heart disease. Long-term study is necessary to understand the function that nutraceuticals play in health and how illnesses are treated. It is also necessary to conduct further study on how nutraceuticals affect disease activity and pathogenesis. In order to turn the majority of possible nutraceuticals into well-established ones and genuinely provide their great advantages, a strong nutraceutical research community must be formed in light of the fast growing interest in the nutraceutical revolution.

KEYWORDS

- **nutraceutical**
- **herbal nutraceutical**
- **diseases**
- **application**

REFERENCES

Agrawal, M.; Tripathi, D. K.; Saraf, S.; Antimisiaris, S. G.; Mourtas, S.; Hammarlund-Udenaes, M.; Alexander, A. Recent Advancements in Liposomes Targeting Strategies to Cross Blood-Brain Barrier (BBB) for the Treatment of Alzheimer's Disease. *J. Control. Release* **2017,** *260*, 61–77.

Anand, P.; Kunnumakkara, A. B.; Newman, R. A.; Aggarwal, B. B. Bioavailability of Curcumin: Problems and Promises. *Mol. Pharm.* **2007,** *4*, 807–818.

Anita, P. S.; Mangesh, S. T.; Prasad, R. V.; Meera, C. S. Nutraceuticals—Global Status and Applications: A Review. *IJPCS* **2012,** *1* (3), 817–832.

Aquila, G.; Marracino, L.; Martino, V.; Calabria, D.; Campo, G.; Caliceti, C.; Rizzo, P. The Use of Nutraceuticals to Counteract Atherosclerosis: The Role of the Notch Pathway. *Oxid. Med. Cell. Lon Gev.* **2019,** *2019*, 5470470.

Ashwlayan, V.; Nimesh, S. Nutraceuticals in the Management of Diabetes Mellitus. *Pharm. Pharmacol. Int. J.* **2018,** *6*, 114–120.

Assadpour, E.; Mahdi Jafari, S. A Systematic Review on Nanoencapsulation of Food Bioactive Ingredients and Nutraceuticals by Various Nanocarriers. *Crit. Rev. Food Sci. Nutr.* **2019,** *59*, 3129–3151.

Basiri, L.; Rajabzadeh, G.; Bostan, A. Physicochemical Properties and Release Behavior of Span 60/Tween 60 Niosomes as Vehicle for α-Tocopherol Delivery. *LWT Food Sci. Technol.* **2017,** *84*, 471–478.

Basnet, P.; Skalko-basnet, N. Curcumin: An Anti-Inflammatory Molecule from a Curry Spice on the Path to Cancer Treatment. *Molecules* **2011,** *16*, 4567–4598.

Bell, S. J.; Goodrick, G. K. A Functional Food Product for the Management of Weight Critical Reviews. *Food Sci. Nutr*. **2002,** *42*, 163–178.

Benzie, I. F. F. Evolution of Antioxidant Defence Mechanisms. *Eur. J. Nutr.* **2000,** *39*, 53–61. DOI: 10.1007/s003940070030

Boozer, C. N.; Nasser, J. A.; Heymsfield, S. B. An Herbal Supplement Containing Ma Huang-Guarana for Weight Loss: A Randomized, Double-Blind Trial. *Int. J. Obes. Relat. Metab. Disord.* **2001,** *25*, 316–324.

Borkar, N.; Saurabh, S. S.; Rathore, K. S.; Pandit, A.; Khandelwal, K. R. An Insight on Nutraceuticals. *PharmaTutor* **2015,** *3* (8), 13–23.

Calfio, C.; Gonzalez, A.; Singh, S. K.; Rojo, L. E.; Maccioni, R. B. The Emerging Role of Nutraceuticals and Phytochemicals in the Prevention and Treatment of Alzheimer's Disease. *J. Alzheimer's Dis* **2020,** *77* (1), 33–51.

Cristiano, M. C.; Froiio, F.; Mancuso, A.; Cosco, D.; Dini, L.; Di Marzio, L.; Fresta, M.; Paolino D. Oleuropein-Loaded Ufasomes Improved the Nutraceutical Efficacy. Nanomaterials **2021,** 11, 105.

Cristiano, M. C.; Froiio, F.; Mancuso, A.; De Gaetano, F.; Ventura, C. A.; Fresta, M.; Paolino D. The Rheolaser Master™ and Kinexus Rotational Rheometer®to Evaluate the Influence of Topical Drug Delivery Systems on Rheological Features of Topical Poloxamer Gel. *Molecules* **2020,** *25*, 1979.

Cui, C.; Zhang, Y.; Wang, L.; Liu, H.; Cui G. Enhanced Anticancer Activity of Glutamate Prodrugs of All-Trans Retinoic Acid. *J. Pharm. Pharm*. **2009,** *61*, 1353–1358.

Dahiya, K.; Nutraceuticals and Their Impact on Human Health. *J. Plant Biochem. Physiol.* **2013,** *1* (4), 1–3.

Dahlen, E. M.; Tengblad, A.; Lanne, T.; Clinchy, B.; Ernerudh, J.; Nystrom, S. H.; Ostgren, C. J. Abdominal Obesity and Low-Grade Systemic Inflammation as Markers of Subclinical Organ Damage in Type 2 Diabetes. *Diab. Metab*. **2014,** *40*, 76–81.

Daly, P. A.; Khrieger, D. R.; Dulloo, A. G. Ephedrine, Caffeine and Aspirin: Safety and Efficacy for the Treatment of Human Obesity. *Int. J. Obes. Relat. Metab. Disord.* **1993,** *17*, S73–S78.

Das, L.; Bhaumik, E.; Raychaudhuri, U.; Chakraborty, R. Role of Nutraceuticals in Human Health. *J. Food Sci. Technol.* **2012,** *49*, 173–183.

DeFelice, S. L. FIM Rationale and Proposed Guidelines for the Nutraceutical Research & Education Act—NREA, 10 Nov 2002. Foundation for Innovation in Medicine.

Defrates, K.; Markiewicz, T.; Gallo, P.; Rack, A.; Weyhmiller, A.; Jarmusik, B.; Hu, X. Protein Polymer-Based Nanoparticles: Fabrication and Medical Applications. *Int. J. Mol. Sci.* **2018,** *19*, 1717.

Desfrançois, C.; Auzély, R.; Texier I. Lipid Nanoparticles and Their Hydrogel Composites for Drug Delivery: A Review. *Pharmaceuticals* **2018,** *11*, 118.

Dillard, C. J.; German, J. B. Phytochemicals: Nutraceuticals and Human Health. *J. Sci. Food Agric.* **2000,** *80*, 1744–1756.

Dima, ¸S.; Dima, C.; Iordăchescu, G. Encapsulation of Functional Lipophilic Food and Drug Biocomponents. *Food Eng. Rev*. **2015,** *7*, 417–438.

Dulloo, A. G.; Duret, C.; Rohrer, D. Efficacy of a Green Tea Extract Rich in Catechin Polyphenols and Caffeine in Increasing 24-H Energy Expenditure and Fat Oxidation in Humans. *Am. J. Clin. Nutr*. **1999,** *70*, 1040–1045.

El Sohaimy, S. A. Functional Foods and Nutraeuticals—Modern Approach to Food Science. *WASJ* **2012,** *20* (5), 691–708.

Elizabeth, A. C. Over the Counter Products: Nonprescription Medications, Nutraceuticals, and Herbal Agents. *Clin. Obstet. Gynecol.* **2002,** *45* (1), 89–98.

Food and Nutrition Board, Institute of Medicine. *Dietary Reference Intakes*; National Academy Press: Washington, DC, 2002.

Fox, C. B.; Sivananthan, S. J.; Duthie, M. S.; Vergara, J.; Guderian, J. A.; Moon, E.; Coblentz, D.; Reed, S. G.; Carter D. A Nanoliposome Delivery System to Synergistically Trigger TLR4 and TLR7. *J. Nanobiotechnol.* **2014,** *12*, 17.

Fresta, M.; Mancuso, A.; Cristiano, M. C.; Urbanek, K.; Cilurzo, F.; Cosco, D.; Iannone, M.; Paolino, D. Targeting of the Pilosebaceous Follicle by Liquid Crystal Nanocarriers: In Vitro and In Vivo Effects of the Entrapped Minoxidil. *Pharmaceutics* **2020,** *12*, 1127.

Frydoonfar, H. R.; McGrath, D. R.; Spigelman, A. D. The Variable Effect on Proliferation of a Colon Cancer Cell Line by the Citrus Fruit Flavonoid Naringenin. *Colorectal Dis.* **2003,** *5*, 149–152.

Gbassi, G. K.; Vandamme, T. Probiotic Encapsulation Technology: From Microencapsulation to Release into the Gut. *Pharmaceutics* **2012,** *4*, 149–163.

German, J. B.; Walzem, R. L. The Health Benefits of Wine. *Annu. Rev. Nutr.* **2000,** *20*, 561–593.

Gumezescu, A.; Oprea, A. E. *Nanotechnology Applications in Food: Flavor, Stability, Nutrition and Safety*, 1st ed.; Academic Press: Cambridge, MA, 2017.

Hardy, G.; Hardy, I.; Ball, P. A. Nutraceuticals—A Pharmaceutical Viewpoint: Part II: Micronutrients, Nutraceuticals and Functional Foods. *Curr. Opin. Clin. Nutr. Med. Care* **2003,** *6* (6), 661–671.

Heyland, D. K. In Search of the Magic Nutraceuticals: Problems with Current Approaches. *J. Nutr.* **2001,** *131* (9), 2591–2595.

Hiramatsu, M.; Yoshikava, T.; Packer, L. *Molecular Interventions in Lifestyle-Related Diseases*; CRC Taylor and Francis: London, 2005.

Houghton, C. A. Sulforaphane: Its "Coming of Age" as a Clinically Relevant Nutraceutical in the Prevention and Treatment of Chronic Disease. *Oxid. Med. Cell. Longev.* **2019,** *2019*, 2716870.

Houston, M. C. Nutraceuticals, Vitamins, Antioxidants, and Minerals in the Prevention and Treatment of Hypertension. *Progr. Cardiovasc. Dis.* **2005,** *47*, 396–449.

http://www.aboutbioscience.org/pdfs/Nutraceuticals.pdf (accessed on 12 Dec 2011).

Israel, O.; Auguster, O.; Edith O. A. Antioxidant and Antimicrobial Activities of Polyphenols from Ethanomedicinal Plants of Nigeria. *Afr. J. Biotech.* **2010,** *9*, 2289–2293.

Jain, A.; Singh, S. K.; Arya, S. K.; Kundu, S. C.; Kapoor, S. Protein Nanoparticles: Promising Platforms for Drug Delivery Applications. *ACS Biomater. Sci. Eng.* **2018,** *4*, 3939–3961.

Kalra, E. K. Nutraceutical Definition and Introduction. *AAPS PharmSci.* **2003,** *5* (3), Article25.

Kathleen, C.; Stephen, D. Nutraceuticals: What Are They and Do They Work? Kentucky Equine Research, Inc. Versailles, KY **2013,** *7* (4), 1–50.

Keservani, R. K.; Sharma, A. K. Flavonoids: Emerging Trends and Potential Health Benefits. *J. Chin. Pharm. Sci.* **2014,** *23* (12), 815.

Keservani, R. K.; Sharma, A. K.; Kesharwani, R. K. Medicinal Effect of Nutraceutical Fruits for the Cognition and Brain Health. *Scientifica* **2016b,** Article ID 3109254, 10 pages. DOI: 10.1155/2016/3109254.

Keservani, R. K.; Sharma, A. K.; Kesharwani, R. K. Nutraceutical and Functional Foods for Cardiovascular Health. In: *Food Process Engineering*; Apple Academic Press, CRC Press, 2016a; pp 257–278, 291–312.

Klein, C.; Sato, T.; Meguid, M. M.; Miyata, G. From Food to Nutritional Support to Specific Nutraceuticals: A Journey Across Time in the Treatment of Disease. *J. Gastroenterol.* **2000,** *35*, 1–6.

Kokate, C. K.; Purohit, A. P.; Gokhale, S. B.; *Pharmacognosy*, 42nd edn.; Nirali Prakashan, 2008.

Kumari, A.; Yadav, S. K.; Yadav, S. C. Biodegradable Polymeric Nanoparticles Based Drug Delivery Systems. *Colloids Surf. B Biointerfaces* **2010,** *75*, 1–18.

Lim, S. B.; Banerjee, A.; Onyuksel, H. Improvement of Drug Safety by the Use of Lipid-Based Nanocarriers. *J. Control. Release* **2012,** *163*, 34–45.

Limer, J. L.; Speirs, V. Phyto-Oestrogens and Breast Cancer Chemoprevention. *Breast Cancer Res.* **2004,** *6*, 119–127.

Liu, L.; Liu, P.; Li, X.; Zhang, N.; Tang C. Novel Soy β-Conglycinin Core Shell Nanoparticles As Outstanding Ecofriendly Nanocarriers for Curcumin. *J. Agric. Food Chem.* **2019,** *67*, 6292–6301.

Livney, Y. D. Nanostructured Delivery Systems in Food: Latest Developments and Potential Future Directions. *Curr. Opin. Food Sci.* **2015,** *3*, 125–135.

Mancuso, A.; Cristiano, M. C.; Fresta, M.; Paolino, D. The Challenge of Nanovesicles for Selective Topical Delivery for Acne Treatment: Enhancing Absorption Whilst Avoiding Toxicity. *Int. J. Nanomed.* **2020,** *15*, 9197–9210.

McCarty, M. F. Toward Practical Prevention of Type 2 Diabetes. *Med. Hypoth.* **2005,** *64*, 151–158.

McClements, D. J. Nutraceuticals: Superfoods or Superfads? In: *Future Foods*; Copernicus, Chem, 2019; pp 167–201.

McClements, D. J.; Li, Y. Structured Emulsion-Based Delivery Systems: Controlling the Digestion and Release of Lipophilic Food Components. *Adv. Colloid Interface Sci.* **2010,** *159*, 213–228.

Meenambal, R; Bharath, M. M. S. Nanocarriers for Effective Nutraceutical Delivery to the Brain. *Neurochem. Int.* **2020,** *140*, 104851.

Mozafari, M. R.; Reed, C. J.; Rostron, C.; Kocum, C.; Piskin, E. Formation and Characterisation of Non-Toxic Anionic Liposomes for Delivery of Therapeutic Agents to the Pulmonary Airways. *Cell. Mol. Biol. Lett.* **2002,** *7*, 243–244.

Na, H. K.; Surh, Y. J. Modulation of Nrf2-Mediated Antioxidant and Detoxifying Enzyme Induction by the Green Tea Polyphenol EGCG. *Food Chem. Toxicol.* **2008,** *46*, 1271–1278.

Natural Health Products Regulations. Pursuant to Subsection 30 (1) of the "Food and Drugs Act". Canada Gazette 137; 2003.

Nelson, N. J. Purple Carrots, Margarine Laced with Wood Pulp? Nutraceuticals Move into the Supermarket. *J. Natl. Cancer Inst.* **1999,** *91*, 755–757.

Pandey, M.; Verma, R. K.; Saraf, S. A. Nutraceuticals: New Era of Medicine and Health. *Asian J. Pharm. Clin. Res.* **2010,** *3*, 11–15.

Rajam, P. R.; Brij, R. M. S.; Sivakumar, T. Nutraceuticals: An Area of Tremendous Scope. *IJRAP* **2011,** *2* (2), 410–415.

Rane, B. R.; Bharath, M. S.; Patil, R. R.; Keservani, R. K.; Jain, A. S. Novel Approaches in Nutraceuticals. In: *Enhancing the Therapeutic Efficacy of Herbal Formulations Through*

Novel Drug Delivery Systems; Kesharwani, R. K., Keservani, R. K., Sharma, A. K., Eds.; IGI Global International Publisher: Pennsylvania, 2020b; Chapter 11, pp 241–266.

Rane, B. R.; Patil, A. S.; Keservani, R. K.; Jain, A. S. Novel Approaches in Herbal Formulation. In: *Enhancing the Therapeutic Efficacy of Herbal Formulations through Novel Drug Delivery Systems*; Kesharwani, R. K., Keservani, R. K., Sharma, A. K., Eds.; IGI Global International Publisher: Pennsylvania, 2020a; Chapter 2, pp 43–68.

Rane, B. R.; Tadavi, S. A.; Keservani, R. K. Naturopathy. In: *Herbal Product Development*; Sharma, A. K., Keservani, R. K., Prakash Gautam, S., Eds.; Apple Academic Press, CRC Press, Taylor & Francis Group, 2020c; Chapter 12, pp 321–347.

Ranzato, E.; Martinotti, S.; Calabrese, C. M.; Giorgio, C. Role of Nutraceuticals in Cancer Therapy. *J. Food Res.* **2014,** *3*, 1–8.

Rasti, B.; Erfanian, A.; Salamat, J. Novel Nanoliposomal Encapsulated Omega-3 Fatty Acids and Their Applications in Food. *Food Chem.* **2017,** *230*, 690–696.

Rathore Kamal, S.; Nema, R. K.; Sulekha, M. Nutraceuticals and Diabetes—An Update. *Planta Indica* Oct–Dec, **2008,** *4* (4), 45–47.

Ratnam, D. V.; Ankola, D. D.; Bhardwaj, V.; Sahana, D. K.; Kumar, M. N. Role of Antioxidants in Prophylaxis and Therapy: A Pharmaceutical Perspective. *J. Control. Release* **2006,** *113*, 189–207.

Rinaldi, A. A Nutraceutical a Day May Keep the Doctor Away. *EMBO* **2005,** *6* (8), 708–711.

Rissanen, T. H.; Voutilainen, S.; Virtanen, J. K.; Venho, B.; Vanharanta, M.; Mursu, J.; Salonen, J. T. Low Intake of Fruits, Berries and Vegetables is Associated with Excess Mortality in Men: The Kuopio Ischaemic Heart Disease Risk Factor (KIHD) Study. *J. Nutr.* **2003,** *133*, 199–204.

Sarris, J.; Byrne, G. J.; Stough, C.; Bousman, C.; Mischoulon, D.; Murphy, J.; Macdonald, P.; Adams, L.; Nazareth, S.; Oliver, G.; Cribb, L.; Savage, K.; Menon, R.; Chamoli, S.; Berk, M.; Ng, C. H. Nutraceuticals for Major Depressive Disorder—More Is Not Merrier: An 8-Week Double-Blind, Randomised, Controlled Trial. *J. Affect. Disord.* **2019,** *245*, 1007–1015.

Schiller, R. N.; Barranger, E.; Schauss A. G.; Nichols, E. Lifestyle Management of Obesity. *JAMA* **2001,** *4*, 34–41.

Sengupta, A.; Ghosh, S.; Das, S. Tomato and Garlic Can Modulate Azoxymethane—Induced Colon Carcinogenesis in Rats. *Eur. J. Cancer Prev.* **2003,** *12*, 195–200.

Sharma, A. K.; Keservani, R. K.; Gautam Surya, P. *Herbal Product Development*; Apple Academic Press, CRC Press, Taylor & Francis Group, 2020.

Silva, A. D.; Lanerolle, P. Nutraceuticals: Concept and Controversies. *CME* **2011,** *56*, 171–173.

Singh, J.; Sinha, S. Classification, Regulatory Acts and applications of Nutraceuticals for Health. *IJPBS* Jan–Mar **2012,** *2* (1), 177–187.

Sosnowska, B.; Penson, P.; Banach, M. The Role of Nutraceuticals in the Prevention of Cardiovascular Disease. *Cardiovasc. Diagn. Ther.* **2017,** *7* (Suppl. 1), S21–S31. http://dx.doi.org/10.21037/cdt.2017.03.20 PMID: 2852991

Sumi, Y. Research and Technology Trends of Nutraceuticals. *Sci. Tech. Trends* **2008,** *28*, 10–21.

Tank Dharti, S.; Gandhi, S.; Shah, M. Nutraceuticals. *Portmanteau Sci. Nat.* **2010,** *5* (3).

Temple, W. J.; Gladwin, K. K. Fruits, Vegetables, and the Prevention of Cancer: Research Challenges. *Nutrition* **2003,** *19*, 467–470.

Trottier, G.; Boström, P. J.; Lawrentschuk, N.; Fleshner, N. E. Nutraceuticals and Prostate Cancer Prevention: A Current Review. *Nat. Rev. Urol.* **2010,** *7*, 21–30.

U. S. Food and Drug Administration. *Liposome Drug Products: Chemistry, Manufacturing, and Controls; Human Pharmacokinetics and Bioavailability; and Labeling Documentation; FDA-2016-D-2817*; Center for Drug Evaluation and Research: Rockville, 2018.

Vargas-Mendoza, N.; Morales-González, A.; Madrigal-Santillán, E. O.; Madrigal-Bujaidar, E.; Álvarez-González, I.; Garcia-Melo, R. F.; Anguiano-Robledo, L.; Fregoso-Aguilar, T.; Morales-Gonzalez, J. A. Antioxidant and Adaptative Response Mediated by Nrf2 During Physical Exercise. *Antioxidants* **2019,** *8*, 196.

Verma, D.; Gulati, N.; Kaul, S.; Mukherjee, S.; Nagaich, U. Protein Based Nanostructures for Drug Delivery. *J. Pharm.* 2018, 285854.

Verma, S.; Utreja, P. Vesicular Nanocarrier Based Treatment of Skin Fungal Infections: Potential and Emerging Trends in Nanoscale Pharmacotherapy. *Asian J. Pharm. Sci.* **2019,** *14*, 117–129.

Wargovich, M. J.; Morris, J, Brown, V.; Ellis, J.; Logothetis, B.; Weber, R. Nutraceutical Use in Late-Stage Cancer. *Cancer Metastasis Rev.* **2010,** *29*, 503–510.

Wei, Y.; Yang, S.; Zhang, L.; Dai, L.; Tai, K.; Liu, J.; Mao, L.; Yuan, F.; Gao, Y.; Mackie, A. Fabrication, Characterization and In Vitro Digestion of Food Grade Complex Nanoparticles for Co-delivery of Resveratrol and Coenzyme Q10. *Food Hydrocoll.* **2020,** *105*, 105791.

Whitman, M. Understanding the Perceived Need for Complementary and Alternative Nutraceuticals: Lifestyle Issues. *Clin. J. Oncol. Nurs.* **2001,** *5*, 190–194.

Yamini, B. T.; Pratibha, T.; Arjmandi, B. H. Nutraceuticals and Cancer Management. *Front. Biosci.* **2005,** *10*, 1607–1618.

Yates, A. A.; Erdman, J. W. Jr; Shao, A.; Dolan, L. C.; Griffiths, J. C. Bioactive Nutrients—Time for Tolerable Upper Intake Levels to Address Safety. *Regul. Toxicol. Pharmacol.* **2017,** *84*, 94–101.

Zhang, Y. J.; Gan, R. Y.; Li, S.; Zhou, Y.; Li, A. N.; Xu, D. P.; Li, H. B. Antioxidant Phytochemicals for the Prevention and Treatment of Chronic Diseases. *Molecules* **2015,** *20*, 21138–21156.

Zheng, J.; Zhou, Y.; Li, Y.; Xu, D. P.; Li, S.; Li, H. B. Spices for Prevention and Treatment of Cancers. *Nutrients* **2016,** *8*, 495. DOI: 10.3390/nu8080495.

CHAPTER 14

PROBIOTIC ROLE OF NUTRACEUTICALS

BHUSHAN R. RANE, NIDHI S. KATE, SAMALI S. RAUT, PRIYANKA J. CHAUHAN, NIMISHA N. PATIL, HRITIK A. PATIL and ASHISH S. JAIN

Shri D.D. Vispute College of Pharmacy & Research Center, Panvel, Raigad, Maharashtra, India

ABSTRACT

There are various foods available on the market today that have adverse effects on one's health, the environment, and society. Consuming them is hazardous to your health. People are battling numerous disorders like diabetes, obesity, leukaemia, osteoporosis, and other severe health conditions as a result of all these factors. Probiotics are therefore crucial in the prevention and treatment of numerous diseases as well as the enhancement of dietary, immune, and physiological functions. It aids digestion and improves intestinal health. Probiotics have both nutritional and therapeutic properties. It improves physical health, increases longevity, and alleviates the symptoms of numerous ailments. People are also becoming more interested in nutraceuticals. They are sold in the market as supplements and a variety of food products. People lead an active lifestyle by eating probiotics as supplements and dietary products. This chapter highlights the impact of probiotics on intestinal health as well as their potential as nutraceuticals.

14.1 NUTRACEUTICALS

Nutraceuticals are nutritional supplements that are used to improve health, postpone ageing, prevent disease, and support the healthy functioning of the

Herbals As Nutraceuticals: Their Role in Healthcare.
Raj K. Keservani, Rajeshwar Kamal Kant Arya, and Rajesh K. Kesharwani (Eds.)

human body (Keservani et al., 2010a, 2010b). Nutraceuticals are currently gaining popularity due to their nutritional and medicinal properties. They are classified as dietary supplements or herbal bioactive substances based on their source. The global nutraceutical market is enormous, at around USD 117 billion. Herbal nutraceuticals aid in the maintenance of health and the promotion of better health, lifespan, and happier life (Sharma et al., 2020; Rane et al., 2020a, 2020b, 2020c). Nutraceuticals have demonstrated encouraging results in the treatment of a variety of ailments, including cancer, neurological diseases, cardiovascular disease, and others. The current review provides an overview of several bioactive substances that operate as nutraceuticals and their involvement in health benefits (Sachdeva et al., 2020; Keservani et al., 2017, 2020). The current interest and continuous global study in nutraceuticals, which are obtained from foods of either vegetal or animal origin, strives to thoroughly elucidate their mechanism of action, their safety, and their efficacy by supporting their function with clinical data (Santini et al., 2017; Daliu et al., 2018). Clarifying their mechanisms of action will pave the way for the development of a new generation of therapeutic agents that can: (i) prevent metabolic syndrome, which is a group of conditions that can occur together and include heart disease, stroke, and type 2 diabetes; (ii) supplement pharmacological therapy, especially for those people who do not meet the criteria for conventional pharmacological therapy (Santini et al., 2018; Daliu et al., 2019).

14.2 HERBAL NUTRACEUTICALS

Herbs have been utilized as all-natural treatments for a variety of physiological ailments since the dawn of time. Traditional medical literature recognized their worth as a gift from nature to humanity for the treatment of ailments. Additionally, several of the herbs have been utilized in food preparation, and a select number have been incorporated into cheese manufacturing as a flavoring and coagulating agent and also as flavor ingredients. Scientific studies on the biological effects and toxicity of chemical compounds found in various herbs have been completed over several years. As a result, literature on the use of plants or their active components in foods and how they interact with food ingredients have emerged recently (Hussian et al., 2014). In a market where demand for nutraceutical goods is rising at a faster rate than ever before, herbal nutraceuticals are gaining importance. Beyond basic nutrition, herbal products with phytochemicals like carotenoids, lutein, phenolic compounds, etc., offer several health advantages to people. Our ancestors employed herbs like garlic, ginkgo Biloba, ginger, ginseng, onion, turmeric, Brahmi, and ashwagandha to treat a variety of illnesses (Bishnoi et al., 2022).

Herbs can be broadly divided into three categories based on their usage and toxicity. The first category consists of "food herbs," which have a mild action, fairly low toxicity, and a low likelihood of producing a negative reaction when taken. Food herbs can be consumed in large amounts for a long time without experiencing any acute or chronic toxicity like. It is conceivable for meals to cause allergic reactions, unusual idiosyncratic reactions, and even popular foods like grapefruit juice, broccoli, and okra might affect how some medications work. The second type is "medicinal herbs," which must be taken as a medication for certain medical disorders and often for a short period as directed by a healthcare professional. These herbs may occasionally interact with medications and frequently produce unpleasant responses. The final category is "poisonous herbs," which must only be used under the strict supervision of qualified clinicians due to their high potential for both acute and chronic toxicity (Hussian et al., 2014).

The market for herbal nutraceuticals is analyzed based on geographies, important players, and market segments. This market has benefited from the COVID-19 epidemic. As a result, there is now a higher demand worldwide for herbal nutraceuticals, which is helping the market for these products grow internationally. The market for herbal nutraceuticals is divided into segments based on product type, form, nature, distribution method, and geography. It is divided into pharmacies, internet stores, supermarkets, and others based on the sales channel. It naturally falls into two categories: conventional and organic. Herbs grown traditionally may contain additions, GMOs, and chemicals. Global demand for conventional herbal nutraceuticals is driven by their easy accessibility and lower cost when compared to alternative organic herbal nutraceuticals. However, the organic segment is anticipated to experience significant growth throughout the forecast period as a result of growing consumer awareness of the negative effects of chemicals, pesticides, and fertilizers, which is causing consumers to turn to organic products and herbal nutraceuticals. The market for herbal nutraceuticals, which had a value of USD 28,329.7 million in 2019, is anticipated to grow to USD 48,446.5 million by 2027, with a CAGR of 7.55 % between 2021 and 2027 (Priya et al., 2022).

14.2.1 AYURVEDIC HERBS AS NUTRACEUTICALS

Sr. no.	Compound	Botanical name	Therapeutic effects	References
1	Aloe vera gel	*Aloe barbadensis*	Anti-inflammatory, anti-allergic	Bishnoi et al. (2022); Mohd et al. (2020); Ansari et al., (2013); Gupta et al. (2010)
2	Brahmi	*Bacopa monnieri*	Nervine tonic, spasmolytic, anti-anxiety	
3	Chamomile	*Allium sativum*	Hay fever, inflammation, muscle spasms, menstrual disorders, insomnia	
4	Garlic	*Allium sativum*	Anti-inflammatory, antibacterial, antigout, nervine tonic	
5	Ginger	*Zingiber officinale*	Stimulant, hyperglycemia, and throat ache	
6	Ginseng	*Panax ginseng*	Adaptogen, stimulating the immune and nervous system	
7	Licorice	*Glycyrrhiza glabra*	Boosts immunity, improve vision, boosts immunity, improves vision	
8	Turmeric	*Curcuma longa*	Anti-inflammatory, anti-septic, antiarthritic, anticancer	
9	Valerian	*Valerian officinalis*	A spasmolytic, mild sedative	

14.3 PROBIOTICS

Elie Metchnikoff, who discovered that consuming live microbes in fermented yoghurt or dairy products improved various GIT qualities, created the probiotics concept (De Simone et al., 2019). The Food and Agriculture Organization of the United Nations and the World Health Organization currently define probiotics as "live microorganisms that, when administered in adequate amounts, confer a health benefit on the host." This definition was reviewed by the International Association for Scientific Prebiotics and Probiotics (ISAPP) (Hill et al., 2014). ISAPP (2018) defines probiotic microorganisms as having enough information about the genus, species, and strain identities; being titled using a valid scientific nomenclature for strain designation; being deposited in a global culture collection, and having favorable health attributes affirmed by at least one human trial. Furthermore, a probiotic-containing product must contain a sufficient number of live microorganisms to offer the stated health benefit until the expiration date. Furthermore, probiotics must be adaptable to adverse GIT conditions, exhibit antagonistic action against infections, stimulate the immune system, have proven positive impacts on the host, and maintain their potency when subjected to certain food processing conditions (Plaza-Diaz et al., 2019). The primary probiotic category consists of lactic acid bacteria (LAB), and the most popular and well-known probiotic bacteria are Lactobacilli and Bifidobacteria. However, additional genera, including Enterococcus, Streptococcus, Bacillus, and Saccharom, have also been commercialized (Sarao et al., 2017). In contrast to genetically modified organisms, there is currently an increasing need for the use of natural commensal microbes, which have specific roles in the GIT. However, bacteria like Akkermansia muciniphila and Faecalibacterium prausnitzii, which are too sensitive to oxygen, stomach pH, and bile salts, as well as problems restricting their widespread multiplication, are viable options (Marcial-Coba et al., 2019). According to several studies, probiotics can act as antibiotics, reduce the likelihood of infections, reduce the incidence of diarrhoea, prevent certain inflammatory diseases and allergy disorders (including atopic dermatitis and rhinitis), and prevent colorectal and bladder cancer (De Prisco and Mauriello, 2016; Plaza-Diaz et al., 2019). The following disorders are prevented or treated by probiotics most consistently: severe respiratory tract infections, necrotizing enterocolitis, antibiotic-associated diarrhoea, acute infectious diarrhoea, and infant colic (Liu et al., 2018). A number of initiatives have been made to increase the bioavailability of probiotics in a wider range of diets. First, in order for probiotics to provide the user with the claimed health advantages, the best strains, culture conditions, and product manufacturing are crucial (Fenster et al., 2019).

14.4 CLASSIFICATION OF PROBIOTICS

Species	Strain	Method of action/target/diseases condition	References
Bacillus spp.	Coagulans	Antibiotic-associated diarrhea	Georgieva et al. (2014); Khaligi et al. (2016); Zhao et al. (2019); Trush et al. (2020); Lopez-Santamarina et al. (2021); Lee et al. (2018)
Bifidobacterium spp.	Breve	Conjugated linoleic acid and conjugated linolenic acid production in culture media	
Enterococcus spp.	Faecium	Atopic dermatitis	
Lactobacillus spp.	Acidophilus	Riboflavin production in foods	
Saccharomyces spp.	Cerevisiae	Irritable bowel syndrome	
Streptococcus spp.	Thermophiles	Eczema, atopic sensitization, food allergy, and wheezing	
	Plantarum	Conjugated linoleic acid production in foods	
	Rhamnosus	Atopic dermatitis, atopic eczema	
	Casei	Antimicrobial ability against Gram-positive and Gram-negative pathogenic and antibiotic-resistant bacteria	
	Lactis	Cow's milk allergy (CMA)	
	Animalis	Atopic dermatitis and asthma	

14.5 RECENT TRENDS

The definition of a probiotic provided by the Food and Agriculture Organization of the United Nations and the World Health Organization (FAO/WHO)—"live microorganisms which when provided in suitable proportions provide a health advantage on the host"—was reaffirmed as pertinent and suitably acceptable for both the uses that are currently in use and those that are on the wanting list. It is generally known and documented that the gut microbiota and health impacts are related. This overall effect on the host has helped to shed light on how interdependent humans and their gut microbes are. Both passive and active methods can be used to influence the gut microbiome. While the latter consists of antibiotics, pre-and probiotics, the former encompasses nutrition, lifestyle, and environment. A phylogenetically constrained diversity of some yeast strains and bacteria were once part of traditional probiotic techniques. However, with the introduction of fecal microbiota transplant (FMT), which was successfully used to treat CDI, IBD, and other disorders, biotherapeutic approaches emerged in recent years (Hage et al., 2017). The creation of functional meals with probiotics has received a lot of attention in recent years. It's because probiotic microorganisms, when consumed in appropriate amounts, offer a variety of health benefits. However, inherent food characteristics like pH, acidity, packaging, and storage requirements of some food types limit the stability and viability of probiotics in diverse food products that can be used as carriers. The survival of probiotic bacteria is also impacted by the gastrointestinal tract's acidic environment. As a result, probiotic carriers or specific delivery systems have been created using microencapsulation and other enrichment techniques. For boosting the viability and bioactivity of probiotics, microencapsulation procedures like emulsion, freeze-drying, extrusion, and others have been demonstrated to be promising (Misra et al., 2022; Niamah et al., 2021). Some of them are discussed below.

14.5.1 ELECTRO-HYDRODYNAMIC PROCESS

Probiotic encapsulation has been targeted for the electro-hydrodynamic (EHD) processes. Probiotics were more likely to survive in the digestive system after being encapsulated using EHD. EHD also improves the probiotics' viability and stability in the food matrixes. The process's inexpensive costs and good reproducibility in a single step make it appealing. Pharmaceutical and food items can both use encapsulated probiotics (Niamah et al., 2021). Through EHD technologies like electrospinning and electrospray, probiotics can be encapsulated. EHD procedures sustain the functionality of live probiotics while working in aqueous solutions at room temperature. EHD is a quick drying method that works with a range of food ingredients and enables effective encapsulation over time. High cell encapsulation and long-term survival are made possible by the qualities of the polysaccharides, proteins, or hybrid compounds used for probiotics encapsulation and the features of the custom matrices created by EHD. The capacity of electrospun fibers and electrospray capsules to contain additional functional substances such as prebiotics, as well as their efficiency to support the targeted administration and survival of the encapsulated probiotic cells under gastrointestinal track conditions makes it a versatile method (Mendes et al., 2021).

14.5.2 VACUUM DRYING

Pharmaceutical formulations containing probiotics must be stable during long-term storage; choosing the best drying technology and protection plan require an awareness of the basic processing steps during drying. Vacuum drying appears to be a promising alternative in theory because it uses more mild process conditions. However, in comparison to freeze drying, viability findings following vacuum drying are less favorable in practice (Broeckx et al., 2016). In a study, a novel method for the preservation of starter cultures and probiotics microwave-vacuum drying was investigated. The effects of chamber pressure (7, 15, and 30 mbar), microwave power (1–5 W/g), and the maximum product temperature (30–50 °C) on the drying time, drying rate, microwave input efficiency, and survival of the model strain *Lactobacillus paracasei ssp. paracasei* F19 (L. paracasei F19) were investigated. Microwave drying times ranged from 50 to 240 min, with microwave power being the primary determinant of drying behavior and effectiveness. The results showed the highest product quality in terms of survival rate, a very quick drying time of 64 min, a high degree of energy efficiency (27%, and a combination of 3 W/g microwave power input and 35 °C maximum temperature at a low chamber pressure of 7 mbar (33%). Therefore, a microwave-vacuum drying procedure is an alternative to traditional freeze or vacuum drying that may be used (Ambros et al., 2018).

14.5.3 SPRAY-DRYING OF PROBIOTICS

Due to the obvious change in human microflora, there is an increase in the population of pathogenic bacteria, which contributes to the development of gastric diseases and disorders. In this way, adding probiotics to food products may help clear harmful bacteria from adhesion sites and control the immune response by activating certain genes in the gastrointestinal tract (GIT) of humans. Probiotic products with a desired shelf life that maintain the viability of the probiotic cells are difficult to develop, however, because probiotics are sensitive to the environmental factors during food manufacture/storage. A sustainable method that allows for mass manufacturing at a cheaper energy cost is spray drying bacteria. In order to ensure that bacteria have enhanced resilience during storage, technological procedures, and digestive stressors, this is also a viable method for encapsulating bacteria within different protective matrices (Sharma et al., 2022). Spray drying involves atomizing feed fluid containing probiotic organisms and dispersed or suspended protectant materials into droplets, which are then rapidly transformed into particles by drying in a heated airflow. The preservation of cell viability is a difficult endeavor because of the severe circumstances and interrelated stresses. Several strategies have been tried to improve cell survival in dried powders, including enhancing the intrinsic stress tolerance of cells, adjusting the composition of protectants, and optimizing the production process and drier settings (Fu et al., 2018; Huang et al., 2017).

14.5.4 ULTRASONICATION

A newer method of encapsulating both hydrophilic and lipophilic ingredients in foods and pharmaceuticals is ultrasonication. With little expense or energy use, it can create stable microbubbles with tiny droplets of narrow size distribution. It is possible to create microcapsules with little or no surfactant. The enormous shear stress generated by ultrasonication, however, may be harmful to the bacterium. Therefore, it is necessary to optimize the ultrasonication parameters, that is, ultrasonication power, time, and before usage (Misra et al., 2022). Dextran and whey protein were used as the encapsulating materials in a study that created double emulsion (W1/O/W2) microcapsules containing probiotic *Lactobacillus Plantarum*. It was shown that the viability of the bacteria remained stable (> 109 CFU/mL) up to the treatment time of 200 s. The existence of all microcapsules allowed living bacteria to endure successive digestion. During sequential digestion, dextran-based capsules seem to be more resilient than whey protein (Pandey et al., 2021).

14.6 HEALTH BENEFITS OF PROBIOTICS IN VARIOUS DISEASES

14.6.1 PROBIOTICS AS ANTICANCER AGENTS

Probiotics may have anticancer properties, according to several studies. By preventing the spread of pro-cancerous bacteria and boosting intestinal immunity, probiotics may lower the risk of developing colon cancer. Probiotic bacteria achieve this suppression through the inhibition of toxic components and a reduction in the permeation of mutagenic substances from intestinal cellular membranes. Apoptosis, which aids in the production of short chains of fatty acids and inhibits the tyrosine-kinase signaling pathway, is achieved to achieve this suppression. Exopolysaccharide and lipopolysaccharide, two probiotic metabolites, aid in the treatment and prevention of cancer. LPS is a substance that is present in Gram-negative bacteria's outer membrane. It stimulates the immune system's T-cell defence mechanism against malignant cells. By activating the apoptotic pathway, the Exopolysaccharide generated by lipopolysaccharide species exerts an anticancer activity on colon carcinoma cells. Cancer cells have a direct cytotoxic effect (Gill et al., 2004; Paulos et. al 2007; Tukenmez et al., 2019).

14.6.2 PROBIOTICS AS ANTIOXIDANT AGENTS

Natural antioxidants can be produced by antioxidant supplements or probiotics isolated from the LAB family. Bioactive peptides such as glutathione, GPX, and CAT are produced by probiotic bacteria. They have antioxidative properties and give protection against fatty acid peroxidation. The main cause of disease development is oxidative stress. As a result, these antioxidants aid in reducing oxidative damage, foraging rate, and antioxidative enzyme activity (Mishra et. al 2015; Kumar et al., 2010; Saiga et al., 2003; Valko et al., 2007; Liu et. al 2010).

14.6.3 PROBIOTICS AS ANTIPLAQUE AGENTS

As the main component of the digestive tract, the oral cavity's good bacteria help to prevent dental caries and can also boost good mucosal immunity. Dental caries is mostly caused by Streptococcus mutans. L. rhamnosus, a probiotic bacterium, reduces the incidence of caries or tooth decay, especially in youngsters, by inhibiting the proliferation of S. mutans. Additionally, daily consumption of probiotics Lactobacilli or Bifidobacteria helps to lower S. mutans levels in plaque or saliva (Jindal et al., 2012; Goel et al., 2014; Twetman et al., 2012).

14.6.4 INFLAMMATORY BOWEL DISEASE (IBD)

Inflammation of the gastrointestinal tract is caused by immune system dysfunction (GIT). Pouchitis, ulcerative colitis, and Crohn's disease are all examples of IBD. Crohn's disease affects the small intestine. Inflammation occurs, which reduces nutrition absorption and affects permeability. Lactobacillus strain (acidophilus and johnsonii), Bifidobacterium strain (bifidum, animalis), Saccharomyces boulardii, and ECN have all been researched for their usefulness in the treatment of IBD. Several strains of VSL3 have been demonstrated to be effective in the prevention of pouchitis. S. boulardii has been demonstrated to be effective in the treatment of Crohn's disease and in reducing relapses (Abraham et al., 2017; Elahi et al., 2007; Maurya et al., 2014).

14.6.5 IRRITABLE BOWEL SYNDROME (IBS)

It's a chronic condition with almost no cure. The treatment aims to alleviate the symptoms. The combination of B. longum, L. acidophilus, L. lactis, and S. thermophilus has shown promising benefits in alleviating IBS symptoms. The usage of bacterial lysates of Enterococcus faecalis and *Escherichia coli* (*E.coli*) has also been demonstrated to be effective in lowering IBS symptoms (Fontana et al., 2013; Drouault et al., 2008; Enck et al., 2008).

14.6.6 LACTOSE INTOLERANCE

Lactose intolerance is brought on by a beta-galactosidase or lactase deficiency. Due to a genetic weakness, lactose cannot be digested into the monosaccharide's glucose and galactose. An individual may experience symptoms including diarrhoea and stomach discomfort after consuming milk or other milk-related food. Due to the utilization of these strains, which have increased Beta-galactosidase activity, in the treatment of lactose intolerant individuals, the combination of Streptococcus thermophilus and L. delbrueckii has demonstrated excellent results (Kechagia et al., 2013).

14.6.7 ROTAVIRUS AND TRAVELER'S DIARRHOEA

The most prevalent cause of acute infantile diarrhoea is rotavirus. When a person goes to a warmer area, they may get symptoms of traveler's diarrhoea. S. boulardii and a combination of L. acidophilus and Bifidum can reduce the mean duration when compared to the traditional one. This probiotic blend has been demonstrated to be an effective treatment for travelers' diarrhoea. In addition, clinical investigations have demonstrated that a combination of L. rhamnosus, B. animalis, and L. reutri can reduce the duration of rotavirus diarrhoea (Kechagia et al., 2013; Goldin et al., 2008; Guérin et al., 2001).

14.6.8 URINARY TRACT INFECTIONS (UTI)

This is very frequent in women and tends to reoccur. Urinary tract infections are caused by *E. coli*, Klebsiella, and Staphylococcus. UTI is associated with a reduction in vaginal lactobacilli. Lactobacillus has been demonstrated to be useful in preventing UTIs. However, the authors acknowledged that randomized clinical studies are necessary (Fontana et al., 2013; Grin et al., 2013).

14.7 MARKETED FORMULATIONS OF PROBIOTICS

Sr. no.	Product/brand name	Type	Contents	Manufacturer
1	Florasante	Capsule	*Bacillus subtilis, Bacillus coagulans,* and *Streptococcus thermophilus*	Himalayan Wellness Company
2	PROBIOTICS 20 BILLION CFU	Capsule	*L. plantarum, L. fermentum, L. acidophilus, B. infantis, L. casei, L. rhamnosus, B. lactis, L. reuteri, L.salivarius, L. paracasei, L. gasseri, S. thermophilus, B. bifidum & B. breve.*	WOW Life Science
3	Acidophilus Probiotic	Tablet	*Lactobacillus Acidophilus*	NATURE'S BOUNTY, INC
4	Probiotic Gummy Vitamins	Chewable candies	*Bacillus coagulans,* and *Bacillus subtilis*	vitafusion™
5	Yakult	Milk drink	*Lactobacillus casei*	Yakult Honsha Co., Ltd.

14.8 CONCLUSION

Probiotics, and nutraceuticals are in high demand worldwide, and during the past ten years, efforts have been made continuously to accommodate that demand. Although other sites are treated with probiotics, the GIT is thought to have the most therapeutic potential. In traditional medicine, probiotics have been utilized to improve the health of gut flora. It improves epithelial barrier performance and stops pathogenic bacteria development. They can be found in a variety of fermented foods, including yakult. It has been studied successfully about the outcomes in treating antibiotic associated diarrhoea, irritable bowel syndrome (IBS), irritable Bowel disease (IBD), and urinary tract infections by utilizing a blend of certain microorganism strains. Furthermore, because living microbes are used in the production process, assuring the quality of such a product during industrial mass manufacturing is the biggest problem. The stability of products in various climatic regions of the world should receive more attention in research. Developed nations ought to step forward and create a global standard for probiotic manufacturing best practices.

14.9 FUTURE PERSPECTIVES

Probiotics are a type of microbe that has been shown to offer health advantages. Those are not taken as medicine, but rather as a dietary supplement. Because most probiotics are unicellular bacteria, they can be easily cultured or grown by supplying the proper media and circumstances such as ideal temperature, pH, nutrients, and minerals. Probiotics can be manufactured on a big scale to meet the needs of the cells utilized by functional foods. The circumstances required for their growth can be simply generated in a medium that does not require a large expenditure. In industries, they are produced in enormous tanks known as bioreactors. Different cells use various resources; thus, they are cultivated in various bioreactors that have been specially built for them. If the probiotic grows more than once in the medium it is being utilized, its effectiveness will improve. This is only achievable if these cultures are consistently given the ideal conditions. Probiotics provide a plethora of health advantages, but there are some questions as well. It has been noted that probiotic cells can penetrate the bloodstream through a process known as bacteraemia, which can result in sepsis, in the case of newborn infants with compromised immunity or severe illnesses. This illness causes the body to have a very strong immunological reaction, which can most often be lethal and includes heavy breathing. Later immune responses are only reactions, with no more major synthesis of antibodies in response to the bacterial activity, because the immunity is already compromised or battling the sickness.

KEYWORDS

- **probiotic**
- **nutraceutical**
- **herbal**
- **classification**
- **uses**
- **recent trends**

REFERENCES

Abraham, B. P.; Quigley, E. M. M. Probiotics in Inflammatory Bowel Disease. *Gastroenterol. Clin. North Am.* **2017,** *46* (4), 769–782. https://doi.org/10.1016/j.gtc.2017.08.003.

Alves Garcia, S. L.; Monteiro da Silva, G.; Svendsen Medeiros, J. M.; Rocha de Queiroga, A. P.; Brito de Queiroz, B.; Bezerra de Farias, D. R.; Correia, J. O.; Florentino, E. R.; Alonso Buriti, F. C. Correction: Influence of Co-Cultures of *Streptococcus Thermophilus* and Probiotic Lactobacilli on Quality and Antioxidant Capacity Parameters of Lactose-Free Fermented Dairy Beverages Containing *Syzygium Cumini* (L.) Skeels Pulp. *RSC Adv.* **2020,** *10* (29), 16905–16905. https://doi.org/10.1039/d0ra90046j.

Amara, A. A.; Shibl, A. Role of Probiotics in Health Improvement, Infection Control and Disease Treatment and Management. *Saudi Pharm. J.* **2015,** *23* (2), 107–114. https://doi.org/10.1016/j.jsps.2013.07.001.

Ambros, S.; Foerst, P.; Kulozik, U. Temperature-Controlled Microwave-Vacuum Drying of Lactic Acid Bacteria: Impact of Drying Conditions on Process and Product Characteristics. *J. Food Eng.* **2018,** *224*, 80–87. https://doi.org/10.1016/j.jfoodeng.2017.12.025.

Ansari, S.; Chauhan, B.; Kalam, N.; Kumar, G. Current Concepts and Prospects of Herbal Nutraceutical: A Review. *J. Adv. Pharm. Technol. Res.* **2013,** *4* (1), 4. https://doi.org/10.4103/2231-4040.107494.

Bampi, G. B.; Backes, G. T.; Cansian, R. L.; de Matos, F. E.; Ansolin, I. M. A.; Poleto, B. C.; Corezzolla, L. R.; Favaro-Trindade, C. S. Spray Chilling Microencapsulation of Lactobacillus Acidophilus and Bifidobacterium Animalis Subsp. Lactis and Its Use in the Preparation of Savory Probiotic Cereal Bars. *Food Bioprocess Technol.* **2016,** *9* (8), 1422–1428. https://doi.org/10.1007/s11947-016-1724-z.

Bharti, N.; Kaur, R. Health Benefits of Probiotic Bacteria as Nutraceuticals. *Eur. J. Mol. Clin. Med.* **2020,** *7* (7), 4797–4807.

Bishnoi, S.; Mudgil, D. Current Concepts, and Prospects of Herbal Nutraceutical. In: *Handbook of Nutraceuticals and Natural* Products; **2022**; pp 189–204. https://doi.org/10.1002/9781119746843.ch10.

Brinques, G. B.; do Carmo Peralba, M.; Ayub, M. A. Z. Optimization of Probiotic and Lactic Acid Production by Lactobacillus Plantarum in Submerged Bioreactor Systems. *J. Ind. Microbiol. Biotechnol.* **2010,** *37* (2), 205–212. https://doi.org/10.1007/s10295-009-0665-1.

Broeckx, G.; Vandenheuvel, D.; Claes, I. J. J.; Lebeer, S.; Kiekens, F. Drying Techniques of Probiotic Bacteria as an Important Step towards the Development of Novel Pharmabiotics. *Int. J. Pharm.* **2016,** *505* (1–2), 303–318. https://doi.org/10.1016/j.ijpharm.2016.04.002.

Daliu, P.; Santini, A.; Novellino, E. A Decade of Nutraceutical Patents: Where Are We Now in 2018? *Expert Opin. Therap. Patents* **2018,** *28* (12), 875–882. https://doi.org/10.1080/13543776.2018.1552260.

Daliu, P.; Santini, A.; Novellino, E. From Pharmaceuticals to Nutraceuticals: Bridging Disease Prevention and Management. *Expert Rev. Clin. Pharmacol.* **2018,** *12* (1), 1–7.https://doi.org/10.1080/17512433.2019.1552135

Das, T.; Das, T.; Ghosh, G. Prebiotics and Probiotics as Functional Foods: Prospect and Promises in Metabolic Diseases. *Prebiotics Probiotics Nutraceuticals* **2022,** 155–186. https://doi.org/10.1007/978-981-16-8990-1_9.

De Prisco, A.; Mauriello, G. Probiotication of Foods: A Focus on Microencapsulation Tool. *Trends Food Sci. Technol.* **2016,** *48*, 27–39. https://doi.org/10.1016/j.tifs.2015.11.009.

de Simone, C. The Unregulated Probiotic Market. *Clin. Gastroenterol. Hepatol.* **2019,** *17* (5), 809–817. https://doi.org/10.1016/j.cgh.2018.01.018.

Dianawati, D.; Mishra, V.; Shah, N. P. Stability of Microencapsulated Lactobacillus Acidophilus and Lactococcus Lactis Ssp. Cremoris during Storage at Room Temperature at Low Aw. *Food Res. Int.* **2013,** *50* (1), 259–265. https://doi.org/10.1016/j.foodres.2012.10.023.

Dinkçi, N.; Akdeniz, V.; Akalin, A. S. Survival of Probiotics in Functional Foods during Shelf Life. *Food Qual. Shelf Life* **2019,** 201–233. https://doi.org/10.1016/b978-0-12-817190-5.00006-9.

Drouault-Holowacz, S.; Bieuvelet, S.; Burckel, A.; Cazaubiel, M.; Dray, X.; Marteau, P. A Double Blind Randomized Controlled Trial of a Probiotic Combination in 100 Patients with Irritable Bowel Syndrome. *Gastroentérologie Clinique et Biologique* **2008,** *32* (2), 147–152. https://doi.org/10.1016/j.gcb.2007.06.001.

El Hage, R.; Hernandez-Sanabria, E.; Van de Wiele, T. Emerging Trends in "Smart Probiotics": Functional Consideration for the Development of Novel Health and Industrial Applications. *Front. Microbiol.* **2017,** *8*. https://doi.org/10.3389/fmicb.2017.01889.

Elahi, B.; Nikfar, S.; Derakhshani, S.; Vafaie, M.; Abdollahi, M. On the Benefit of Probiotics in the Management of Pouchitis in Patients Underwent Ileal Pouch Anal Anastomosis: A Meta-Analysis of Controlled Clinical Trials. *Digest. Dis. Sci.* **2007,** *53* (5), 1278–1284. https://doi.org/10.1007/s10620-007-0006-z.

Enck, P.; Zimmermann, K.; Menke, G.; Müller-Lissner, S.; Martens, U.; Klosterhalfen, S. A Mixture of *Escherichia coli* (DSM 17252) and Enterococcus Faecalis (DSM 16440) for Treatment of the Irritable Bowel Syndrome-a Randomized Controlled Trial with Primary Care Physicians. *Neurogastroenterol. Motility* **2008,** *20* (10), 1103–1109. https://doi.org/10.1111/j.1365-2982.2008.01156.x.

Fenster, K.; Freeburg, B.; Hollard, C.; Wong, C.; Rønhave Laursen, R.; Ouwehand, A. The Production and Delivery of Probiotics: A Review of a Practical Approach. *Microorganisms* **2019,** *7* (3), 83. https://doi.org/10.3390/microorganisms7030083.

Fischbach, M. Microbiome: Focus on Causation and Mechanism. *Cell* **2018,** *174* (4), 785–790. https://doi.org/10.1016/j.cell.2018.07.038.

Fontana, L.; Bermudez-Brito, M.; Plaza-Diaz, J.; Muñoz-Quezada, S.; Gil, A. Sources, Isolation, Characterisation and Evaluation of Probiotics. *Br. J. Nutr.* **2013,** *109* (S2), S35–S50. https://doi.org/10.1017/s0007114512004011.

Fu, N.; Huang, S.; Xiao, J.; Chen, X. D. Producing Powders Containing Active Dry Probiotics with the Aid of Spray Drying. *Adv. Food Nutr. Res.* **2018,** *85*, 211–262. https://doi.org/10.1016/bs.afnr.2018.02.003.

Georgieva, M.; Andonova, L.; Peikova, L.; Zlatkov, A. Probiotics—Health Benefits, Classification, Quality Assurance and Quality Control—Review. *Pharmacia* **2014,** *61*, 22–31

Gill, H. S.; Guarner, F. Probiotics and Human Health: A Clinical Perspective. *Postgraduate Med. J.* **2004,** *80* (947), 516–526. https://doi.org/10.1136/pgmj.2003.008664.

Goel, R.; Vedi, A.; Goyal, P. Probiotics: Contribution to Oral Health. *J. Orofacial Health Sci.* **2014,** *5* (1), 12. https://doi.org/10.5958/2229-3264.2014.00193.2.

Goldin, B. R.; Gorbach, S. L. Clinical Indications for Probiotics: An Overview. *Clin. Infect. Dis.* **2008,** *46* (s2), S96–S100. https://doi.org/10.1086/523333.

Grin, P. M., Kowalewska, P. M., Alhazzan, W., Fox-Robichaud, A. E. Lactobacillus for Preventing Recurrent Urinary Tract Infections in Women: Meta-Analysis. *Can. J. Urol.* **2013,** *20* (1), 6607–6614.

Guérin-DananC.; Meslin, J.-C.; Chambard, A.; Charpilienne, A.; Relano, P.; Bouley, C.; Cohen, J.; Andrieux, C. Food Supplementation with Milk Fermented by Lactobacillus Casei DN-114 001 Protects Suckling Rats from Rotavirus-Associated Diarrhea. *J. Nutr.* **2001,** *131* (1), 111–117. https://doi.org/10.1093/jn/131.1.111.

Gupta, S. Chamomile: A Herbal Medicine of the Past with a Bright Future (Review). *Mol. Med. Rep.* **2010,** *3* (6). https://doi.org/10.3892/mmr.2010.377.

Hill, C.; Guarner, F.; Reid, G.; Gibson, G. R.; Merenstein, D. J.; Pot, B.; Morelli, L.; Canani, R. B.; Flint, H. J.; Salminen, S.; Calder, P. C.; Sanders, M. E. Expert Consensus Document. The International Scientific Association for Probiotics and Prebiotics Consensus Statement on the Scope and Appropriate Use of the Term Probiotic. *Nat. Rev. Gastroenterol. Hepatol.* **2014,** *11* (8), 506–514. https://doi.org/10.1038/nrgastro.2014.66.

Hussain, S. A.; Panjagari, N. R.; Singh, R. R. B.; Patil, G. R. Potential Herbs and Herbal Nutraceuticals: Food Applications and Their Interactions with Food Components. *Crit. Rev. Food Sci. Nutr.* **2014,** *55* (1), 94–122. https://doi.org/10.1080/10408398.2011.649148.

Jangra, M.; Belur, P. D.; Oriabinska, L. B.; Dugan, O. M. Multistrain Probiotic Production by Co-Culture Fermentation in a Lab-Scale Bioreactor. *Eng. Life Sci.* **2016,** *16* (3), 247–253. https://doi.org/10.1002/elsc.201500069.

Javanmard, A.; Ashtari, S.; Sabet, B.; Davoodi, S. H.; Rostami-Nejad, M.; Akbari, M. E.; Niaz, A.; Mortazavian, A. M. Probiotics and Their Role in Gastrointestinal Cancers Prevention and Treatment; an Overview. *Gastroenterol. Hepatol. From Bed Bench* **2018,** *11* (4).

Jindal, G.; Pandey, R. K.; Singh, R. K.; Pandey, N. Can Early Exposure to Probiotics in Children Prevent Dental Caries? A Current Perspective. *J. Oral Biol. Craniofacial Res.* **2012,** *2* (2), 110–115. https://doi.org/10.1016/j.jobcr.2012.05.001.

Kechagia, M.; Basoulis, D.; Konstantopoulou, S.; Dimitriadi, D.; Gyftopoulou, K.; Skarmoutsou, N.; Fakiri, E. M. Health Benefits of Probiotics: A Review. *ISRN Nutr.* **2013,** 1–7. https://doi.org/10.5402/2013/481651.

Keservani, R. K.; Kesharwani, R. K.; Vyas, N.; Jain, S.; Raghuvanshi, R.; Sharma, A. K. Nutraceutical and Functional Food as Future Food: A Review. *Der Pharmacia Lettre* **2010a,** *2* (1), 106–116.

Keservani, R. K.; Kesharwani, R. K.; Sharma, A. K.; Vyas, N.; Chadoker, A. Nutritional Supplements: An Overview. *Int. J. Curr. Pharm. Rev. Res.* **2010b,** *1* (1), 59–75.

Keservani, R. K.; Sharma, A. K.; Kesharwani, R. K. An Overview and Therapeutic Applications of Nutraceutical and Functional Foods. *Rec. Adv. Drug Deliv. Technol.* **2017,** 160–201.

Keservani, R. K.; Sharma, A. K.; Kesharwani, R. K. (Eds.). *Nutraceuticals and Dietary Supplements: Applications in Health Improvement and Disease Management*; CRC Press, 2020.

Khalighi, A.; Behdani, R.; Kouhestani, S. Probiotics: A Comprehensive Review of Their Classification, Mode of Action and Role in Human Nutrition. In: *Probiotics and Prebiotics in Human Nutrition and Health* 2016. https://doi.org/10.5772/63646.

Kumar, M.; Kumar, A.; Nagpal, R.; Mohania, D.; Behare, P.; Verma, V.; Kumar, P.; Poddar, D.; Aggarwal, P. K.; Henry, C. J. K.; Jain, S.; Yadav, H. Cancer-Preventing Attributes of Probiotics: An Update. *Int. J. Food Sci. Nutr.* **2010,** *61* (5), 473–496. https://doi.org/10.3109/09637480903455971.

Kumar, V.; Naik, B.; Kumar, A.; Khanduri, N.; Rustagi, S.; Kumar, S. Probiotics Media: Significance, Challenges, and Future Perspective - a Mini Review. *Food Product. Process. Nutr.* **2022,** *4* (1). https://doi.org/10.1186/s43014-022-00098-w.

Lee, E.-S.; Song, E.-J.; Nam, Y.-D.; Lee, S.-Y. Probiotics in Human Health and Disease: From Nutribiotics to Pharmabiotics. *J. Microbiol.* **2018,** *56* (11), 773–782. https://doi.org/10.1007/s12275-018-8293-y.

Liu, J.; Huang, S. S.; Zhao, Z. Research on Antioxidative Capacity of Lactic Acid Bacteria. *China Dairy Ind.* **2010,** *38* (5), 38−41.

Liu, Y.; Tran, D. Q.; Rhoads, J. M. Probiotics in Disease Prevention and Treatment. *J. Clin. Pharmacol.* **2018,** *58*, S164–S179. https://doi.org/10.1002/jcph.1121.

Lopez-Santamarina, A.; Gonzalez, E. G.; Lamas, A.; Mondragon, A. del C.; Regal, P.; Miranda, J. M. Probiotics as a Possible Strategy for the Prevention and Treatment of Allergies: A Narrative Review. *Foods* **2021,** *10* (4), 701. https://doi.org/10.3390/foods10040701.

Marcial-Coba, M. S.; Knøchel, S.; Nielsen, D. S. Low-Moisture Food Matrices as Probiotic Carriers. *FEMS Microbiol. Lett.* **2019,** *366* (2). https://doi.org/10.1093/femsle/fnz006.

Maurya, P.; Mogra, R.; Bajpai, P. Probiotics: An Approach Towards Health and Disease. *Trends Biosci.* **2014,** *7* (20), 3107–3113.

Mendes, A. C.; Chronakis, I. S. Electrohydrodynamic Encapsulation of Probiotics: A Review. *Food Hydrocoll.* **2021,** *117*, 106688. https://doi.org/10.1016/j.foodhyd.2021.106688.

Mishra, V.; Shah, C.; Mokashe, N.; Chavan, R.; Yadav, H.; Prajapati, J. Probiotics as Potential Antioxidants: A Systematic Review. *J. Agric. Food Chem.* **2015,** *63* (14), 3615–3626. https://doi.org/10.1021/jf506326t.

Misra, S.; Pandey, P.; Dalbhagat, C. G.; Mishra, H. N. Emerging Technologies and Coating Materials for Improved Probiotication in Food Products: A Review. *Food Bioprocess Technol.* **2022,** *15* (5), 998–1039. https://doi.org/10.1007/s11947-021-02753-5.

Mohd, S.; Kaur, J. A Review on List of Herbal Nutraceuticals Having Health Benefits. *Eur. J. Mol. Clin. Med.* **2020,** *7* (7), 2552–2557.

Niamah, A. K.; Gddoa Al-Sahlany, S. T.; Ibrahim, S. A.; Verma, D. K.; Thakur, M.; Singh, S.;Patel, A. R.; Aguilar, C. N.; Utama, G. L. Electro-Hydrodynamic Processing for Encapsulation of Probiotics: A Review on Recent Trends, Technological Development, Challenges and Future Prospect. *Food Biosci.* **2021,** *44*, 101458. https://doi.org/10.1016/j.fbio.2021.101458.

Obafemi, Y. D.; Oranusi, S. U.; Ajanaku, K. O.; Akinduti, P. A.; Leech, J.; Cotter, P. D. African Fermented Foods: Overview, Emerging Benefits, and Novel Approaches to Microbiome Profiling. *NPJ Sci. Food* **2022,** *6* (1), 15. https://doi.org/10.1038/s41538-022-00130-w.

Pandey, P.; Mettu, S.; Mishra, H. N.; Ashokkumar, M.; Martin, G. J. O. Multilayer Co-Encapsulation of Probiotics and γ-Amino Butyric Acid (GABA) Using Ultrasound for Functional Food Applications. *LWT* **2021,** *146*, 111432. https://doi.org/10.1016/j.lwt.2021.111432.

Paulos, C. M.; Wrzesinski, C.; Kaiser, A.; Hinrichs, C. S.; Chieppa, M.; Cassard, L.; Palmer, D. C.; Boni, A.; Muranski, P.; Yu, Z.; Gattinoni, L.; Antony, P. A.; Rosenberg, S. A.; Restifo, N. P. Microbial Translocation Augments the Function of Adoptively Transferred Self/Tumor-Specific CD8+ T Cells via TLR4 Signaling. *J. Clin. Investig.* **2007,** *117* (8), 2197–2204. https://doi.org/10.1172/JCI32205.

Pedroso, D. L.; Dogenski, M.; Thomazini, M.; Heinemann, R. J. B.; Favaro-Trindade, C. S. Microencapsulation of Bifidobacterium Animalis Subsp. Lactis and Lactobacillus Acidophilus in Cocoa Butter Using Spray Chilling Technology. *Braz. J. Microbiol.* **2013,** *44* (3), 777–783. https://doi.org/10.1590/s1517-83822013000300017.

Plaza-Diaz, J.; Ruiz-Ojeda, F. J.; Gil-Campos, M.; Gil, A. Mechanisms of Action of Probiotics. *Adv. Nutr.* **2019,** *10* (suppl_1), S49–S66. https://doi.org/10.1093/advances/nmy063.

Rane, B. R.; Patil, A. S.; Keservani, R. K.; Jain, A. S. Novel Approaches in Herbal Formulation. In: *Enhancing the Therapeutic Efficacy of Herbal Formulations through Novel Drug Delivery Systems*; Kesharwani, R. K., Keservani, R. K., Sharma, A. K., Eds.; IGI Global International Publisher: Pennsylvania, 2020a; Chapter 2, pp 43–68.

Rane, B. R., Bharath, M. S., Patil, R. R., Keservani, R. K., Jain, A. S. Novel Approaches in Nutraceuticals. In: *Enhancing the Therapeutic Efficacy of Herbal Formulations through Novel Drug Delivery Systems*; Kesharwani, R. K., Keservani, R. K., Sharma, A. K.; IGI Global International Publisher: Pennsylvania, 2020b; Chapter 11, pp 241–266. ISBN13: 9781799844532.

Rane, B. R., Tadavi, S. A., Keservani, R. K. Naturopathy. In: *Herbal Product Development.* Sharma, A. K., Keservani, R. K., Prakash Gautam, S., Eds.; Apple Academic Press, CRC Press, Taylor & Francis Group, 2020c; Chapter 12, pp 321–347. ISBN: 9781771888776.

Sachdeva, V.; Roy, A.; Bharadvaja, N. Current Prospects of Nutraceuticals: A Review. *Curr. Pharm. Biotechnol.* **2020,** *21* (10), 884–896. https://doi.org/10.2174/1389201021666200130113441.

Saiga, A.; Tanabe, S.; Nishimura, T. Antioxidant Activity of Peptides Obtained from Porcine Myofibrillar Proteins by Protease Treatment. *J. Agric. Food Chem.* **2003,** *51* (12), 3661–3667. https://doi.org/10.1021/jf021156g.

Santini, A.; Novellino, E. Nutraceuticals—Shedding Light on the Grey Area between Pharmaceuticals and Food. *Expert Rev. Clin. Pharmacol.* **2018,** *11* (6), 545–547.

Santini, A.; Novellino, E. To Nutraceuticals and Back: Rethinking a Concept. *Foods* **2017,** *6* (9), 74. https://doi.org/10.3390/foods6090074.

Sarao, L. K.; Arora, M. Probiotics, Prebiotics, and Microencapsulation: A Review. *Crit. Rev. Food Sci. Nutr.* **2015,** *57* (2), 344–371. https://doi.org/10.1080/10408398.2014.887055.

Sharma, A. K.; Keservani, R. K.; Gautam Surya, P. *Herbal Product Development*; Apple Academic Press, CRC Press, Taylor & Francis Group, 2020.

Sharma, R.; Rashidinejad, A.; Jafari, S. M. Application of Spray Dried Encapsulated Probiotics in Functional Food Formulations. *Food Bioprocess Technol.* **2022**. https://doi.org/10.1007/s11947-022-02803-6.

Singhi, S. C.; Kumar, S. Probiotics in Critically Ill Children. *F1000Research* **2016,** *5*, 407. https://doi.org/10.12688/f1000research.7630.1

N

O

P

Q

R

S